# Lecture Notes in Computer Science 12428

More information about this series at http://www.springer.com/series/7409

Constantine Stephanidis ·
Jessie Y. C. Chen · Gino Fragomeni (Eds.)

# HCI International 2020 – Late Breaking Papers

## Virtual and Augmented Reality

22nd HCI International Conference, HCII 2020
Copenhagen, Denmark, July 19–24, 2020
Proceedings

Springer

*Editors*
Constantine Stephanidis
University of Crete and Foundation
for Research and Technology – Hellas
(FORTH)
Heraklion, Crete, Greece

Jessie Y. C. Chen
U.S. Army Research Laboratory
Aberdeen Proving Ground, MD, USA

Gino Fragomeni
U.S. Army Combat Capabilities
Development Command Soldier Center
Orlando, FL, USA

ISSN 0302-9743          ISSN 1611-3349  (electronic)
Lecture Notes in Computer Science
ISBN 978-3-030-59989-8          ISBN 978-3-030-59990-4  (eBook)
https://doi.org/10.1007/978-3-030-59990-4

LNCS Sublibrary: SL3 – Information Systems and Applications, incl. Internet/Web, and HCI

This Springer imprint is published by the registered company Springer Nature Switzerland AG
The registered company address is: Gewerbestrasse 11, 6330 Cham, Switzerland

# Foreword

The 22nd International Conference on Human-Computer Interaction, HCI International 2020 (HCII 2020), was planned to be held at the AC Bella Sky Hotel and Bella Center, Copenhagen, Denmark, during July 19–24, 2020. Due to the COVID-19 pandemic and the resolution of the Danish government not to allow events larger than 500 people to be hosted until September 1, 2020, HCII 2020 had to be held virtually. It incorporated the 21 thematic areas and affiliated conferences listed on the following page.

A total of 6,326 individuals from academia, research institutes, industry, and governmental agencies from 97 countries submitted contributions, and 1,439 papers and 238 posters were included in the volumes of the proceedings published before the conference. Additionally, 333 papers and 144 posters are included in the volumes of the proceedings published after the conference, as "Late Breaking Work" (papers and posters). These contributions address the latest research and development efforts in the field and highlight the human aspects of design and use of computing systems.

The volumes comprising the full set of the HCII 2020 conference proceedings are listed in the following pages and together they broadly cover the entire field of human-computer interaction, addressing major advances in knowledge and effective use of computers in a variety of application areas.

I would like to thank the Program Board Chairs and the members of the Program Boards of all Thematic Areas and Affiliated Conferences for their valuable contributions towards the highest scientific quality and the overall success of the HCI International 2020 conference.

This conference would not have been possible without the continuous and unwavering support and advice of the founder, conference general chair emeritus and conference scientific advisor, Prof. Gavriel Salvendy. For his outstanding efforts, I would like to express my appreciation to the communications chair and editor of HCI International News, Dr. Abbas Moallem.

July 2020                                                                 Constantine Stephanidis

# HCI International 2020 Thematic Areas and Affiliated Conferences

Thematic Areas:

- HCI 2020: Human-Computer Interaction
- HIMI 2020: Human Interface and the Management of Information

Affiliated Conferences:

- EPCE: 17th International Conference on Engineering Psychology and Cognitive Ergonomics
- UAHCI: 14th International Conference on Universal Access in Human-Computer Interaction
- VAMR: 12th International Conference on Virtual, Augmented and Mixed Reality
- CCD: 12th International Conference on Cross-Cultural Design
- SCSM: 12th International Conference on Social Computing and Social Media
- AC: 14th International Conference on Augmented Cognition
- DHM: 11th International Conference on Digital Human Modeling & Applications in Health, Safety, Ergonomics & Risk Management
- DUXU: 9th International Conference on Design, User Experience and Usability
- DAPI: 8th International Conference on Distributed, Ambient and Pervasive Interactions
- HCIBGO: 7th International Conference on HCI in Business, Government and Organizations
- LCT: 7th International Conference on Learning and Collaboration Technologies
- ITAP: 6th International Conference on Human Aspects of IT for the Aged Population
- HCI-CPT: Second International Conference on HCI for Cybersecurity, Privacy and Trust
- HCI-Games: Second International Conference on HCI in Games
- MobiTAS: Second International Conference on HCI in Mobility, Transport and Automotive Systems
- AIS: Second International Conference on Adaptive Instructional Systems
- C&C: 8th International Conference on Culture and Computing
- MOBILE: First International Conference on Design, Operation and Evaluation of Mobile Communications
- AI-HCI: First International Conference on Artificial Intelligence in HCI

# Conference Proceedings – Full List of Volumes

**http://2020.hci.international/proceedings**

# HCI International 2020 (HCII 2020)

The full list with the Program Board Chairs and the members of the Program Boards of all thematic areas and affiliated conferences is available online at:

**http://www.hci.international/board-members-2020.php**

# HCI International 2021

The 23rd International Conference on Human-Computer Interaction, HCI International 2021 (HCII 2021), will be held jointly with the affiliated conferences in Washington DC, USA, at the Washington Hilton Hotel, July 24–29, 2021. It will cover a broad spectrum of themes related to human-computer interaction (HCI), including theoretical issues, methods, tools, processes, and case studies in HCI design, as well as novel interaction techniques, interfaces, and applications. The proceedings will be published by Springer. More information will be available on the conference website: http://2021.hci.international/

General Chair
Prof. Constantine Stephanidis
University of Crete and ICS-FORTH
Heraklion, Crete, Greece
Email: general_chair@hcii2021.org

**http://2021.hci.international/**

# Contents

# Virtual, Augmented and Mixed Reality
# Design and Implementation

# Haptic Helmet for Emergency Responses in Virtual and Live Environments

Florian Alber[✉], Sean Hackett[✉], and Yang Cai[✉]

Carnegie Mellon University, 4720 Forbes Avenue, Pittsburgh, PA 15213, USA
{falber,shackett}@andrew.cmu.edu, ycai@cmu.edu

**Abstract.** Communication between team members in emergency situations is critical for first responders to ensure their safety and efficiency. In many cases, the thick smoke and noises in a burning building impair algorithms for navigational guidance. Here we present a helmet-based haptic interface with eccentric motors and communication channels. As part of the NIST PSCR Haptic Interfaces for Public Safety Challenge, our helmet with an embedded haptic interface in the headband enables communication with first responders through haptic signals about direction, measurements, and alerts. The haptic interface can be connected over LoRa for live communication or via USB to VR simulation system. With our affordable, robust, and intuitive system we took victory in the Haptic Challenge after the VR and live trials at a firefighter training facility.

**Keywords:** Hyper-reality · Haptics · Haptic interface · Wearable sensors · Sensor fusion · Augmented reality · Helmet · First response · Firefighter

## 1 Introduction

Emergency responses such as fire fighting takes place in an extremely hazardous environment where visibility and communication are often very poor. Communicating between team members and recognizing firefighters's activities are critical to ensure their personal safety and mission status. It also helps to assess firefighters' health condition and location.

Haptic interfaces are useful in first response scenarios where regular perceptual channels such as visual and audio are not possible. Haptic interfaces are also valuable in advanced virtual reality training in order to bring physical touches into the digital world. The goal of the haptic interface is to integrate available sensors into first responder's gear with haptic sensing, and stimulating capabilities. Our haptic interface contains haptic vibration motors. The haptic actuators are embedded into the helmet, which provides left, right, forward and backward directional signals. The helmet has an embedded computing and a wireless communication system.

We have developed a prototype of the haptic helmet as shown in Fig. 1. The haptic interface on the helmet enables the user to communicate with haptic signals over LoRa for live testing or connect to the VR terminal through a USB cable. In contrast to prevailing augmented reality technologies, which overlays virtual reality objects on top

© Springer Nature Switzerland AG 2020
C. Stephanidis et al. (Eds.): HCII 2020, LNCS 12428, pp. 3–11, 2020.
https://doi.org/10.1007/978-3-030-59990-4_1

of the real world scene, this haptic helmet enables mission critical information from the firefighter to be shared with other first responders monitoring the scene so they can make more informed decisions during missions and improve the safety of the firefighters and victims.

## 2 Related Work

Haptic communication is a primitive way to perceive the world by touching. Haptic interfaces have been widely used on mobile phones and video games. To embed haptic interfaces into the first responders' equipment such as helmets, gloves, or shoes are challenging because individual bodies and individual areas of the body have different vibration perception thresholds [1]. Mobile phones normally use eccentric motors to enable vibrations. There are cylindrical or disk shapes [2], providing different intensity of buzzes. There are also piezo-based haptic actuators [3]. E-fibers are growing technology for affordable and high-resolution sensing that can be seamlessly integrated into clothes [4], in contrast to traditional pressure or resistance sensors, such as Flex sensors [5]. Besides, photometric stereo optical imaging technology enables extremely high-fidelity microscopic 2D surface pressure sensing [6].

In 2019, National Standardization and Technology (NIST)'s Public Safety Communication Research (PSCR) Division launched Haptic Interfaces for Public Safety Challenge to assess the potential of using virtual reality environments as a development tool to prototype and iterate on designs for public safety technologies [7]. The prototypes are evaluated in two key areas: how the prototypes impact a first responder's performance in three virtual reality (VR) scenarios [law enforcement, emergency medical services (EMS), and fire service]; and once embedded into firefighter personal protective equipment (PPE), how the prototypes assist firefighters in a live realistic scenario as they navigate and conduct a search and rescue task at a firefighter training facility. The haptic interfaces are tested in VR and physical environments. The VR environments are created with Unreal Engine™ that can be connected to the haptic interfaces in three VR scenarios created by NIST VR team:

The Firefight scenario is to use the haptic interface to navigate through the office space filled with smoke and fire, locate and rescue the virtual victims, and exit the building as quickly as possible. Given a rough location of the victim and the floor layout in a virtual world using Unreal, we developed a navigation tracker that can assist in directing the user through the building. Due to limited visibility the user must rely on the haptic actuators on the helmet to provide directions and navigation while in the environment.

The EMS scenario is to assess the blood pressure levels for triage of multiple victims, using a simulated haptic pressure sensor array. In our case, we use haptic interfaces on the helmet to estimate the systolic blood pressure in the following critical categories in bands: $<80$, 80–90, and $>90$ mm-Hg. The data are updated on a tablet and then the responder moves to guide the responder to the next patient in line.

The Law Enforcement scenario is to enable a SWAT member to locate the active shooters through a haptic interface. The information is sent to the haptic interface to assist situation awareness by directing the user's gaze. In this case, we use the haptic actuators hidden in the user's helmet, which provides four directional signals: front, left, right, and back. We can also modulate the vibration frequencies and strengths to enrich the haptic vocabulary. The NIST VR team created the virtual underground parking lot surrounded by multiple assailants using tools such as Unreal. The user would take cover behind some concrete barriers as the "first-person shooter". The simulated SWAT commander, who can see the overview of the parking lot, will send the haptic signal through the four-button gamepad to assist the user to locate and mitigate the threats. The haptic interface and the average time to locate and mitigate the threats.

## 3   Haptic Helmet and its Connectivity's

For the NIST Haptic Challenge VR scenarios the helmet is connected via USB to the PC running Unreal Engine. In Unreal the appropriate output response (e.g. direction to target/victim, or blood pressure reading) is calculated and sent to the microprocessor in the haptic helmet. A USB connection is used here as the VR headset already requires a wired connection and avoids the connectivity complexity of a wireless system.

However for the real life scenarios a wireless system becomes a necessity, as first responders go deep inside burning buildings. A LoRa module in the helmet receives commands from an external LoRa transmitter. Information from floor plans, drone footage, and helmet sensor data can be used by a human outside the building to manually control the haptic helmet to assist the first responder. The haptic helmet connections for the VR and real life scenarios are shown in Fig. 1 and 2.

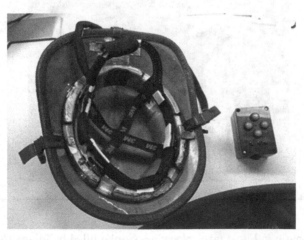

**Fig. 1.** The haptic helmet prototype with onboard processor and communication channels

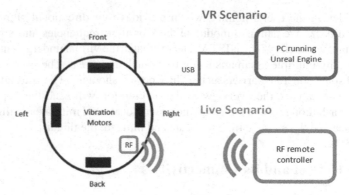

**Fig. 2.** Diagram of the haptic helmet connected to VR simulator terminal and live communication over LoRa RF

## 4 Haptic Interface for VR Training Scenarios

The Haptic Helmet is compatible with common VR headsets. For example, HTC Vive headset and hand controllers can be used alongside the Helmet and work seamlessly. For the NIST Haptic Interfaces for Public Safety Challenge, there were three scenarios to tackle; Law, Fire and EMS. A HTC Vive VR headset was used alongside the first version of the haptic helmet prototype shown in Fig. 3.

**Fig. 3.** The haptic helmet used with a HTC Vive VR system

The Fire scenario in Fig. 4 takes place in a smoke filled building. The area around the user was split into 90-degree sections. The haptics indicate the direction of a victim to be rescued relative to the user; *forward, left, right, or back*. The direction updated every two seconds. The EMS scenario in Fig. 5 involved measuring the blood pressure of a victim using a haptic interface. Medical equipment was applied to the victim and the helmets back motor buzzed corresponding to their blood pressure: 1 buzz = Low, Below 80 mmHg, 2 buzz = Medium, 80–90 mmHg, and 3 buzz = High, Above 90 mmHg. A wrist mounted display was used to log the blood pressure values.

**Fig. 4.** Fire scenario with the haptic helmet to navigate through a smoky building

**Fig. 5.** EMS scenario where the blood pressure readings are transmitted through the haptic interface and logged into the tablet.

**Fig. 6.** SWAT scenario where the direction to a target is passed by the haptic interface

The SWAT scenario in Fig. 6 took place in an indoor parking lot. Like the Fire scenario, the area around the user was split into 90-degree sections. The haptic actuators indicated the direction of the target relative to the user whether it is *front, left, right,* or *back*. By turning towards the direction of buzz the target could be easily found and then engaged.

## 5  Haptic Interface for Live Scenarios

For the live scenario a firefighter wearing the helmet would be instructed towards the location of a dummy placed in the test building. This location would be known by a team member who would follow and direct the firefighter using a remote controller.

The haptic helmet from the VR challenge had to go through a number of developments to make it suitable for the live challenge. As the firefighter would be wearing a flame hood and SCBA mask, the contact with their head would be reduced. Therefore higher power haptic motors were embedded in the headband of the helmet with additional electronics to drive them.

The haptic motors and wiring were concealed in the headband and connected to a small enclosure mounted on the helmet containing the control electronics. Having all the components embedded in the headband meant that the haptic system could be swapped between helmets. This provides a low cost implementation inside the high cost helmet shell.

A controller containing a LoRa module was used to communicate the haptic instructions to the helmet. LoRa was chosen due to the low data rate of this application and its long range, which would be useful for sending signals deep into buildings. The protocol for using the controller and the helmet is shown in Table 1.

**Table 1.**  Protocol of the controller and helmet

| Controller | Helmet |
| --- | --- |
| -Setup LoRa Tx/Rx module, addressing, freq, signal power | -Setup LoRa Tx/Rx module, addressing, freq, signal power |
| -Directional button press<br>-Send packet | |
| | -Receive packet, RSSI<br>-Haptic vibration<br>-Return ACK |
| -Receive ACK, RSSI<br>-Ready for next button press | |

At the firefighter training facility in Fig. 7, the teams were given a short time to dress their firefighters with their haptic system. The helmet, which was already a required piece of firefighter PPE, could be donned quickly by the firefighter and required no additional gear. The dial adjustment mechanism of the headband also meant it could fit a variety of head sizes.

The fully equipped firefighter wearing the haptic helmet, shown in Fig. 7, were instructed on how to interpret the haptic signals. The navigation approach was refined

**Fig. 7.** Firefighter training facility (left) and instructing firefighter with the navigation approach of the haptic helmet (right)

from the VR scenarios. Here a Stop and Go approach was used to minimise the chance of mistakes occurring which could be fatal. A front buzz meant go forward, a back buzz meant stop, and left and right buzzes meant to turn 90° left/right on the spot. This approach was very effective in the tight rooms and hallways of the training facility. The floor plan and route map of the facility is shown in Fig. 8

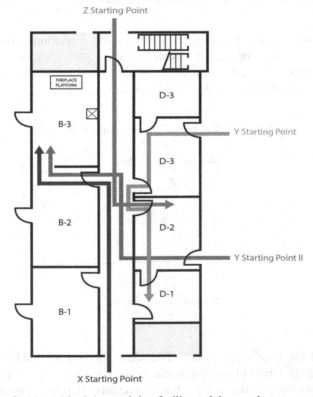

**Fig. 8.** Map of firefighter training facility and the search rescue routes

The search and rescue task took place in a pitch black smoke filled environment. The firefighters were told to let the haptics guide them and not use traditional navigation techniques. The team member used a thermal camera, shown in Fig. 9, to follow the movement of the firefighter and guide them to the dummy location in the building. Finally the team member had to guide the firefighter back out of the building along the same route. Each team followed their respective firefighters through the building during the trials to remove any possible communication and tracking issues which could occur from guiding the firefighter from outside the building.

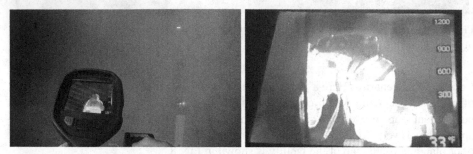

**Fig. 9.** Live footage from the haptic challenge when entering the training facility (left) and image of thermal camera of the firefighter inside the building (right)

The search and rescue task took place six times with a rotation of firefighters and judges to provide a fair comparison between all the competing teams. A base time of 3:00 min was given, with a cut-off time of 6:00 min which would count as a fail. The team results from competition are shown in Table 2.

**Table 2.**  Search & rescue routes and times achieved

| Round | Route | Time |
|-------|-------|------|
| 1 | Z | ** |
| 2 | Y1 | ** |
| 3 | Y2 | ** |
| 4 | X | 1:56 |
| 5 | Z | ** |
| 6 | Y1 | 1:15 |

*** currently waiting on final time from NIST.*

The NIST Haptic Challenge was judged on scores from the VR and Live challenges. The haptic integration, ease of use, haptic clarity, and time taken to complete the trials were all taken into account. After completion of the competition our team was announced as the winners and also received an award for the most commercially promising system.

# 6 Conclusions

In this paper, we presented a hyper-reality helmet with built in haptics to guide first responders, as well as activity recognition to gain real-time information from the first responders.

The haptic helmet intuitively guided users through NISTs VR scenarios intuitively with little instructions. The final haptic system embedded in the firefighter helmet for the haptic challenge live trials proved highly effective. The firefighter judges were extremely enthusiastic about our system which was low cost, had good haptic clarity and completed the live trials rapidly. As a result our team were the winners of the haptic challenge, proving the value of our haptic system.

**Acknowledgement.** We are grateful to awards from the NIST PSCR/PSIA Program's Haptic Interfaces for Public Safety Challenge and the encouragement of the public safety SME team. This project is also funded in part by Northrop Grumman Corporation's SOTERIA Program and Carnegie Mellon University's Mobility21 National University Transportation Center, which is sponsored by the US Department of Transportation. We would like to thank Professors Roberta L. Klatzky, Mel Siegel and Lenny Weiss for their expertise and guidance during the course of the haptic interface development.

# References

1. Parsons, K.C., Griffin, M.J.: Whole-body vibration perception thresholds. J. Sound Vib. **121**, 237–258 (1988). https://www.sciencedirect.com/science/article/pii/S0022460X88800270
2. Vibrating Mini Motor Disc. https://www.adafruit.com/product/1201?gclid=Cj0KCQjwwODl BRDuARIsAMy_28V5fb3mi-TVeX-R5mG-k-Q1yAc3TzsacWUku0P6ZGQPT9NvEcztdIY aAjrpEALw_wcB
3. Piezo Speaker. https://leeselectronic.com/en/product/49611.html
4. Sundaram, S., Kellnhofer, P., Li, Y., Zhu, J.Y., Torralba, A., Matusik, W.: Learning the signatures of the human grasp using a scalable tactile glove. Nature **569**, 698–702 (2019)
5. http://cfg.mit.edu/sites/cfg.mit.edu/files/Sundaram_et_al-2019-Nature.pdf
6. Flex Sensor. https://learn.sparkfun.com/tutorials/flex-sensor-hookup-guide
7. Yuan, W., Dong, S., Adelson, E.H.: GelSight: high-resolution robot tactile sensors for estimating geometry and force. Sensors **17**(12), 2762 (2017) https://www.ncbi.nlm.nih.gov/pmc/art icles/PMC5751610/
8. NIST Haptics Interfaces Challenge (2019). https://www.nist.gov/communications-technology- laboratory/pscr/funding-opportunities/open-innovation-prize-challenges-0

# eTher – An Assistive Virtual Agent for Acrophobia Therapy in Virtual Reality

Oana Bălan[1], Ștefania Cristea[1], Gabriela Moise[2], Livia Petrescu[3], Silviu Ivașcu[1(✉)], Alin Moldoveanu[1], Florica Moldoveanu[1], and Marius Leordeanu[1]

[1] Faculty of Automatic Control and Computers, University POLITEHNICA of Bucharest, 060042 Bucharest, Romania
ivascu.silviu10@gmail.com

[2] Department of Computer Science, Information Technology, Mathematics and Physics (ITIMF), Petroleum-Gas University of Ploiesti, 100680 Ploiesti, Romania

[3] Faculty of Biology, University of Bucharest, Bucharest, Romania

**Abstract.** This paper presents the design, a pilot implementation and validation of eTher, an assistive virtual agent for acrophobia therapy in a Virtual Reality environment that depicts a mountain landscape and contains a ride by cable car. eTher acts as a virtual therapist, offering support and encouragement to the patient. It directly interacts with the user and changes its voice parameters – pitch, tempo and volume – according to the patient's emotional state. eTher identifies the levels of relaxation/anxiety compared to a baseline resting recording and provides three modalities of relaxation - by determining the user to look at a favorite picture, listen to an enjoyable song or read an inspirational quote. If the relaxation modalities fail to be effective, the virtual agent automatically lowers the level of exposure. We have validated our approach with a number of 10 users who played the game once without eTher's intervention and three times with assistance from eTher. The results showed that the participants succeeded to finish the game quicker in the last gameplay session where the virtual agent intervened. Moreover, their biophysical data showed significant improvements in terms of relaxation state.

**Keywords:** Acrophobia · Virtual agent · Exposure therapy

## 1 Introduction

Phobia is a type of anxiety disorder manifested through an extreme and irrational fear towards objects and situations. According to the newest statistics, 13% of the world's population suffer from a certain type of phobia [1]. They are classified into social phobias (agoraphobia, fear of public speaking) and specific phobias (triggered by specific objects or situations). At world level, 15–20% people [2] experience specific phobias at least once in their lifetime. Acrophobia (fear of heights) is in the top, affecting 7.5% of people worldwide, followed by arachnophobia (fear of spiders) with 3.5% and aerophobia (fear of flying) with 2.6%. The recommended treatment in case of phobia is either medical (with pills) or psychological – in-vivo exposure to the fear-provoking stimuli and

© Springer Nature Switzerland AG 2020
C. Stephanidis et al. (Eds.): HCII 2020, LNCS 12428, pp. 12–25, 2020.
https://doi.org/10.1007/978-3-030-59990-4_2

Cognitive Behavioral Therapy (CBT), a modality that determines the patient to change his thoughts towards the objects generating fear. Nearly 80% of phobics find relief in medicines and CBT. However, treatment should be provided continuously, as in more than 50% of cases, the disorder tends to relapse. The medication prescribed includes anti-anxiety and anti-depressive drugs that alleviate anxiety symptoms [3], with side effects such as impaired cognition and tendency to create dependence [4].

Virtual Reality (VR) has emerged in recent years and a series of systems for phobia therapy have been tested and validated, either as commercial products or research items. They are called Virtual Reality Exposure Therapy (VRET) systems and are preferred by more than 80% of patients over the classical in-vivo exposure therapy [5].

VR phobia therapy has the advantages of providing a safe exposure environment, with a various range of modifiable stimuli and immediate intervention from the therapist. Our approach replaces the human therapist with a virtual one, in the form of a female software avatar, called eTher, that evaluates the patient's level of anxiety based on recorded biophysical signals, provides guidance and encouragement, changes its voice parameters (pitch, tempo and volume) according to the user's emotional state and automatically adjusts the levels of game exposure. We continued the previous research [6–8] and enriched the software by adding a virtual agent with the appearance of a virtual avatar.

The paper is structured as follows: Sect. 2 presents related work, Sect. 3 introduces emotions and biophysical signals, Sect. 4 describes the virtual environment for acrophobia therapy, Sect. 5 details eTher's implementation and Sect. 6 outlines its validation – experimental procedure and results. Finally, in Sect. 7 we provide conclusions and future research directions.

## 2 Related Work

Virtual Reality (VR) has been successfully used for 25 years in mental disorders treatment [9, 10]. One of the first studies on virtual environments and acrophobia treatment is presented in [11]. Three virtual environments have been created to be used for exposure therapy. They contain an elevator, balconies and bridges. 17 subjects were randomly divided in two groups: a treatment group and a control group. The subjects from the treatment group have been gradually exposed in the VR environments and the results indicated that virtual height situations generated the same experience as physical world height exposure.

A large debate about the potential of VRET in acrophobia treatment is highlighted in [12]. VR can elicit phobic stimuli and the patients may feel like being in situations that are difficult to access in the real-world environment. Also, VR offers a solution to treat the patients who cannot imagine an acrophobic environment. VRET can be used by anyone including those who do not have the courage to recognize the phobia and seek treatment. The technological advances made possible the development of cheap VR applications and devices to be used in therapies. The VR-based triggers of acrophobic behavior are identified and described in [12]: visuo-vestibular triggers, postural triggers, visual and motion triggers. The authors concluded that VR can be used both as a tool for treating acrophobia, but also for investigating and understanding it [12].

Recently, Freeman et al. [13] showed that VR has not been used at its maximum potential for mental healthcare therapy. They conducted a systematic review of the field, where both benefits and issues have been identified. The main benefit is the increased access to treatment, while the main concern refers to the therapy's quality control. The highlighted benefits of VR for mental healthcare are: the usage of VR enables accurate therapeutic strategies implementations; helpful situations for therapy can be created; the treatment can be repeated without additional costs and can be delivered to patients' homes [13]. VR offers the possibility to implement various therapeutic techniques, more real situations can be simulated and patients can gradually and repeatedly experiment them until they overcome their problems.

In 2018 the results of an automated treatment for fear of heights were presented in [14]. A software application called Now I Can Do Heights with a virtual coach was involved in the treatment of 49 subjects. 51 subjects were allocated to the control group. The VR-based treatment was designed during 6 sessions of 30 min each, over a period of 2 weeks. The results proved the efficiency of the VR-based automated psychological therapy: fear of heights has decreased for all the participants from the VR group.

The efficiency of VR cognitive behavior therapy was proved in a large experiment in [15]. 193 subjects, aged between 18 and 65, were divided in two groups. One group was exposed to VR-CBT applications and gamified VR environments and the other was a control group. A significant reduction of acrophobia was recorded for the VR-CBT group after three months of therapy, compared to the control group [15]. Also, the authors concluded that VR-based treatment can be performed even in the absence of the therapist.

A pilot study was performed in [16] in order to evaluate e-virtual reality exposure for acrophobia treatment. 6 subjects were exposed for six sessions during three weeks in VR-based therapy: three of them participated in e-VRET sessions (without a therapist) and three were involved in traditional p-VRET sessions (in the physical presence of the therapist). The results showed that there is no significant difference between e-VRET and p-VRET sessions regarding the anxiety level recorded. The authors made a first step in proving that VRET can be used over Internet for phobia treatment [16].

Most of the automated applications for acrophobia treatment use artificial intelligence to estimate the subjects' fear level in acrophobic situations. Various biophysical data were collected and learning models have been used to make predictions about the patients' fear level. In the experiment presented in [17], the EEG data of 60 participants was acquired and fed to a deep learning model to detect the subjects' acrophobia level.

In [6–8], we investigated the efficiency of different machine learning classifiers in a VR system for treating acrophobia. The system automatically estimated fear level based on multimodal sensory data – EEG, pulse, electrodermal activity and a self-reported emotion evaluation. The results showed classification accuracies ranging from 42.5% to 89.5%, using the Support Vector Machine, Random Forest and k Nearest Neighbors techniques. The most important features for fear level classification were GSR, HR and the EEG in the beta frequency range.

VRET gained the status of effective therapy for various mental disorders according to the results of the meta-analyses presented in [18]. 30 studies about VRET in different disorders have been analyzed: 14 studies with specific phobias, 8 with social anxiety disorder or performance anxiety, 5 with posttraumatic stress disorder and 3 with panic disorder. The authors concluded that VRET is an equal medium for exposure therapy [18].

## 3 Emotions and Biophysical Data

Accurate recognition of emotions allows the appropriate adjustment of the therapeutic attitude in phobias. Phobic behavior is a defense-like overreaction to a certain category of stimuli [19]. As a manifestation of autonomic nervous system activation, emotions can be identified by measuring and analyzing physiological reactivity. Nowadays a wide variety of modern neurophysiological methods are used for biophysical signal sensing of emotions: Electromyography - EMG, Electrodermal Activity – EDA, Electroencephalography - EEG, Heart Rate Variability – HRV.

In response to phobic conditioning, we chose to record Galvanic Skin Response (GSR) and Heart Rate (HR). It is expected that GSR and HR amplitudes would increase as a result of phobic stimuli without habituation and decrease as a result of the virtual agent's therapeutic intervention.

## 4 VRET Game for Acrophobia

The designed system is based on a VR game for acrophobia therapy, to which we added a virtual agent acting as a virtual therapist.

The acrophobia game depicts a mountain landscape, with three possible scenarios that can be selected from the start menu: a ride by cable car, one by ski lift and one by foot. The game is rendered over the HTC Vive Head Mounted Display and the interaction is ensured by pressing the buttons from the HTC Vive's controllers. In the current version, eTher is implemented only for the cable car ride (Fig. 1). Throughout the ride, there are 10 stops where the cable car gets blocked and the user's emotional state is evaluated considering the biophysical data (GSR and HR) recorded during the previous level. A level is defined as the ride segment between two consecutive stops. Each level takes approximately 10 s, time in which the cable car moves slowly, so that the user can look on the window, move his head, rotate and have a full realistic experience of the environment. The ride takes a semi-ascending path, so that the stops are set at the following altitudes: starting point – 28 m, Stop1 – 138 m, Stop2 – 264 m, Stop3 – 327 m, Stop4 – 213 m, Stop5 – 388 m, Stop6 – 460 m, Stop7 – 395 m, Stop8 – 470 m, Stop9 – 607 m, Stop10 – 640 m.

Before the start of the game, we store the user's profile – name, age, level of acrophobia (low, medium, high) determined by completing 3 questionnaires: Heights Interpretation Questionnaire [20], Visual Height Intolerance Severity Scale [21], Acrophobia Questionnaire [22]. We also store each user's favorite song/image/quote that will represent the relaxation modalities.

At start, we also measure the baseline HR and GSR during a relaxation period of 3 min. The difference between the average baseline relaxation HR and GSR values and the average HR and GSR values recorded during a game level will determine the current anxiety level. According to the current anxiety level, eTher will either allow the user to continue the game (if the anxiety level is low) or provide relaxation modalities by randomly presenting the user his favorite image, song or quote. The game ends when the user reaches the final stop or after a predefined number of game epochs – for the current implementation we chose 20. He may also leave the game anytime he feels uncomfortable or experiences motion sickness. A full description of eTher's workflow is detailed in Sect. 5.

## 5   eTher for Acrophobia Therapy

### 5.1   Description

eTher (Fig. 2) is a therapeutic agent providing assistance to the user through social interactions in order to increase the efficiency of the VR game in acrophobia therapy. It is actually a virtual agent equipped with conversational service capabilities in order to maintain a dialogue with the users. The agent aims to keep the users in a comfortable zone defined by their baseline relaxation HR and GSR values during exposure to various acrophobia anxiety-producing stimuli in the VR environment.

**Fig. 1.** Virtual environment seen from the cable car     **Fig. 2.** eTher female game avatar

The main capabilities of eTher are:

- Provides assistance in acrophobia treatment by monitoring patients' biophysical data during the therapy.
- Offers encouragement, motivation, corrective feedback and support to the patients.
- Provides personalized guidance during the VR game.
- Automatically adjust the level of exposure.

The architecture of eTher is inspired by INTERRAP [23] and MEBDP [24] (Fig. 3).

The therapeutic agent consists of three modules: a World Interface, which provides information exchange with the environment, a Knowledge Base consisting of facts and rules and a Control Unit. The Environment is defined by patients and the VR game. The World Interface contains user's data acquisition systems, sensors systems, systems for describing game scenarios and actuators, elements through which the agent modifies the environment. The Knowledge Base and Control Unit are accordingly structured in

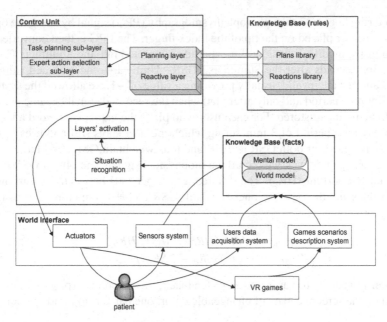

**Fig. 3.** eTher architecture

two layers: one for reactive behavior and one for planned behavior. The Knowledge Base (facts) of the agent contains the beliefs about the environment (world model), and beliefs about itself (mental model). For example, the world model contains: relaxing songs; motivational, encouraging, congratulating expressions; conversation topics; music with changeable parameters; patients' profiles; games scenarios descriptions; etc. The Knowledge Base (rules) contains a plans library and a reactions library.

A plan is a tuple $p = <condition, procedure>$
If the condition is true, then the procedure is run.

The Control Unit contains algorithms for situations recognition (a reactive or planning situation), a module which controls layers activation and two layers for each type of situation. All layers contain Reinforcement Learning techniques to decide which rules, plans or cooperation protocols will be selected at any given state of the environment. An environment state is defined by anxiety levels, patients' profiles and game scenarios. The planning layer is divided in two sub-layers similar to MEBDP [24]: task planning sub-layer to decompose a global task in subtasks and expert action selection sub-layer to select action for each sub-task. The reactive layer and planning layer generate actions which are transferred to actuators to be performed.

## 5.2 Implementation

For measuring HR and GSR, we used the Shimmer3 GSR+ Unit [25]. It captures the skin's electrical conductance via two electrodes placed on the medial phalanges of the middle and ring fingers and the photoplethysmography (PPG) signal by using the optical pulse probe sensor placed on the tip of the index finger. The PPG is then converted into HR by an in-lab modified version of the module integrated into the Shimmers C# API. As in the first seconds after the Shimmers3 GSR+ Unit starts, the recorded values for both GSR and HR are invalid, usually providing a value of $-1$, we allowed the system to have a calibration period and only after that, when the recorded signal became valid, the data could be used and stored. For each user who plays the game, we record a baseline HR and GSR for a period of 3 min during which the user stays still and tries to relax. The data is averaged for both GSR and HR and thus we obtain $GSR_b$ and $HR_b$.

During each game level, we record the biophysical signals and obtain GSR current ($GSR_c$) and HR current ($HR_c$). The differences (in percent) between the current biophysical values and the baseline ones (for both GSR and HR) are computed using the following formula:

$$P_{HR} = 100 * (HR_c - HR_b) / HR_b$$
$$P_{GSR} = 100 * (GSR_c - GSR_b) / GSR_b$$
(1)

In order to provide the user a form of feedback, these percents ($p_{HR}$ and $p_{GSR}$) are translated on the screen as bars of changeable color, one bar for $p_{HR}$ and one for $p_{GSR}$, thus:

$$P_{HR/GSR} \leq 10\%, \ color = green$$
$$P_{HR/GSR} > 10\% \ and \ P_{HR/GSR} \leq 40\%, \ color = yellow$$
$$P_{HR/GSR} > 40\% \ and \ P_{HR/GSR} \leq 70\%, \ color = orange$$
$$P_{HR/GSR} > 70\%, \ color = red$$
(2)

$p_{HR/GSR}$ refers to either $p_{HR}$ or $p_{GSR}$.

The objective of eTher is to keep the user within the green and yellow area of comfort. ETher is a humanoid software agent with the shape of a female game avatar. It is designed using the Unity game engine, starting from a realistic human face. The appearance of the agent needs to be according the users' preference. The agent must have a positive voice that inspires trustworthiness, competence and warmth. After a preliminary analysis of the gender dimension, we chose to design the therapeutic agent with a female voice for several reasons: (i) it is perceived as helping, not commanding [26]; (ii) we are biologically set from intrauterine life to prefer the female voice, to identify the mother's voice, not necessarily that of the father [27]; (iii) the female voice is clearer and more melodious, having a calming and soothing effect (the processing of the female voice is done in the same auditory area dedicated to musical information) [28] (iv) the female voice offers greater confidence than the male voice due to a higher pitch [29]. By default, the female avatar has a normal, neutral voice in terms of voice parameters – pitch, tempo and volume. The voice parameters are afterwards modified in Audacity [30] in order to create the four Agent Interactions (AIs) described in Table 1. For AI1, pitch, tempo and volume are amplified with 10%. This means that the agent's voice is more energetic, dynamic and lively, appropriate for giving encouragements and motivating the user to go on with the game. For AI2, eTher's voice is rather neutral. For AI3, the parameters are decreased with 10%, meaning that the voice is graver, the words are pronounced slower and the volume is lower in order to make the user relax, detach and diminish his level of anxiety. In the case of AI4, all three parameters are decreased with 20%. This will be played when the subject is extremely tense and the voice needs to be sober, serious, with a staccato rhythm and quieter so that he may completely understand his emotional state and relax.

**Table 1.**  Virtual agent interactions

| Interaction | Phrase | Voice parameters | | |
|---|---|---|---|---|
| | | Pitch | Tempo | Volume |
| AI1 | "Good job! Keep going!" | +10% | +10% | +10% |
| AI2 | "Enjoy and relax for a while" | 0% | 0% | 0% |
| AI3 | "Calm down and relax" | −10% | −10% | −10% |
| AI4 | "You are too tense. Take a deep breath and try to relax more" | −20% | −20% | −20% |

We identified the following situations or tuples *<condition, procedure>* which compose the Therapy Plan Library (Table 2):

**Table 2.**  Therapy plan library

| Situation no. | Condition | | Procedure | |
|---|---|---|---|---|
| | GSR color | HR color | Agent interaction | Change game level (only after the relaxation modalities are provided) |
| Situation1 | Green | Green | AI1 | No |
| Situation2 | Green | Yellow | AI1 | No |
| Situation3 | Yellow | Green | AI1 | No |
| Situation4 | Yellow | Yellow | AI1 | No |
| Situation5 | Green | Orange | AI2 | −1 level |
| Situation6 | Orange | Green | AI2 | −1 level |
| Situation7 | Yellow | Orange | AI2 | −1 level |
| Situation8 | Orange | Yellow | AI2 | −1 level |
| Situation9 | Orange | Orange | AI3 | −2 levels |
| Situation10 | Orange | Red | AI3 | −2 levels |
| Situation11 | Red | Orange | AI3 | −2 levels |
| Situation12 | Green | Red | AI3 | −2 levels |
| Situation13 | Red | Green | AI3 | −2 levels |
| Situation14 | Yellow | Red | AI3 | −2 levels |
| Situation15 | Red | Yellow | AI3 | −2 levels |
| Situation16 | Red | Red | AI3 | −3 levels |

A diagram that details the therapy workflow is presented in Fig. 4.

In Situation1–Situation4, $GSR_{color}$ and $HR_{color}$ are both either green or yellow. This means that the user is relaxed or rather relaxed and he can continue to the next game level. eTher in the form of AI1 appears on the screen and encourages him to go on. In Situation5–Situation8, one of $GSR_{color}$ or $HR_{color}$ are Green/Yellow and the other is Orange, suggesting that the subject tends to become anxious. eTher plays AI2 and then presents randomly the user's favorite image, song or quotation for 20 s. After these 20 s, the subject's emotional state is evaluated again. If it falls into Situation1–Situation4, AI1 appears and he may continue the game from there. If it falls into Situations 5–8, the player is taken to the previous level, so the level of exposure decreases with 1, if it exists. For Situations 9–15, the exposure decreases with 2 game levels, while for Situation16, the user is automatically taken to 3 game levels behind the current one. In Situation9–Situation15, $GSR_{color}$ and $HR_{color}$ are both orange or one is red. In all these cases, AI3 is played and if the relaxation modalities fail to work, the exposure is lowered with 2 levels. In Situation15, both $GSR_{color}$ and $HR_{color}$ are red. AI4 is played by eTher and if the relaxation means are inefficient, the user is taken 3 game levels behind the current one.

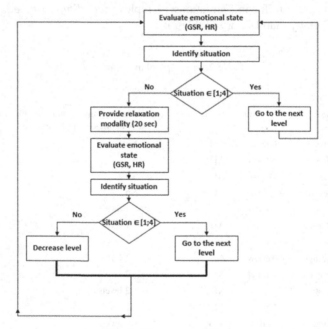

**Fig. 4.** Therapy workflow

# 6  eTher Validation

## 6.1  Experiment Validation

In order to validate our pilot implementation of eTher, we have performed a series of tests with a number of 10 acrophobic subjects, 5 men and 5 females, aged 24–50. The experiment consisted in obtaining a baseline GSR and HR recording during a relaxation period of 3 min, followed by a gameplay session without intervention from eTher, other 3 gameplay sessions spanned over a period of 3 days assisted by eTher and a subjective evaluation where the participants were asked to share their opinion about the human-agent interaction by answering the following questions:

1. How was your experience with the agent?
2. Was it easy to be used?
3. Did you feel safe during gameplay?
4. Have you felt an improvement of your acrophobic (comfort) state during the game?

As our research is in incipient phase, we designed and evaluated the functionality of the therapeutic agent simultaneously, by adjusting its capabilities during the current development phase. As we wanted to find out more about the human-agent interaction, at the end of our experiment, the subjects took part in an interview with the researchers and verbally expressed their opinions and shared their experience.

## 6.2  Results

In the last session of the experiment, for all users, the most frequent situation was Situation1 (44%), followed by Situation2 (21%), Situation3 (17%), Situation6 (5%), Situation13 (9%) and Situation15 (4%). The participants succeeded to reach the final stop of the game after completing 11.75 game levels on average.

There were 14% exposure changes with 1 level and no exposure changes with 2 or 3 levels. Of the 14% exposure changes with 1 level, 9% were due to the fact that a high level of anxiety appeared in Stop1 and Level1 had to be repeated. As Level0 does not exist, the user had to replay Level1. Probably they experienced anxiety at the start of the game, due to the fact that they knew they are monitored or felt uneasy in the presence of the tester.

Both skin conductance parameters (Fig. 5) and heart rate (Fig. 6) decreased at the end of the three days of gameplay, indicating a reduction of phobic anxiety.

In order to assess the efficiency of the proposed model, we applied a paired-samples t-test that compared the average of the recordings at the final session for the same group of subjects before and after therapy. We started from the hypothesis that the average values decrease after therapy for both GSR and HR. A $p < 0.05$ value was considered statistically significant. The average GSR before therapy was 1.68 uS and after therapy, 0.9 uS. In what concerns HR, before therapy the average was 77.34 bpm and after therapy, 75.17 bpm.

According to Table 3, we reject the null hypothesis and support the research hypothesis stating that virtual agent-assisted therapy has a positive effect on acrophobia alleviation.

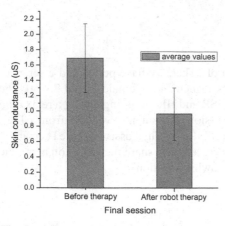

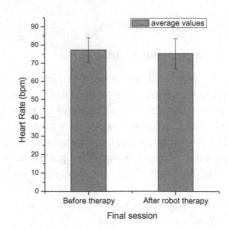

**Fig. 5.** Reduction of GSR after virtual agent therapy

**Fig. 6.** Reduction of HR after virtual agent therapy

**Table 3.** Compared averaged recordings before and after therapy

| N = 10 | Before therapy | | After eTher-based therapy | | | |
|--------|------|----------------|------|----------------|---|----|
| df = 9 | Mean | Std. deviation | Mean | Std. deviation | t | p* |
| GSR (uS) | 1.689 | 0.450 | 0.961 | 0.341 | −6.822 | 0.000 |
| HR (bpm) | 77.342 | 6.681 | 75.173 | 8.168 | −3.814 | 0.004 |

For exemplification, we present the analysis of one of the subjects both before and after therapy.

The data obtained from the electrodermal recording was processed using the ledalab software [31, 32]. Its processing includes electrical noise reduction, the detection and measurement of artifacts, as well as signal decomposition into two components: Skin Conductance Level – SCL and Skin Conductance Response – SCR.

SCL is a component that varies slowly as a result of general changes in autonomous arousal, while SCR is a rapidly varying component which represents the phasic response to a succession of stimuli provided in the environment. These two processes have different neurological mechanisms.

Figure 7 presents the evolution of GSR during gameplay before (a) and after (b) therapy.

Figure 8 presents individual responses to stimuli obtained after processing raw data (before therapy – a and after therapy – b) using the Discrete Decomposition Analysis (DDA) method.

In what concerns the qualitative questionnaire, we hereby express the opinion of the user with the lowest level of VR experience: "Nice and interesting experience, although it was the first time I used the VR glasses; it wasn't easy for me to use them, as I had to stay focused and alert. Yes, I felt secure and I enjoyed knowing that a virtual therapist

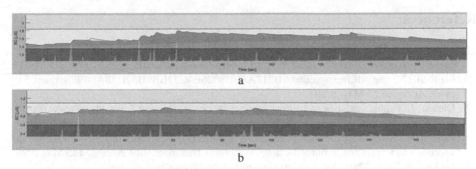

**Fig. 7.** The evolution of GSR during gameplay before (a) and after (b) therapy

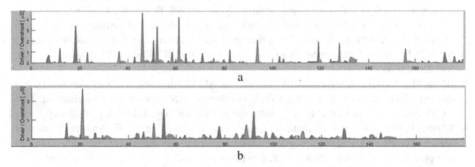

**Fig. 8.** Individual responses to stimuli before (a) and after (b) therapy.

takes care of me. I felt an improvement of my acrophobic condition, but also a stress because I do not have necessary abilities for interacting with the virtual environment".

## 7   Conclusions

In this paper, we presented the pilot implementation and validation of an assistive virtual agent which identifies emotional states and provides guidance by adjusting exposure levels in a virtual environment for acrophobia therapy. The therapeutic agent automatically adjusts its speech and voice parameters according to the player's anxiety level and supplies relaxation modalities. In an experiment with 10 acrophobic users, we have showed a significant improvement of the anxiety levels, observable in the GSR and HR rates in the final session of the experiment. As future directions, we plan to perform more experiments and to refine the system so that the patients will be able to use it in safety conditions.

**Acknowledgements.** The work has been funded by the Operational Programme Human Capital of the Ministry of European Funds through the Financial Agreement 51675/09.07.2019, SMIS code 125125, UEFISCDI proiect 1/2018 and UPB CRC Research Grant 2017.

# References

1. World Health Organization: Depression and other common mental disorders. Global health estimates. https://apps.who.int/iris/bitstream/handle/10665/254610/WHO-MSD-MER-2017.2-eng.pdf;jsessionid=4AB217A8ED8026E4C51BB9B57D58ADEE?sequen ce=1. Accessed 25 May 2020
2. Olesen, J.: Phobia statistics and surprising facts about our biggest fears. http://www.fearof. net/phobia-statistics-and-surprising-facts-about-our-biggest-fears/. Accessed 25 May 2020
3. Fadden, H.: Acrophobia (definition, causes, symptoms and treatment). https://www.thehealth yapron.com/acrophobia-definition-causes-symptoms-treatment.html. Accessed 25 May 2020
4. Berkley Wellness: The risks of anti-anxiety pills. https://www.berkeleywellness.com/healthy-mind/mood/sleep/article/risks-anti-anxiety-pills. Accessed 25 May 2020
5. Garcia-Palacios, A., Hoffman, H.G., See, S.K., Tsay, A., Botella, C.: Redefining therapeutic success with virtual reality exposure therapy. CyberPsychol. Behav. **4**, 341–348 (2001)
6. Bălan, O., Moise, G., Moldoveanu, A., Leordeanu, M., Moldoveanu, F.: Automatic adaptation of exposure intensity in VR acrophobia therapy, based on deep neural networks. In: Twenty-Seventh European Conference on Information Systems (ECIS2019), Stockholm-Uppsala, Sweden (2019)
7. Bălan, O., Moise, G., Moldoveanu, A., Leordeanu, M., Moldoveanu, F.: Does automatic game difficulty level adjustment improve acrophobia therapy? Differences from baseline. In: 24th ACM Symposium on Virtual Reality Software and Technology, Tokyo, Japan (2018)
8. Bălan, O., Moise, G., Moldoveanu, A., Leordeanu, M., Moldoveanu, F.: An investigation of various machine and deep learning techniques applied in automatic fear level detection and acrophobia virtual therapy. Sensors **20**, 496 (2020)
9. Moldoveanu, A., et al.: The TRAVEE system for a multimodal neuromotor rehabilitation. IEEE Access **7**, 8151–8171 (2019). https://doi.org/10.1109/access.2018.2886271
10. Ferche, O., et al.: From neuromotor command to feedback: a survey of techniques for rehabilitation through altered perception. In: 2015 E-Health and Bioengineering Conference (EHB), Iasi, pp. 1–4 (2015). https://doi.org/10.1109/ehb.2015.7391454
11. Hodges, L.F., et al.: Virtual environments for treating the fear of heights. Computer **28**(7), 27–34 (1995). https://doi.org/10.1109/2.391038. Accession Number: 5033494
12. Coelho, C.M., Waters, A.M., Hine, T.J., Wallis, G.: The use of virtual reality in acrophobia research and treatment. J. Anxiety Disord. **23**(5), 563–574 (2009). https://doi.org/10.1016/j. janxdis.2009.01.014
13. Freeman, D., et al.: Virtual reality in the assessment, understanding, and treatment of mental health disorders. Psychol. Med. **47**, 2393–2400 (2017). https://doi.org/10.1017/s00332917 1700040x
14. Freeman, D., et al.: Automated psychological therapy using immersive virtual reality for treatment of fear of heights: a single-blind, parallel-group, randomised controlled trial. Lancet Psychiatry **5**, 625–632 (2018)
15. Donker, T., et al.: Effectiveness of self-guided app-based virtual reality cognitive behavior therapy for acrophobia: a randomized clinical trial. JAMA Psychiatry **76**(7), 682–690 (2019). https://doi.org/10.1001/jamapsychiatry.2019.0219
16. Levy, F., Leboucher, P., Rautureau, G., Jouvent, R.: E-virtual reality exposure therapy in acrophobia: A pilot study. Journal of telemedicine and telecare **22**(4), 215–220 (2016). https:// doi.org/10.1177/1357633x15598243
17. Hu, F., Wang, H., Chen, J., Gong, J.: Research on the characteristics of acrophobia in virtual altitude environment. In: Proceedings of the 2018 IEEE International Conference on Intelligence and Safety for Robotics, Shenyang, China, August 24–27, 2018 (2018)

18. Carl, E., et al.: Virtual reality exposure therapy for anxiety and related disorders: a meta-analysis of randomized controlled trials. J. Anxiety Disord. **61**, 27–36 (2019)
19. Boucsein, W.: Electrodermal Activity. Springer, Cham (2012). https://doi.org/10.1007/978-1-4614-1126-0
20. Steinman, S.A., Teachman, B.A.: Cognitive processing and acrophobia: validating the heights interpretation questionnaire. J. Anxiety Disord. **25**, 896–902 (2011). https://doi.org/10.1016/j.janxdis.2011.05.001
21. Huppert, D., Grill, E., Brandt, T.: A new questionnaire for estimating the severity of visual height intolerance and acrophobia by a metric interval scale. Front. Neurol. **8**, 211 (2017)
22. Cohen, D.C.: Comparison of self-report and behavioral procedures for assessing acrophobia. Behav. Ther. **8**, 17–23 (1977)
23. Müller, J.P.: The Design of Intelligent Agents. LNCS, vol. 1177. Springer, Heidelberg (1996). https://doi.org/10.1007/BFb0017806. ISBN:978-3540620037
24. Nakano, M., et al.: A two-layer model for behavior and dialogue planning in conversational service robots. In: Proceedings of IEEE-IROS-05, pp. 1542–1547 (2005)
25. Shimmer Sensing GSR Unit. https://www.shimmersensing.com/products/shimmer3-wireless-gsr-sensor. Accessed 25 May 2020
26. Borkowska, B., Pawlowski, B.: Female voice frequency in the context of dominance and attractiveness perception. Anim. Behav. **82**(1), 55–59 (2011)
27. Lee, G.Y., Kisilevsky, B.S.: Fetuses respond to father's voice but prefer mother's voice after birth. Dev. Psychobiol. **56**(1), 1–11 (2014)
28. Sokhi, D.S., Hunter, M.D., Wilkinson, I.D., Woodruff, P.W.: Male and female voices activate distinct regions in the male brain. Neuroimage **27**(3), 572–578 (2005)
29. Re, D.E., O'Connor, J.J., Bennett, P.J., Feinberg, D.R.: Preferences for very low and very high voice pitch in humans. PLoS ONE **7**(3), 32719 (2012)
30. Audacity. https://www.audacityteam.org/. Accessed 25 May 2020
31. Ledalab. http://www.ledalab.de/. Accessed 25 May 2020
32. Benedek, M., Kaernbach, C.: Decomposition of skin conductance data by means of nonnegative deconvolution. Psychophysiology **47**, 647–658 (2010)

# A Color Design System in AR Guide Assembly

Xupeng Cai, Shuxia Wang, Guangyao Xu$^{(\boxtimes)}$, and Weiping He

School of Mechanical Engineering, Northwestern Polytechnical
University, Xi'an, People's Republic of China
2768360933@qq.com

**Abstract.** With the rapid development of human-computer interaction, computer graphics and other technologies, Augmented Reality is widely used in the assembly of large equipment, but in practical applications there are a lot of problems such as unreasonable color design. Based on this, we propose a projection display color design system based on color theory and image processing technology, the system can provide the designer and user with a color design system, solve the problem of color contrast, and realize the distinction of projection interface and projection plane. First of all, using the camera to gather the image of projection plane, and using the image processing to deal with image, so as to determine the color of the projection plane, then based on the color model and the theory of color contrast to choose the appropriate projection interface's color, to ensure that there is a sharp contrast between the projection plane and projection interface, finally, the effectiveness of the proposed system is verified by experiment.

**Keywords:** Color design · Augmented reality · Image processing · Color model first section

## 1  Introduction

Augmented Reality (AR) is an important branch of Virtual Reality (VR) technology [1] and a research hotspot in recent years. The technology can make a virtual object that computer graphics created place in the reality world precisely, and presented in the eyes of users with visualization technology, it can give users a sense of reality and make users feel immersive that has been widely used in maintenance, assembly and other aspects. The principle is shown in Fig. 1.

Tracking and positioning technology is the basis of Augmented reality technology. According to the real-time tracking and precise positioning can carry out the fusion of virtual objects and the real world. Because of the different ways of realization, common tracking and positioning technologies can be divided into three kinds, include the sensor-based, based on machine vision and based on hybrid tracking and positioning. Sensor-based tracking and positioning completely depends on the kinds of sensors, in general, the placement positions of sensors are relatively fixed and the universality is weak. The tracking and positioning based on machine vision is mainly to obtain the images in real time by camera, and to track the feature points of objects by image processing technology, so as to realize the tracking of objects. Hybrid tracking and positioning is

© Springer Nature Switzerland AG 2020
C. Stephanidis et al. (Eds.): HCII 2020, LNCS 12428, pp. 26–39, 2020.
https://doi.org/10.1007/978-3-030-59990-4_3

to obtain the direction of the object that tracked through sensors such as inertia sensors and gravity tilt sensors, and to obtain the position of the object through machine vision, so as to realize positioning and tracking, which will be applied more in the future [2].

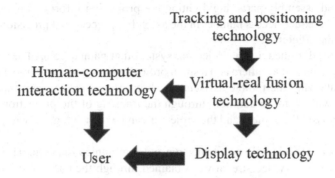

**Fig. 1.** The principle of augmented reality

Display technology is the core of Augmented reality technology. When it is clearly and accurately displayed in front of users that can achieve its the maximum value. At present, most of displays are presented to the user through the wearable helmet. However, due to the large volume and certain weight of this head-mounted display device, and the fact that it is worn on the head, prolonged wearing will aggravate the fatigue of users. With the rapid development of projector display technology, more and more scholars focus on the projection display, thus producing projection augmented reality. All kinds of virtual information are projected in front of the user through the projector, which greatly reduces the user's physical burden, and has been successfully applied in the assembly and maintenance of large equipment such as aircraft and ships.

Servan J [3] using the laser projector to establish the airbus A400M assembly guidance system, as shown in Fig. 2, in the computer generated virtual aircraft assembly line structure, and through the projector projection to the actual assembly scenario, it can avoid workers to refer to paper manual operation frequently, so as to improve the work efficiency, and is verified by experiment, the results show that the assembly guidance system can effectively improve the efficiency of assembly workers.

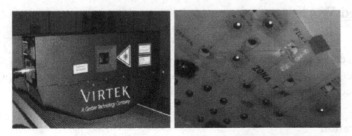

**Fig. 2.** Wing laser projection guidance

American FIILS team applied Augmented reality to fasteners assembly of aircraft in F35, to establish 3D optical projection guidance system, it can display clearly all assembly instruction information on the assembly parts surface, and through the precise tracking and positioning technology that can achieve the perfect fusion of assembly information and assembly parts, highlighting the projection information, components and annotations, reduce access number in the assembly process and precision assembly, thus improve the efficiency.

Gurevich [4] designed a new projection system that named TeleAdviso for remote assistance. It consists of a camera and a small projector mounted on a remote-controlled robotic arm, allowing the remote assistant to observe the workers' current status and communicate with them. In addition, through the tracking of the projection space, the connection between the camera and the projector can be realized, greatly improving the availability.

Matt Adcock [5] designed a new projection and communication system by using laser projector and camera. Video stream was obtained through the camera and transmitted to the remote computer, enabling remote experts to observe the actual working scene of workers. Through the miniature projector displays the annotations of the experts and get the good results.

Ashish Doshi [6] applied the projection augmented reality to the automotive industry in the operation of manual welding, through two projector displayed the position of spot welding accurately, and verified its effectiveness in GM Holden, the results showed that when we used the projection augmented reality, the standard deviation of artificial spot welding was reduced by 52%, it also showed the practicability of the projection augmented reality.

But there are also some problems with the above research, the main problem is the projection interface design, only simple displays the assembly information, no professional interface design, especially in color design, resulted in the contrast deficiencies between projection interface and projection plane that affects the applications of projection augmented reality system. Therefore, this paper proposes a projection interface color design method, so as to improve the practicability of projection augmented reality system.

The chapter arrangement of this paper is as follows: the first part is related background introduction, the second part is related work, the third part is system design, the fourth part is user experiment, and the last part is conclusion and future work.

## 2   Related Works

Since projectors and cameras are needed in the projection augmented reality system, it is necessary to calibrate the projector camera system before use, conduct color correction of the projector and understand the properties of color, so as to provide support for the subsequent color design. The details are as follows:

### 2.1   Calibration

The purpose of calibration is to obtain the internal parameter matrix and external parameter matrix of the projector and camera, and the internal parameter matrix represents its internal fixed parameters, the external parameter matrix represents its spatial position information. In general, the internal parameter only needs to be calibrated once, the external parameter is often calibrated due to the change of position. Specific steps of calibration are as follows:

Step1: print a checkerboard and stick it on a plane board as the calibration board.
Step2: adjust the direction of the calibration board to obtain images in different directions.
Step3: extract checkerboard corner points from the images.
Step4: estimate the internal parameter matrix and external parameter matrix without distortion.
Step5: the least square method is applied to estimate the distortion coefficient in case of actual radial distortion.
Step6: according to the maximum likelihood method, the results are optimized to improve the accuracy, so as to obtain internal parameters matrix and external parameters matrix.

## 2.2 Projector Color Correction

In projection augmented reality system, because of the influence of projection plane's color, texture and light environment, the projected image's color usually distorted and lead to the distortion of camera image, it can cause great influence to images and influence the later image processing, so we need to study the projector camera photometric model, through the color correction to eliminate the color distortion.

Nayar [7] proposed a complete projector photometric camera model, based on this, Fujii and Harish simplified and optimized the projector photometric model, and established a new projector camera photometric model, which was represented by a matrix

$$C = A(F + VP) \tag{1}$$

In the above equation, the matrix $C$ is the image color information of camera capture, the matrix $P$ is the image color information of projection, the matrix $F$ is external environment light, the matrix $A$ is the surface albedo of projection plane, the matrix $V$ is color mixed matrix that represents the mutual interference between the three channels in a pixel, such as $V_{BR}$ is the influence of the blue channel to red channel, its size represents the strength of the interference, the matrix form is as follows:

$$C = \begin{bmatrix} C_R \\ C_G \\ C_B \end{bmatrix}, P = \begin{bmatrix} P_R \\ P_G \\ P_B \end{bmatrix}, F = \begin{bmatrix} F_R \\ F_G \\ F_B \end{bmatrix} \tag{2}$$

$$A = \begin{bmatrix} A_R & 0 & 0 \\ 0 & A_G & 0 \\ 0 & 0 & A_B \end{bmatrix}, V = \begin{bmatrix} V_{RR} & V_{GR} & V_{BR} \\ V_{RG} & V_{GG} & V_{BG} \\ V_{RB} & V_{GB} & V_{BB} \end{bmatrix} \tag{3}$$

By projection black, red, green, blue, and any a color image, and use the camera to capture, according to matrix operation, the above parameters are obtained, and achieves the image color correction, the correction results as shown in Fig. 3, figure (a) is the projection plane image that camera captures with concave and convex decorative pattern of the curtain that with a certain degree of fold, figure (b) is the projection image on white wall that captured by the same camera, figure (c) is the uncompensated image, figure (d) is the compensated image, the results show that the compensated camera image is closer to the projected image than uncompensated camera image. However, before and after compensation, the image appears a certain degree of folding phenomenon, mainly caused by the concave and convex material of the curtain itself, which involves the elimination of projection concave and convex texture, has little to do with this paper, so it is not discussed.

(a) projection plane image          (b) projection image

(c) uncompensated image          (d) compensated image

**Fig. 3.** Color collecting results

## 2.3  Color Attributes

Color is the visual perception caused by light waves of different wavelengths. Although there are all kinds of colors, they generally have three attributes, namely hue, saturation and value.

Hue, that is the color of display, and can be expressed more clearly. The common hue is the seven colors of red, orange, yellow, green, blue, indigo and purple obtained by optical dispersion decomposition, while black, white and gray belong to the colorless system with no hue, which mainly depends on the spectrum of human eyes. Different wavelengths represent different hue. Common hue rings include six hue rings, etc., as shown in Fig. 4.

Purity, as the name implies, refers to the degree of color purity, also known as saturation, the higher the purity, the brighter the color, otherwise it is dim, where the ultimate purity is the primary color, when the continuous addition of black, white or other colors, the purity gradually reduced, will become dim, until colorless, as shown in Fig. 5.

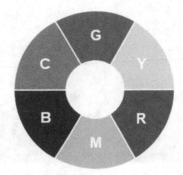

**Fig. 4.** Six hue rings (Color figure online)

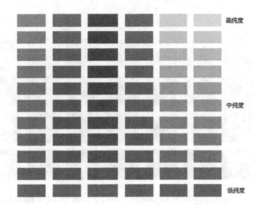

**Fig. 5.** Gradient diagram of purity

Value refers to the bright degree of color, not only depends on the intensity of light source and the influence of the surface albedo. The value generally has two kinds of circumstances, first, different hue has different value, this is the inherent nature of the different hue, generally are not influenced by the light source and the surface of the object, the yellow value is the highest, violet lowest; second, the same hue has different value in different light sources or objects, in general, the value in strong light is higher than that in weak light. The gradient diagram is shown in Fig. 6.

In addition to light by change the value, you can add color to change its value, saturation will change, but the change is not a linear correlation, need to be determined according to specific circumstances, such as in red to add white and black respectively, due to add other color, saturation must have been declining, but on the contrary, the value of the former is rising, which declined value, as shown in Fig. 7.

**Fig. 6.** Gradient diagram of value (Color figure online)

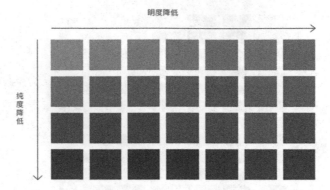

**Fig. 7.** Gradient diagram of saturation and value (Color figure online)

## 3  System Design

Based on the above theory, a projection interface color design method for projection augmented reality is proposed in this paper. The algorithm flow chart is shown in Fig. 8, the main steps can be summarized into four stages, called image acquisition stage, image processing stage, color judgment stage and color selection stage.

### 3.1  Image Acquisition Stage

This stage includes two processes: open camera and acquisitive image, in which the camera is a calibrated camera, and image acquisition is conducted by the function of OpenCV Library. Before camera acquisition, many of parameters such as camera frame rate and image size can be set, then save and display the acquisitive images.

### 3.2  Image Processing Stage

This stage deals with the image that acquired in the image acquisition stage, which is divided into two stages of pretreatment and reprocessing. In the preprocessing stage, we need to perform operations such as color conversion on the original image to achieve the

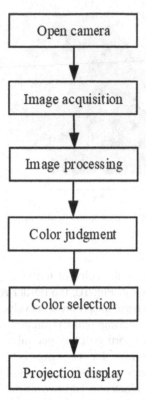

**Fig. 8.** Projection interface color design flow

purpose of noise reduction and make the image more realistic; in the reprocessing stage, the image is decomposed from three channels into a single channel through channel separation, which provides a basis for judgment. Above processing, color conversion is the most important processing, the following detailed introduction.

The images collected by the camera are saved in the RGB color model, but the HSV model can better set the judgment thresholds and obtain accurate judgment results when obtaining the projection plane color through image processing. The following is an introduction to the RGB and HSV color models.

**RGB Color Model**
This model is an additive color model that adds the three color components of red, green, and blue in different proportions. A maximum of each color component is 1, the minimum value is 0, resulting in a variety of colors, commonly used in display devices. With three components as the axis, we can build up a three-dimensional color space, as shown in Fig. 9, in which point A is blue, blue components is 1, red and green component is 0; Point B is magenta, the blue and red components are 1, the green component is 0; Point C is red, the red component is 1, the green and blue components are 0; D is yellow, the green and red component is 1, and the blue component is 0.

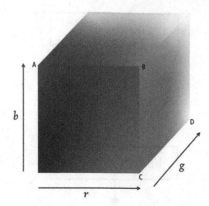

**Fig. 9.** RGB color space (Color figure online)

## HSV Color Model

The model is created according to the color of the visual color space, also known as hexagonal cone model, as shown in Fig. 10, the border represents hue, the horizontal axis represents saturation, the vertical axis represents value, the range of hue is 0°–360°, the hue of red is 0°, in red as the starting point, counterclockwise, green is 120°, blue is 240°; saturation represents the proportion of the spectral color, the higher the proportion, the closer the color is to the spectral color. The range is 0%–100%, the larger the value, the higher the saturation; value represents the brightness of the color, with values ranging from 0% to 100%, where 0% is black and 100% is white.

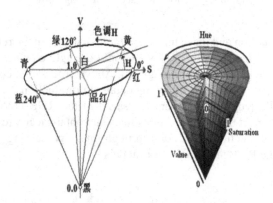

**Fig. 10.** HSV color space (Color figure online)

**Color Conversion**

Assume that the value of a certain color in the RGB color model is $(r, g, b)$, and the value in the corresponding HSV model is $(h, s, v)$, and set $max$ and $min$ is the maximum and minimum of $r, g, b$, then the value can be calculated by the following formula:

$$h = \begin{cases} 0°, & \text{if } max = min \\ 60° \times \frac{g-b}{max-min} + 0°, & \text{if } max = r \text{ and } g \geq b \\ 60° \times \frac{g-b}{max-min} + 360°, & \text{if } max = r \text{ and } g < b \\ 60° \times \frac{b-r}{max-min} + 120°, & \text{if } max = g \\ 60° \times \frac{r-g}{max-min} + 240°, & \text{if } max = b \end{cases} \tag{4}$$

$$s = \begin{cases} 0, & \text{if } max = 0 \\ \frac{max-min}{max}, & \text{otherwise} \end{cases} \tag{5}$$

$$v = max \tag{6}$$

The above formula, the value range of $h$ is 0°–360°, the value range of $s$ is 0–1, the value range of $v$ is 0–1. However, in order to better adapt to various types of data, the above value range is further processed in OpenCV Library, the $h$ fall to 180° from 360°, the value $h$ and $v$ increase to 255, but this does not affect the judgement, we can realize the transformation of color by the above conversion formula.

### 3.3 Color Judgment Stage

This stage is to judge the projection plane image color and pass the judgment result to the color selection stage. According to the values of each component in Table 1, and set the judgment threshold to determine the color of the projection plan image.

**Table 1.** HSV range of common colors

|      | red | | orange | yellow | green | cyan | blue | purple | black | white | gray |
|------|-----|-----|--------|--------|-------|------|------|--------|-------|-------|------|
| hmin | 0 | 155 | 11 | 26 | 35 | 78 | 100 | 125 | 0 | 0 | 0 |
| hmax | 10 | 180 | 25 | 34 | 77 | 99 | 124 | 155 | 180 | 180 | 180 |
| smin | 43 | | 43 | 43 | 43 | 43 | 43 | 43 | 0 | 0 | 0 |
| smax | 255 | | 255 | 255 | 255 | 255 | 255 | 255 | 255 | 30 | 43 |
| vmin | 46 | | 46 | 46 | 46 | 46 | 46 | 46 | 0 | 220 | 46 |
| vmax | 255 | | 255 | 255 | 255 | 255 | 255 | 255 | 46 | 255 | 220 |

According to the data in the Table 1, for each color, the most obvious judgment is based on the $h$ value, that is respective hue, no overlapping range, just from 0°–180°, where attention needs to be paid to the red hue range, it includes two ranges, and the saturation and value range are almost exactly the same without any distinction; for black, white and gray, the biggest judgment basis is respective value, without overlapping range, and their hue range is exactly the same, black saturation range is the maximum, white minimum.

Therefore, when judging the color of the projection interface image, it is necessary to judge according to the following steps.

Step1: judge whether it is a color image or a black and white image
Step2: if it is a black and white image, just judge its value
Step3: if it is a color image, only judge the hue. In order to improve the accuracy of judgment, judge the range of the other two values, so as to accurately judge the color of the image.

### 3.4  Color Selection Stage

This stage is based on the judgment results of the color judgment stage to select the color of the projection interface and display it on the projection plane through the projector, which involves color selection [8].

There are two types of color selection methods in this paper. The first is contrast color selection and projection interface is highlighted, which mainly includes medium difference hue and contrast hue; the second is harmonious color selection [9], which mainly includes selection of same hue and adjacent hue. Among them, contrast color selection is applied to the color of the projection interface, while harmonious color selection is mainly applied to the visual elements on the projection interface.

In terms of contrast color selection, the medium difference hue refers to the color difference of 90° on the hue ring, the contrast between the two is not very strong, not only to ensure a strong contrast, to give a sense of comfort, so in the color selection, the priority level is set as the first; the contrast hue refers to the color difference of 120° on the hue ring, the contrast is relatively strong and gives people a sense of pleasure, its priority is set as the second.

In harmonious color selection, the selection of same hue is to choose in the same color, which makes integral color keep consistent, by adjusting the saturation and value that makes the whole more harmonious and sets priorities for the second; adjacent hue selection refers to adjacent hue in hue ring, there are different angles in different hue ring, for example, in twelve color ring, the angle of adjacent hue is 30°, in twenty-four hue ring, the angle is 15°, the adjacent hue can make whole more colorful, the priority level is set as the first. Through the harmonious color selection that makes more harmonious visual elements displayed in the projection interface.

## 4  User Study

### 4.1  Experiment Design

In order to verify the feasibility and comfort of the system, we look for 20 volunteers for our experiment. The proportion of men and women is 1:1, and the age is between 20 and 30 years, there are no people with color blindness and no exposure to similar systems.

Each volunteer needs to be done two times assembly task, using the traditional projection augmented reality system for the first time, that is not used in this paper's color design system, the second time to use the new projection augmented reality system, which uses the paper's color design system, every time task contains two assembly tasks, the biggest difference between the two assembly task is the projection plane color, the rest is same, the first assembly task projection plane is white, the second assembly task of projection plane is blue. In order to reduce the influence of learning effect, it is necessary to disrupt the experiment order of volunteers. If the first volunteer uses the traditional system for experiment, the second volunteer needs to use the new system first, and so on, until all volunteers complete the experiment.

Before the experiment, the assembly tasks are explained to the volunteers to minimize the impact of insufficient familiarity on the experimental results. The assembly tasks are to assemble two kinds fasteners of different sizes and quantities on the aircraft panel. Recording the assembly time and error rate in the process of experiment, at the end of the experiment, we require each volunteer to fill in the questionnaire, which uses Likert scale, as shown in Table 2, where each question contains five options, respectively is very disagree, disagree, neutral, agree, very agree, the corresponding scores respectively 1, 2, 3, 4 and 5 points.

**Table 2.** Survey questions

| Number | Question |
| --- | --- |
| 1 | You like projection augmented reality system |
| 2 | Projection augmented reality system is easy |
| 3 | You can easily complete the assembly task |
| 4 | The interaction is very convenient |
| 5 | New projection augmented reality system is more comfortable |
| 6 | You prefer the new projection augmented reality system |

### 4.2 Results Analysis

Efficiency is the requirement of assembly, which is mainly reflected by time and accuracy. In this paper, the assembly time and error rate of each experimenter are recorded in the experiment, and the statistical results are shown in Table 3.

As can be seen from Table 3, the traditional system needs 150.65 s with an error rate of 8.13%, while the new system only needs 120.43 s with an error rate of 1.65%. The main reason is that the traditional system does not carry out color design, in the second assembly task, because the color of the projection plane and the projection interface are very close, the recognition degree is reduced, which leads to longer assembly time and higher assembly error rate.

**Table 3.** Assembly time and error rate

| Projection augmented reality system | Assembly time (s) | Error rate (%) |
|---|---|---|
| Traditional | 150.65 s | 8.13% |
| New | 120.43 s | 1.65% |

Finally, qualitative statistical analysis is conducted through questionnaires after the experiment, and the results are shown in Fig. 11.

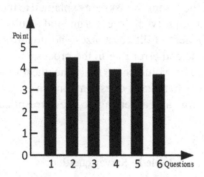

**Fig. 11.** The statistical results of survey questions

As can be seen from Fig. 11, the recognition degree of the new projection augmented reality system is very high. From the statistical results of questions 1, 2 and 3, it can be clearly seen that projection augmented reality system is very popular, not only because it is easy to learn, but also because it can help workers to complete tasks faster and better. The statistical results of question 4 show that users do not find it difficult to interact, indicating that the interaction mode of projection augmented reality system is reasonable and effective. The results of questions 5 and 6 show the superiority of this new system. Compared with the traditional projection augmented reality system, the new projection augmented reality system is more comfortable and favored by users.

## 5 Conclusion and Future Work

For more than two decades, Augmented reality has got an increasing number of attentions in many fields especially the manufacturing. Lots of scholars confirm that the projection Augmented reality is better than traditional Augmented reality.

In this paper, we propose a projection display color design system which can provide the designer and user with a color design system, and solve the problem of color contrast, and realize the distinction of projection interface and projection plane. Through the analysis of the experimental results, we can draw the conclusion that the color design system in this paper can serve the projection augmented reality system well, it can improve the assembly efficiency, and improve the practicability and applicability of the projection augmented reality system.

There are some problems in this paper that we will solve to improve the efficiency and accuracy of system it in the future as follow:

(1) Improve the robustness of image processing. At present, we only achieve the simple color judgment, and will explore more other ways to improve the judgment accuracy.

(2) Create a color selection model. Now, only realize the qualitative selection according to the color contrast theory, in the future, the algorithm will be further optimized.

**Acknowledgement.** We would especially like to thank volunteers and related scholars for their special contribution to this research. Besides we would also like to appreciate the anonymous reviewers for their invaluable criticism and advice for further improving this paper.

**Funding Information.** This research is partly sponsored by the civil aircraft special project (MJZ-2017-G73), National Key R&D Program of China (Grant No. 2019YFB1703800), Natural Science Basic Research Plan in Shaanxi Province of China (Grant No. 2016JM6054), the Programme of Introducing Talents of Discipline to Universities (111 Project), China (Grant No. B13044).

# References

1. Milgram, P., Kishino, F.: A taxonomy of mixed reality visual displays. IEICE Trans. Inf. Syst. **77**(12), 1321–1329 (1994)
2. Azuma, R., Leonard, J., Neely, H., et al.: Performance analysis of an outdoor augmented reality tracking system that relies upon a few mobile beacons. In: IEEE/ACM International Symposium on Mixed & Augmented Reality (2006)
3. Serván, J., Mas, F., Menéndez, J., et al.: Using augmented reality in AIRBUS A400M shop floor assembly work instructions. In: AIP Conference Proceedings, pp. 633–640 (2012)
4. Gurevich, P., Lanir, J., Cohen, B., et al.: TeleAdvisor: a versatile augmented reality tool for remote assistance (2012)
5. Adcock, M., Gunn, C.: Using projected light for mobile remote guidance. Comput. Support. Coop. Work **24**(6), 1–21 (2015)
6. Doshi, A., Smith, R.T., Thomas, B.H., et al.: Use of projector based augmented reality to improve manual spot-welding precision and accuracy for automotive manufacturing. Int. J. Adv. Manuf. Technol. **89**, 1–15 (2017)
7. Nayar, S.K., Peri, H., Grossberg, M.D., et al.: A projection system with radiometric compensation for screen imperfections. In: ICCV Workshop on Projector-Camera Systems (PROCAMS) (2003)
8. Khan, R.Q., Khan, W.Q., Kaleem, M., et al.: A proposed model of a color harmonizer application. In: 2013 5th International Conference on Information and Communication Technologies, pp. 1–7 (2013)
9. Fei, L., Lan, L.: Brief analysis about the color sequence harmony principle applying in product form design. In: 2009 IEEE 10th International Conference on Computer-Aided Industrial Design & Conceptual Design, pp. 191–193 (2009)

# An Augmented Reality Command and Control Sand Table Visualization of the User Interface Prototyping Toolkit (UIPT)

Bryan Croft$^{(\boxtimes)}$, Jeffrey D. Clarkson$^{(\boxtimes)}$, Eric Voncolln$^{(\boxtimes)}$, Alex Campos$^{(\boxtimes)}$, Scott Patten$^{(\boxtimes)}$, and Richard Roots$^{(\boxtimes)}$

Naval Information Warfare Center Pacific, San Diego, CA, USA
{bryan.croft,jeff.clarkson,eric.voncolln,andres.a.campos,
scott.a.patten,richard.roots}@navy.mil

**Abstract.** The User Interface Prototyping Toolkit (UIPT) constitutes a software application for the design and development of futuristic stylized user interfaces (UIs) for collaborative exploration and validation with regards to decision making and situational awareness by end users. The UIPT toolkit effort is targeted to Command, Control, Communications, Computers, Intelligence, Surveillance and Reconnaissance (C4ISR) and Cyber missions for the U.S. Navy. UIPT enables the Navy to significantly reduce the risks associated with pursuing revolutionary technology such as autonomous vehicles, artificial intelligence and distributed sensing in terms of the design of the user interface. With this in mind, augmented reality is examined to determine if this technology can be useful to the Navy's mission in supporting the warfighter in the future vision of the Navy. This work examines the development and evaluation of an augmented reality virtual sand table presentation overlaid over the same set of information presented in the UIPT displayed on a large touch table. This new way to present the same information allows for a unique collaborative setting to be evaluated as a potential future user interface for the Navy. This is applied in the context where a group of warfighters gather in collaborative virtual space for decision making and situational awareness. The development and experimentation with an augmented reality interface initiates the means to validate a futuristic Navy vision of user interfaces that support the warfighter in a Command and Control environment.

**Keywords:** Command and Control · Command Center of the Future · Information systems · Naval Innovative Science and Engineering · Rapid Prototyping · User interfaces · User Interface Rapid Prototyping Toolkit · Unity3D

## 1 Introduction

A Naval Innovative Science and Engineering (NISE) project titled User Interface Prototyping Toolkit (UIPT) and its follow-on project Information Warfare Commander (IWC) continue to explore the rapid production of high-fidelity user interfaces (UIs) prototypes

C. Stephanidis et al. (Eds.): HCII 2020, LNCS 12428, pp. 40–53, 2020.
https://doi.org/10.1007/978-3-030-59990-4_4

to be utilized by end users, researchers, designers, and engineers working together to develop and validate, new concepts of operations (CONOPS) and emerging technologies. The IWC project was targeted to include exploration and integration of technologies involved with a future vision and function of Naval Command and Control spaces with emphasis on the human-computer interfaces and interaction.

The objective is to examine the ability to rapidly develop a specific human-machine interface prototype for future Navy information systems that address emerging operational and tactical threats and is currently targeting advanced UI designs specifically for a large multitouch display device as well as the exploration of other forms of UIs. UIPT/IWC is a software development effort to explore rapid prototyping of information systems while standardizing forward looking UI interactions, such as multitouch and Augmented Reality (AR). One goal of the project was to examine the combined use of the touch table interface with the AR interface appearing virtually above the same virtual environment. The full benefits of AR is still an unknown with regards to user experience.

## 1.1 Motivation

The US Navy is focused on maintaining a decisive advantage over adversarial forces. The Warfighter as an individual end user and as collective end users plays a critical role in maintaining this advantage. New technologies such as Artificial Intelligence, Machine Learning, and Autonomous Systems grow in importance and applicability. How the user interfaces with these technologies becomes critical, especially in a fast-paced environment where situational awareness and decision making are out pacing the human's ability to not only comprehend but to understand, formulate and then act in the most optimal way possible. The UI between the end user and advanced capabilities such as ambient intelligence, autonomous systems and the overwhelming influx of data from intelligence sources and sensors, is positioned as a key element that allows the end users awareness, understanding and control with regards to these complex systems. How end users will be able to utilized such interfaces and maximize the man-machine performance and relationship is still largely unknown.

The lack of User Experience (UX) design in software development in the past has resulted in software user interfaces which are non-intuitive, difficult to use, non-task oriented, and have a general lack of form and function required for knowledge and decision support end use. The goal for the UI in the UIPT/IWC project was to overcome this trend and provide an easy to use and valid interface for the Warfighters. UI prototypes which focus on innovative concepts continue to progress through the efforts of processes such as User-Centered Design (UCD) [1]. For the domain of interest, it becomes more than just the User Experience (UX). It is the timely and well-designed placement and interaction of information on the UI that can better support the critical functions of situational awareness, decision making and control of the supporting systems. The exploration for the use of AR must follow a similar path for it to become an interface of the future.

The application of AR to the UIPT/IWC project is still in its infancy and even more so when it comes to actual military end users making use of it in operational and tactical situations. The Navy, like many other organizations, has expressed interest in AR as a technology for end use; however, the practical application remains difficult. The current size, weight, fitment and calibration of AR head hear is an operational barrier, though these obstacles are will become less of an issue as technology matures. Current motivation in the evaluation of AR is directed more toward the actual end use for visualization, situational awareness, and scene understanding especially within a collaborative environment shared with multiple end users. AR allows individually tailored perspectives of the interface, along with the option to provide remote collaboration through virtual avatars. This provides some unique features not present in the current common flat display user interface.

## 1.2  Technical Background

The UIPT/IWC projects are based on a technical and design centered approach from years of experience developing user interfaces for the Navy. Previous examples include the Multi-modal Watch Station [2, 3] (see Fig. 1) from the 1990s, a touch-screen application which was user task centric and included a task management system in support of the end users. MMWS provided the concept of man power reduction through a 3 to 1 increase in

**Fig. 1.** Multi-Modal Watch Station (MMWS), 1990s

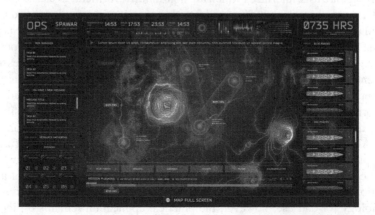

**Fig. 2.** Advanced user interface prototype design, 2015

efficient tasking. From then to today, a 55" multi-touch workstations [4–6] (see Fig. 2) for the NIWC Pacific Command Center of the Future VIP demonstration room has demonstrated a technological increase in the design and implementation of UIs.

NIWC Pacific programs involving mission planning, task-managed designs, decision support aids and visualization have provided a wealth of experience in design and development of user interfaces. This included many user engagements and knowledge elicitation sessions to validate UI/UX concepts and requirements, all of which provided the foundational basis for the UIPT/IWC interface. This knowledge, experience and examples provide sound principles which can be applied to an AR based interface.

UI design and development has found greater emphasis during the last decade plus [1, 2] with changes to UI by such companies as Apple, Google and a variety of new comers. While current AR technologies do not have the display quality and capability such as found in a 55" touch table display [4] these AR based technologies in terms of visualization quality are well underway to becoming sharp and vibrant displays of their own accord. Some of the current research in AR [7–10] has incorporated technologies such as ray tracing that allow AR based virtual objects to be almost indistinguishable from real objects within the AR view. While this is not the intention of AR technology for Warfighter uses, owing to the fact that the augmented objects must be easily distinguishable, it does demonstrate that these displays will at some point provide high quality graphics which can then appropriately integrated for a wide variety of uses.

This effort grew out of a mockup demonstration of a future Navy information system as depicted in a NIWC Pacific Vision video, produced in 2015. NIWC Pacific contracted a multi-touch display manufacturer to work with NIWC designers and subject matter experts to design and manufacture Touch Table hardware and code the mockup application. This mockup (see Fig. 3) was specifically developed for demonstrations to DoD VIP visitors striving to understand the future direction of C4ISR for the U.S. Navy. The initial and current concept for the AR interface is to have it overlaid on top of the UIPT/IWC interface. This is an interesting concept which in of itself covers some interesting human factors questions about overlaying AR visualization on top of a more conventional interface and yet have them interplay in unison.

**Fig. 3.** Touch table used by UIPT

### 1.3  User Interface Display Hardware

Early exploration for AR prior to and during development of UIPT/IWC project was based off of the Microsoft HoloLens device. For this AR interface connections were made to Command and Control (C2) systems with the intent to display data from the C2 system in a virtual sand table and be able to not only view this information but send commands through UI elements back to the C2 system. This AR interface was additionally implemented for collaboration which providing control on classified information between users of the system. This allowed users of different classification levels to collaborate and interact in the same virtual space but were able to visualize information that pertained to their classification but have it filtered from the view of others that do not share the same classification level. This provided an initial foundation of the promise of AR both for visualization and collaborative efforts.

This work will not address the comparison or utility of certain AR headsets over others. This has been subjective with a variety of undefined or unscientific experiments. The general assumption is that AR devices will improve in time as technology improves and better understanding and data from user experience with AR equipment will pave the path going forward.

### 1.4  Interfacing to Information Systems

For the case of UIPT/IWC project the emphasis is placed on the user interface rather than the backend components and thus as a working prototype the data is often supported in the form or hardcoded entries in spreadsheets or as comma separated values. These files are read into support information to be presented as well as timed events to illustrate actively changing scenarios. This ability has been improved through the integration and connection to the Next Generation Threat System (NGTS) [11]. The NGTS system is a synthetic environment generator that provides models of threat and friendly vehicles such as aircraft, ground vehicles, ships, submarines, weapon systems, sensors and other elements commonly found in military systems. Both the simulator and the UIPT/IWC act together as the source for the AR system. The AR system provides a one-to-one matching of the information sets integrated with UIPT/IWC and NGTS.

## 2  Objectives

The UIPT/IWC application allows users to explore the impact of new technology on emerging operational concepts before large program investments are made, the Navy can significantly reduce the risks associated with pursuing revolutionary technology. The Navy is currently investing in new technologies, such as, autonomous vehicles, artificial intelligence, distributed sensing and integrated fires, but the way in which Fleet users will work with and "team" with these new capabilities is largely unknown. Included in this is the use of new human-computer interface devices and methods such as AR.

The objective of the addition of an AR component to the UIPT/IWC project is to provide users, researchers, designers and developers with the means to explore, analyze, develop and test new CONOPs and emerging technology that maximize the human interaction efficiency especially in the realm of collaboration. Currently, there are few, if any, AR tools dedicated to the examination of these specific futuristic CONOPs and technology and the effects on C4ISR decision making through the user interface. The AR component will enable the acceleration, experimentation and evaluation of these concepts through implementation, testing and end use. Prototyping and testing before major investment will save money as Fleet users will already have been informed of and validated the concepts of AR applied to future C2 systems.

### 2.1 Development of the Augmented Reality Interface Overlay

The effort to integrate AR into the UIPT/IWC software application was a collaborative effort amongst several technology groups at NIWC Pacific. The primary effort was a collaboration between the UIPT/IWC and Battlespace Exploration of Mixed Reality (BEMR) teams. This permitted the quick development of UIs to collaboratively explore and validate new CONOPS in AR and the impact on the end user decision making process. Initial development involved the use of the Magic Leap One AR device. The data shared between UIPT/IWC and the AR device included touch table view orientation (position, rotation and camera zoom level) in addition to the position and heading of NGTS air, surface and subsurface tracks.

Critical for this particular development effort and the primary objective was to determine if an AR view could be overlaid on top of an existing touch table view. The AR view would be offset from the touch table view but would align in position and stay aligned with movement of display elements in the touch table display. Size, position, orientation and other factors had to be synchronized correctly and efficiently across two independent and different but shared views. This primary effort then allows a human-computer interface study to determine if this particular technique can be validated for effectiveness under operational use. This turned out to be the main emphasis of the development effort.

### 2.2 Employment of the Toolkit

The AR component of the UIPT/IWC project is still an initial prototype and requires additional development effort to be able to run human evaluation experiments. Preliminary judgements can be made to see if further development and experimentation is warranted. The preliminary conclusion is that continuation down this path is valuable and shows promise as a collaborative environment for improved situational awareness and decision making for the Naval Warfighter end user.

## 3 Approach – Design and Development of an AR Overlay

The technical approach to the AR component of the UIPT/IWC project has two main goals, first the examination of a novel user interface in the exploration of future Navy human-machine interface capabilities, and second the development of a creation of a AR prototype application which can be used collaboratively with other non-AR users of a touch table display of the equivalent information set. Both goals are required to meet the objective of the UIPT/IWC project. Designers and software developers work hand in hand in an agile development process with weekly evaluations of the AR component for both form and functionality.

### 3.1 User Interface Design

The AR component of the UIPT/IWC project utilized a simple user interface design by creating an overlay of the same air, surface and subsurface track data directly above the touch table interface. The elevation of air tracks and subsurface tracks were exaggerated to provide more awareness of the depth of the battlespace. Each air and subsurface AR tack renders a light pillar that anchors it to the paired position of the UIPT/IWC touch table interface. AR headset images from the current prototype have been captured that illustrate the concept. In Fig. 4 the real-world view through the AR devices shows both the surrounding area as well as the outer edge of the touch table device and its display along with several AR tracks overlaid on top of the real-world view.

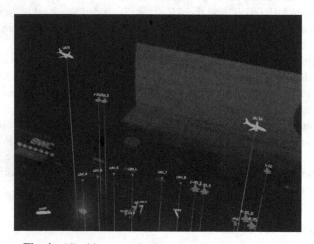

**Fig. 4.** AR objects overlaid atop touch table application

Figure 5 illustrates the use of billboard-based data tags where AR objects are provided with names of the entities. Colorized entities also represent the type of entities presented as either friend, foe or neutral. Note: alignment of AR content is shifted from the UIPT/IWC interface only in the image capture due to the relative position of the recording camera on the device.

**Fig. 5.** AR objects with data tags indicating type of object

Figure 6 demonstrates the correlation and synchronization of the UIPT/IWC display objects with those of the AR prototype objects. Leader lines are drawn down from AR entities to those of touch table entities. This illustrates the position, orientation and correlation between the two user interfaces but provides an AR based visualization of the same dataset. The AR view is synchronized to the touch table view, pan, rotation and zoom actions.

**Fig. 6.** AR objects with leader lines corresponding to touch table object locations

Figure 7 shows the AR virtual terrain occluding the touch table UIPT/IWC's side menus. This depicts one of the potential challenges with the AR overlaid concept which is that the table top menu is seen through the virtual AR terrain.

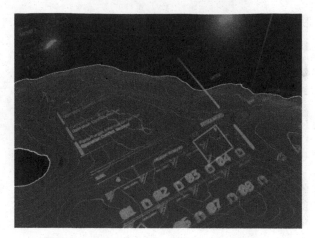

**Fig. 7.** AR image occluding touch table menu

## 3.2 Experimentation Plans and Metrics

The AR component of UIPT/IWC is still in early stages of development and is essentially a prototype application. No formal or rigorous experimentation efforts have been performed. Only internal evaluations have been formulated from which indicate that further development is warranted, which would allow UX sessions with Naval Warfighter feedback. Common technical parameters to be examined in the future will be divided into prototype usability and target concept utility [11]. For prototype usability a combination of the following metrics will be collected during testing as seen in Table 1.

**Table 1.** Usability metrics

| Metric | Description |
|---|---|
| 1. Successful task completion: | Each scenario requires the user to obtain specific data that would be used in a typical task. The scenario is successfully completed when the user indicates they have found the answer or completed the task goal. In some cases, users will be given multiple-choice questions |
| 2. Error-free rate: | Error-free rate is the percentage of user who complete the task without any errors (critical or non-critical errors) |
| 3. Time on task: | The amount of time it takes the user to complete the task |
| 4. Subjective measures: | Ratings for satisfaction, ease of use, ease of finding information, etc. where users rate the measure on a 5 to 7-point Likert scale |
| 5. Likes, dislikes and recommendations: | Users provide what they liked most about the prototype, what they liked least about the site, and recommendations for improving the prototype |

For target concept utility, we will focus on Situational Awareness during collaboration. The final metrics have not been determined at this time, however a modified Situation Awareness Global Assessment Technique (SAGAT) [12] could be used to collect metrics during the course of testing seen in Table 2.

**Table 2.** Utility metrics

| Metric | Technical parameter |
| --- | --- |
| 1. Perception | Object detection time sec |
| 2. Comprehension | Object recognition time (sec) and quality (scale) |
| 3. Projection | Object intention recognition time (sec) and quality (scale) |

### 3.3   The AR Component Network Development

The UIPT/IWC project is built and developed using Unity [12]. Unity is a cross-platform game engine produced by Unity Technologies which is generally used for 2D and 3D game development and simulations. The use of Unity for advanced display systems development was successfully demonstrated as a means to build C2 system UIs. The ability to integrate AR was also an integrated feature of Unity and third-party tools for Unity. Such a third-party tool was the Photon Unity Networking 2 (PUN 2) plugin. This permitted the creation of a broadcast server that could broadcast the UIPT/IWC and NGTS data across a Local Area Network (LAN) instead of having to go through the Photon public servers. This was useful because the receipt of networked data could be provided offline and could easily be integrated with Unity and the Magic Leap One device. This additionally allowed the project to be developed in a standalone unclassified environment and easily be integrated into a classified environment.

Photon is a high-level set of libraries and functions that allow developers to easily convert Unity game objects and prefabrications (prefabs) into objects which can send data over the network to another instance of a Unity build. This was the manner in which key pieces of data was transferred to the companion Magic Leap version of the software to allow the content to be rendered in the AR content correctly. Color coding was utilized to differentiate objects in the scene, commonly referred to as tracks, to be displayed according to type along with a data tag indicating the track name.

Photon uses the lobby room convention found in many online games however, this was modified so that the connection would be both immediate and automatic. This allowed any Magic Leap device to connect and join. In this networked online game concept the first user to enter the lobby room would become the host for subsequent users that join. This provides a frame of reference for subsequent AR users. Built in is the capability that if the host should leave or be disconnected then a randomly chosen current user is then selected as the new host. This preserves continuity of the experience and removal of reliance on the just the first device to join the lobby room.

The main component which is used to send data over the network is an objected called PhotonView. This can be used to send the data of game objects over the network.

This is done by attaching the scripts that contain the data to the list of elements being observed by the PhotonView or, by taking it directly from components like the Transform of a game object which holds the position, rotation and scale. Additionally, method calls can be made over the network using Remote Procedure Calls (RPCs) which can be used in sending chat messages or updating values observed by the game object PhotonView. RPCs however, are more commonly used for communicating infrequent events whereas a OnPhotonSerializeView method call is more appropriate to use. OnPhotonSerializeView allows the opening of a predefined data stream which serialized data is passed from host to client and vice versa. This happens by encoding high level Unity C# code and translating it to a binary file which can then be easily parsed and sent over the network using a PhotonStream class from PUN 2. On the other side this binary file can then be un-serialized, read and reconstructed into the C# code.

### 3.4   The AR Component Camera Synchronization Development

Track data synchronization was the first task which was followed by a more difficult problem of synchronizing the camera between the Magic Leap One device view and the camera view of the UIPT/IWC software on the touch table device. Essentially the way that the camera works for each of these two components is very different. With the Magic Leap One device mounted on an end user's head, the camera view is essentially aligned with the view of the end user. For the UIPT/IWC application the camera is a game object which is independent of anything that can be panned, which equates to movement on a two-dimensional plane in which tracks are laid out, rotated in a line orthogonal to the camera axis of rotation, and zoomed in and out. This equates to moving along a line orthogonal to the camera plane. Furthermore, Unity cameras have two mode; perspective mode which mimics human vision in the sense that objects that are further away will decrease in size and conversely increase in size as they come closer to the camera view. UIPT/IWC application content is displayed in camera perspective view.

In order to emulate the camera behavior in Magic Leap One, the size of the virtual world needed to be scaled so the elements in the AR view would be in the same position, orientation and scale so that they correspond exactly to the UIPT touch table elements upon which they are superimposed. This was accomplished by a using a scaling factor which was calculated to convert from the scale of 1 km to 1 Unity unit of measure in the UIPT/IWC version and 1 m to 1 Unity unit of measure in the Magic Leap AR version. A simple scaling of the map by a factor of 1000 accomplished this. This however, was not the only problem. Magic Leap One view required the UIPT touch table view be in orthographic camera mode to prevent the depth distortion of a perspective view. To accomplish matching views, scaling and switching from perspective to orthographic view, a trial and error process occurred to find the best scaling factor to achieve this. This was set as $0.613771151640216 * OrthographicSize - 1.0032187899731$. OrthographicSize is the current size of the orthographic camera in Unity. The scaling equation is dependent on the real-world size of the touch table display. In future work, an in-application adjustment feature would be helpful to assist with scale calibration.

### 3.5  The AR Component Camera Synchronization Development

With the track data synchronized the next challenge was to have only one touch table as the host as the AR Magic Leap One host and any other touch table would be considered a just standalone hosts that do not share the synchronized camera data. The problem of interest occurs when the touch table is initially the host and then gets disconnected for any reason. The Magic Leap One device then becomes the new host and when the touch table is back up online everything would have to be reset. The solution to this problem was to create an additional widget. This widget is placed in the top right corner of the visor view just out of sight. The widget monitors the host connection such that when it becomes the host because the touch table has disconnected it will then disconnect the Magic Leap One devices and wait with periodic checking to see if the touch table was reconnected and serving as the host. It would then reconnect as the client.

### 3.6  The AR Component Utilizing Simulation Data

More realistic real-world data was simulated using NGTS to generate the scenario that would reflect the desired events. NGTS as a modeling and simulation enterprise tool for Warfighters is used to create scenarios where tracks can be positioned anywhere in the world using the Distributed Interactive Simulation (DIS) version 6 standard protocol to create Protocol Data Units (PDUs) to be exchanged over the network. In NGTS there is also the option to record data and then store it in a chapter10 file. This allows playback of the same scenario which is useful for repeated demonstrations. The receipt of DIS PDUs in UIPT/IWC was achieved via a third party took called Architecture Development Integration Environment (AMIE). This provided a middleware solution to open a connection to NGTS, capture DIS PDU data and then generate from it a data structure usable in Unity to create the actual entities.

### 3.7  The AR Component Graphics and Computational Load Reduction

Since AR devices have limitations in both graphics and computations the 3D model set that was used in UIPT/IWC were too high in the number of polygons used in the models and would needlessly use too much of the resources in the Magic Leap One device an thus hinder the rendered result. The solution was to obtain a set of reduced resolution models and determine new methods for shading operations. A mesh decimation algorithm using Euler operations of vertex clustering, vertex deletion, edge collapsing and half end collapsing were utilized. This provided adequate 3D model representation while allowing the Magic Leap One to not be hindered by expensive resources.

### 3.8  The AR Component Entity Data Tags

An additional display factor was considered for the AR view. Data tags for each entity was required. A data tag is a simple text object that is presented next to the object of interest. The text data is obtained over the networks which is typically the call sign or name of the entity but can include amplifying information. There was a need to have this data tag continue to be placed next to the entity it belongs to as well as face towards the view of the end user of the AR device. With AR the user is free to walk around the sand table display and view it from different angles. In the UIPT/IWC case the AR user can walk around the physical touch table while viewing the scene from different angles but still have the data tags aligned to the user's current view.

## 4   Conclusions and Future Work

The UIPT/IWC project has been in development for about two years while the effort to explore the use of AR for Naval Warfighter applications has just began. Internal evaluation of the prototype to date provide good indicators that further development followed by human-computer evaluation from Navy Warfighters is warranted. The UIPT/IWC project continues to find applicable use in multiple domains. The question still remains if AR technology poses significant advantages in future human-computer interfaces that allow a unique collaboration environment for improved situational awareness and decision making? The question needs to be addressed at least to the determination of its value as well as the proper end use. Future work should continue down the development path until an adequate application prototype exists that will permit the desired end user feedback and thus validate the need to move forward or not move forward with AR technologies for future Navy human-computer interfaces.

## References

1. Van Orden, K.F., Gutzwiller, R.S.: Making user centered design a cornerstone for navy systems. In: Proceedings, vol. 144, pp. 10–1388, October 2018
2. Campbell, N., Osga, G., Kellmeyer, D., Lulue, D., Williams, E.: A human-computer interface vision for naval transformation. In: Space and Naval Warfare Center Technical Document, vol. 3183, June 2013
3. Osga, G.: Building goal-explicit work interface systems. In: Human Systems Integration Symposium, Tysons Corner, VA, June 2003
4. Goldman, K.H., Gonzalez, J.: Open exhibits multitouch table use finding. In: Ideum Open Exhibits Papers, July 2014
5. Hinckley, K., Guimbretiere, F., Agrawala, G., Apitz, G., Chen, N..: Phrasing techniques for multi-stroke selection gestures. In: Ideum Open Exhibits Papers (2006)
6. Schoning. F., Steomocle. F., Kruger, A., Hinrichs, K.: Interscopic multi-touch surfaces: using bimanual interactive for intuitive manipulation of spatial data. In: Ideum Open Exhibits Papers (2009)
7. Cossairt, O., Nayar, S., Ramamoorthi, R.: Light field transfer: global illumination between real and synthetic objects. ACM Trans. Graph. **27**(3), 1–6 (2008). https://doi.org/10.1145/1360612.1360656
8. Agusanto, K., Li, L., Chuangui, Z., Sing, N.W.: Photorealistic rendering for augmented reality using environment illumination. In: Proceedings of the 2nd IEEE and ACM International Symposium on Mixed and Augmented Reality (ISMAR 2003) (2003). https://doi.org/10.1109/ismar.2003.1240704
9. Guo, X., Wang, C., Qi, Y.: Real-time augmented reality with occlusion handling based on RGBD images. In: 2017 International Conference on Virtual Reality and Visualization (ICVRV 2017) (2017). https://doi.org/10.1109/icvrv.2017.00069

10. Lira dos Santos, A., Lemos, D., Falcao Lindoso, J.E., Teichrieb, V.: Real time ray tracing for augmented reality. In: 2012 14th Symposium on Virtual and Augmented Reality (2012). https://doi.org/10.1109/svr.2012.8
11. Next-Generation Threat System (NGTS): Published by NAVAIRSYSCOM on December 9th 2014
12. Unity. https://unity3d.com

# Exploring Augmented Reality as Craft Material

Lauren Edlin[⊠], Yuanyuan Liu, Nick Bryan-Kinns, and Joshua Reiss

School of Electronic Engineering and Computer Science,
Queen Mary University of London, Mile End, London E1 4NS, UK
L.Edlin@qmul.ac.uk

**Abstract.** Craft making is associated with tradition, cultural preservation, and skilled hand-making techniques. While there are examples of digital craft making analyses in the literature, Augmented Reality (AR) applied to craft making practice has not been explored, yet applying AR to craft making practices could bring insight into methods of combining virtual and physical materials. This paper investigates how AR is considered by craft makers. We find that narrative is essentially physically located in craft objects, and while virtual elements may describe and annotate an artefact, it is not considered part of the craft artefact's narrative.

**Keywords:** Craft making in HCI · Augmented Reality · Computational materials

## 1 Introduction

Craft making is associated with traditional practices, cultural preservation and skilled hand-making techniques [38]. Alongside this deep connection to cultural traditions, craft making practices have increasingly incorporated digital technologies, often resulting in new traditional-digital hybrid making processes [17,39]. Craft has thus been of recent interest in the Human-Computer Interaction (HCI) community as a way of investigating 'materiality' in making practices (e.g. [7,23] and the conceptualization of computation as a material [44]).

While attention has been given to analyses of a number of materials and digital tools in craft, the particular technology of Augmented Reality (AR) applied to craft making practices has not been explored. AR is a range of interactive technologies that allow a user to experience virtual content superimposed on the real world [5]. Since AR consists of systems that connect virtual content to physical objects without observable interactions between them, subjecting this technology to craft making practices (and vice versa) can shed light on methods

Supported by AHRC funded Digital Platforms for Craft in the UK and China (AH/S003401/1), and EPSRC+AHRC Centre for Doctoral Training in Media and Arts Technology (EP/L01632X/1).

© Springer Nature Switzerland AG 2020
C. Stephanidis et al. (Eds.): HCII 2020, LNCS 12428, pp. 54–69, 2020.
https://doi.org/10.1007/978-3-030-59990-4_5

of combining virtual and physical materials, and offer further reflection on the nature of 'immaterial' materials in making.

We use *critical making* to investigate AR as craft material. Critical making examines how social understandings are expressed in made objects through intermittent steps of conceptual analysis, exploratory making and prototype construction, and reflective critique [33] p. 253. Critical making stems from *critical design*, which explores the social contexts, ethical assumptions and values that shape making practices and their reflections in the made objects themselves [6,18].

Critical making is an appropriate method to investigate AR in craft making because craft is replete with culturally-embedded values and associations, e.g. Arts and Crafts movement in the UK [2,26]. Thus, any understanding of craft making with virtual components should consider these attributes that shape making processes. In addition, the reflective user-design practice invites the influence and opinions of craft practitioners, which makes craft makers primary stakeholders in developments arising from this research.

This research presents the conceptual analysis stage in a critical making framework, which involves compiling relevant concepts and theories for further work. We do this through an empirical investigation into how craft makers understand their own practices and values associated with these practices and whether AR is believed to fit with these practices. We first interview craft makers about their making processes and conception of material. Then we construct an AR craft prototype and conduct analysis with craft makers to understand whether they perceive AR virtual content as aligned with craft values.

The main over-arching questions investigated in this research are:

- What attributes of craft do expert makers view as important in the context of craft making with AR technologies?
- How do craft makers understand AR technologies as a potential material?
- How do users receive a simple AR for craft prototype based on the findings, what are promising directions to explore in future research?

The overarching theme that emerged is the importance and nature of narrative in a craft object. Narrative is necessarily physically expressed. To understand an object's tradition or history, one must touch and observe the artefact. Since computation is immaterial, it cannot manifest as an inherent narrative. However, the artefact requires additional information to "fill in" the story or wider cultural context the maker wishes to highlight. This leaves room for computation to assist narrative, and there is disagreement as to whether this allows computation to be considered to be a material and have a narrative in itself.

## 2   Background

### 2.1   Attributes of Craft

Traditional craft making is defined as small-scale production of functional or decorative objects [34], and commonly associated with practices involving natural materials such as pottery, ceramics, metal-working, wood turning, weaving, and textiles. However, what is considered craft evolves in conceptualisation through socio-cultural, historical, and technological contexts [10]. Attributes and values of craft making include:

- *Skill and Expertise* [8,38]: to craft is to "participate skillfully in some small-scale process" [2], p. 311. Sennett [38] argues that craftsmanship is an innate desire to do skilled, practiced work with full engagement for its own sake.
- *Embodied Practice through Hand-Making:* craft is knowledge of a process through learned coordination of the body, and hand-making is the exemplar of embodied practice [25,29].
- *Close Proximity with Physical Material* [1]: A material's unique and uncertain 'feedback' to the maker's manipulation shapes the making process. Pye ([2], p. 342) coined this balance between uncertainty and craftsmanship 'the workmanship of risk'.

## 2.2 Computation as a Craft Material or Medium

Recent work in *digital craft* (or hybrid craft [22,47]) bridges the conceptual gap between traditional and modern making practices by "combining emerging technologies with hands-on physical making practices, leading to the production of digital artifacts (such as code)" ([31], p. 720). Thus, digital craft making is investigated to inform the role of material, or 'materiality', in making and interaction design practices e.g. [41,46]. Much of this work involved experimental making with craft materials, including clay pottery [37], leather making[42], and woven basketry [47].

In craft, data and computation [19] interact with instantiating physical objects, which situates computation as 'immaterial'. Vallgaarda and Sokoler coin 'computational composites' to describe the expression of computation through integration with physical materials ([44] p. 514). They argue that since the inner workings of computational processes are not perceivable to the human eye, material properties of computation can only be studied through its composition with directly observable materials and its form-giving abilities [45].

The role of communication lends computation to be conceptualised as a *medium* rather than a material [20]. A medium uses physical properties to convey a message through consistent organization (e.g. film, comics) [28]. Gross et al. [23] identified three main material approaches to making: Tangible User Interfaces (TUIs) present computation in physical form, computation as metaphysical material, and craft applied to HCI extends communicating tradition to digital making. The interconnected role of physical, immaterial, and communication resemble the notion of medium, and highlight the potential of expanding crafts traditional narratives through digital components.

## 2.3 Related Work: Craft and Augmented Reality

**Mobile Augmented Reality:** *Augmented reality* (AR) refers to a range of interactive technology systems that allow a user to experience virtual content superimposed on the real world [5,27]. *Virtual content* can be textual, audio, symbolic, or 2D and 3D graphical visualizations positioned in a user's real world view in a "spatially contextual or intelligent manner" ([4] p. 2) using sophisticated hardware and computer vision techniques.

We limit the scope of our investigation to *marker-based AR* technologies that can be implemented as a mobile phone application. Marker-based AR identifies and tracks a designated object, image or surface in the real world using a camera and overlays content according to the marker position ([3], p. 12-3).

**AR Craft:** The Spyn Toolkit was developed to investigate how the memories and thoughts that makers have while making an object can be virtually embedded into the created object [35,36]. Spyn is a mobile phone application that allows knitters to record audio, visual and textual information throughout the making process, and associate them with locations in the final knitted piece. The authors found that recipients of the craft objects with virtual 'notes' reacted positively to these virtual elements, which made the crafts more personal.

# 3 Understanding Making Processes and the Conceptualisation of Material

We interview craft makers' to gain insight into makers perspectives on their own making practices. We asked makers to identify the attributes and principles they consider integral to craft making, and whether AR can serve as a potential material for their particular type of craft making.

## 3.1 Method

We conducted a qualitative study consisting of semi-structured expert interviews [11] with self-described craft makers. Craft makers are experts of their particular making practices, which includes unique understandings of the affordances of their chosen materials and tools, as well as insights into how these materials may interact with computational or virtual elements.

**Interview Participants:** We selected interviewees from a number of making contexts in order to gain insights into the diverse experience of makers. Craft making is done in a variety of contexts which shape the values associated with the making process. For instance, craft is made and sold through self-employed businesses, makerspaces, designer-maker fairs and galleries [40]. Craft is also a leisure activity [32], and features in political activism against mass consumption [15]. Finally, the characteristics of craft-based businesses in the UK tend to significantly differ between rural and urban areas in production scale and access to materials, tools and networks [9,15].

We approached twelve potential interviewees that between them represent most of the contexts just described, with nine participants agreeing to be interviewed. Table 1 provides a summary of each participant's making context, primary craft type, and location where they make, sell and/or show their work. The interviewees vary in use of computation in their work, with some having digital technology significantly feature in their work and others using only traditional making techniques. Ute was the only participant to have created an AR application. All names have been changed to maintain anonymity, and all interviewees gave written consent to participate in this study.

**Interview Structure:** A semi-structured interview format was chosen to ensure consistency between interviews while allowing flexibility to probe based on interviewees' responses. The questionnaire consisted of twelve questions with additional probes as needed. Questions centered around interviewees' own reflections on their making practices, how they understand materials and whether computation and digital technology fits in that understanding, and their views on incorporating AR into their craft practices. A short demonstration of simple

**Table 1.** Semi-structured interview participant summary

| Interview participants | | | |
|---|---|---|---|
| Name | Making context | Craft type | Location |
| James | Self-Employed Small Business | Digital Visual Art, Digital Illustration, Printing | Urban |
| Sophie | Self-Employed Small Business, Teaching | Knitting, Sewing, Garment and Cushion Design | Urban |
| Rachel | Designer in Residence | Research through Making, Textiles, Ceramics, Jesmonite, 3D Printing and Laser Scanning | Urban |
| Lotte | Self-Employed Small Business and Studio | Weaving, Sewing, Textiles, Garment Making | Urban |
| Christine | Academic Researcher and Repair Activist | Exploratory Making, Textiles | Urban |
| Karina | Self-Employed Small Business | Painted Silk Scarves, Printed Illustration | Urban and Rural |
| Robert | Self-Employed Small Business, Teaching | Ceramics, Pottery, Painting | Rural |
| Elizabeth | Self-Employed Small Business | Narrative Textiles and Costumes | Rural |
| Ute | Academic Researcher and Craft Activist | E-textiles, Urban knitting, 3D printing | Urban |

marker-based AR technologies was also prepared to prompt discussion. Interview lengths varied between 22 min and roughly 1 h. The interviews took place between June 9th–23rd 2019 in various locations in the UK. All interviews were recorded and transcribed by the first author.

**Data Analysis:** All transcripts were analyzed using *thematic analysis* (TA), which is a common method for analysing semi-structured interviews, with the aim of "minimally organising and describing the data set in (rich) detail" ([12], p. 6). In contrast to content analysis, which organizes data by its manifest content and takes the frequency of topic occurrence into account [43], TA takes the manifest and latent meanings in the data as more fundamental to theme organization than frequency. This better reflects a shared sense of meaning through the experiences described in the data ([13] p. 57). As this study is concerned with investigating makers' perceptions and experiences with craft making, we find TA to be a better fit in a critical making framework. An inductive coding strategy was used [24], meaning that categories and themes were not predetermined before analysis but rather induced directly from the data set. The themes discovered are described below.

## 3.2    Results

### Theme 1: Craft Making is Articulating Ideas Physically with some External Constraints using Various Approaches to Hand-Making

*Craft Making is Instantiating a (Vague) Idea into Physical Form through Whatever Means Required:* The most fundamental component of craft making, according both traditional makers and makers who use digital technologies, is instantiating or working out an idea into a physical form. All participants, except the two purely traditional makers, stressed that the idea of what they want to make comes first and making techniques were then chosen or learned as required to actualize the idea. Traditional makers further emphasised that while having an idea to work from is important, the idea was described as vague, undefined, and primarily developed through the making process itself. They impressed that interaction with and understanding the physicality of that material is what primarily shapes the idea of what they are making rather than pure conceptualization.

*Hand-Making is Equated with Thinking or Mediated Interaction with Materials Through Tools:* Stemming from this, there was also a disparity between traditional and digital makers on the role of hand-making: traditional makers emphasised hand-making as embodied knowledge, while the makers more familiar with digital technologies perceived hand-making as only one, albeit fundamental way of interfacing with materials. Traditional makers explicitly described hand-making as the direct link between the conception of what to make into the direct physical manipulation of materials. For instance, Elizabeth is a a traditional maker who using textiles, sewing and other techniques of assemblage to make elaborate costumes and set pieces. She stated: "It's almost, I think through my fingers, they know what to do."

Participants who employ digital technology in their work also perceived hand-making as essential to craft, however they emphasised the potential interactions between tools and materials. Rachel is a designer-maker who investigates combining traditional and digital making techniques, and posited expanding hand-making to include building and maintenance of machines:

> ...but then there's also the interesting argument in if the machines been made by hand. Let's say its a ceramic extruding 3D printer that's fed by hand and most of the time it will be stopping it and starting it and cleaning it... it's very high maintenance, then there's just as much handwork that's going into that. But with craft, it's making something, it has to have an element of hand.

The conception situates hand-making as any part of the process where hands interact with materials or tools. Nonetheless, machine maintenance is not explicitly considered as "thinking" according to traditional makers.

## Theme 2: Craft Materials are Defined by Purposeful Physical Manipulation

*Craft Materials are Anything that is Interacted with Purposefully:* Both makers familiar with digital technologies and traditional makers characterised *craft materials* as anything that is used purposefully. Therefore, a deliberate and reasoned making process defines craft rather than using a specific type of material. Specifically how a material is used purposefully is more elusive. Again traditional makers emphasise direct interaction and experimentation with materials, and "finding" the resulting craft artefact through the making process. Robert, who makes hand-built pots that often incorporate found natural objects such as stones and twigs, described making as fitting the pot to the found object:

> So basically, I look at the stick and just get used to it being around and think, what sort of shape is it wanting. Does it want to be on the top of something, tall and narrow, does it want to be broad, squat, does it want something rough or something really quiet and sleek.

Both traditional and digital practitioners agreed that purpose is found in pushing materials to their limits through considered experimentation. However, makers familiar with digital technology highlighted that purpose is often situated in researching how to use a specific tool, or the properties of material.

*Computation Must Purposefully Interact with Physical Objects to be a Craft Material - Therefore, AR is not a Material:* As evidenced in the previous category, the participants were open to anything being a craft material, including computation. However unlike the traditional and physical materials that are more commonly used in craft, computation was given further stipulations on how and where it could be used specifically in craft making. First, because computation is not physically observable, it may become a material only if it interacts with a physical object. This follows the notion of 'computational composites' described by Vallgaarda and Sokoler [44] as discussed in Sect. 2.2. Interaction with physical objects includes computation that describes and shapes physical materials as found in digital fabrication techniques.

Second, there was skepticism as to whether computation can be used purposefully because is cannot be experimented with or shaped by hand. Computation and other digital technologies were seen to be "cutting corners" and shortening the making process, which takes out the times and deliberation inherent to craft making and reduces its value. Elizabeth stated her disappointment with digital enhancements in photos, which expended to virtual elements incorporated with craft artefacts:

> ... it's a bit depressing. It's amazing that's there but then, the fact that it doesn't exist. I suppose that's the thing that somebody's setting up these incredible shoots and actually making it manifest in the real world. I think there's a big difference between that, and being able to press a few buttons and make something that looks like it is but actually it isn't... I've got a lot of respect for people who actually make things actually happen in the real world.

Third, virtual elements that do not explicitly interact with physical objects in an observable way, like AR, are consequently not materials. Several participants noted that AR technology was not yet sophisticated enough to be used in a "tactile" manner necessary for craft making. In addition, there was again skepticism about the ability to purposefully incorporate AR - it was noted that video and audio can already supplement physical artefacts, and therefore AR was not necessary or needed.

### Theme 3: A Craft Object's Narrative is Fundamentally Physically Situated

*A Craft Object's Narrative Exists on a Spectrum of Being More or Less Physically Bound:* The most salient theme that emerged was the explication of what it means for a craft object to have a *narrative*. Craft is generally characterised as having 'authenticity', history, or a story due to links with cultural tradition. But what does it really mean for a physical object to have and communicate a narrative? While all participants agreed that craft is synonymous with narrative, its comprehension differed between participants.

How a craft object conveys narrative was described either as more or less abstracted from the actual craft object. Traditional makers placed narrative as the physical object itself - therefore, it can only be communicated through perceptual and sensational interactions with that object. A person can access the narrative by touching, observing, and subsequently get impressions of where, how, and why that object was made. For example, Robert described how many of his ceramics pieces 'echo' natural landscapes:

... there's two slates, they were found at the seaside and it was in Cornwall where there was some distant hills, landscape hills, and then cliffs and the sea, so I poured glaze on that just to have vaguely a sort of landscape-y feel to it, just to echo the place where I found them.

In this case, location is central to that artefacts narrative. While an observer may not learn this precise information from looking at the piece, it nevertheless may invoke a response due to the inspiration that shaped it.

Some participants, on the other hand, conceived narrative as situated in the larger social context that additionally informs the artefacts meaning. Narrative therefore not only manifests through observations of the object, but also through affiliated concepts and philosophies. Christine, for instance, works with repair activist groups to support more sustainable consumption practices. She described allowing people to practice sewing on an old sweater before trying to repair their own clothing:

... people don't always want to do their first darn on their precious jumper that they're really upset they've got a hole in... so I take (the sweater) with me, people can have a go on it, they can cut holes they can stitch into it, they can test things out on it. So it's a kind of record of those workshops, its a record of other people's work. Otherwise there would really be no record of all of those kinds of things.

A person observing the sweater will most likely gain any understanding of the meaning behind the craft practices displayed on it without additional access to the associated craft activism workshops. Therefore the artefact alone is insufficient to convey what it most meaningful about it and further social context is required.

*Error Generates Narrative, Reproductive Making Processes Kills Narrative:* Both traditional and digital makers agreed evidence of"trials and errors" that belie an artefact as hand-made create narrative, while digital fabrication and computation erase markers of individuality and narrative. This supports the notion that narrative is primarily physically situated, and the direct manipulation of materials to instantiate an idea are what points to the narrative. Several participants mentioned that there is no purpose in creating an item that appears "perfect" as digital fabrication can accomplish this create multiple identical products. An error-free object is depersonalised therefore devoid of interesting narrative.

*AR may Give Definition to Opaque Narrative but is not Innate Narrative:* Finally, participants broadly agreed that AR elements may present annotations that inform viewers about an object's narrative, but the AR elements themselves are not part of the object's narrative. Most participants were accepting of AR as an additional element to an artefact - AR may be used to present recorded personal statements, or present additional images and video related to the objects historical or cultural context. However, because these are virtual components that are not observed to interact physically, these additions are not conceived as part of an artefacts narrative.

One participant, however, argued that a physical craft object alone cannot communicate a precise meaning. Ute is a researcher who uses both traditional and digital making in her work. In the case of a quilt-making project that involved collaboration, she stated:

> Using actually augmented reality that doesn't affect the physical artefact, but just overlays it with those different experiences in some way. If you think about maybe first giving my voice and overlaying it, and then it's the voice of my co-creator who is telling a different story about it, I think that would be really interesting because it would be closing the gap between the meanings that we as crafters put into our stuff so the audience can understand.

Unlike most other participants, Ute was familiar with AR and other digital technologies and supported the disseminating information about projects through channels such as social media. Traditional makers, on the other hand, found the idea of further defining craft objects meaning through words to be confining and undesirable.

## 4 Craft Maker's Reflections on AR Craft Virtual Content

We investigated whether the type of AR virtual content influences makers perceptions regarding AR as amenable to craft. The results of the semi-structured interviews demonstrated that AR must be incorporated purposefully with the physical artefact, and may at least further define the meanings associated with the object despite not being inherent to the narrative. We build an AR craft prototype that features several types of virtual content, namely audio, 3D models, and representations of hand recordings, and have several makers reflect on the extent to which the type of virtual content fulfills these criteria.

### 4.1 Participants

Craft makers and sellers were approached at a craft market and asked if they wanted to participate in a short study. Five people were approached with four agreeing to participate by signing a consent form. All participants were from a small business context, and therefore only represent one craft context.

Participant 1, James, was also an interviewee in the first study and was asked to participate in the second study after expressing interest. James creates and sells hand-drawn digital illustrations, and therefore has experience with digital tools. Participant 2 fixes, improves, and sells small items such as bespoke watches, and therefore has knowledge of mechanical making tools. Participant 3 is a traditional painter and illustrator who also works at a market stall selling custom knitted scarves. Participant 4 creates and sells sustainable clothing items in a permanent store next to the market, and is familiar with sewing and weaving.

## 4.2   AR Craft Prototype

An experimental prototype was constructed to demonstrate simple AR technologies. The prototype is an AR block jewellery piece, and consists of three wooden blocks on a braided string. The wooden block faces were painted using water colours, acrylic paint, and ink. The theme of the necklace was the 'countryside in the UK Cotswolds', and the scenes on the block faces represent common rural landscapes, farm animals and products representative of the area. Figure 1 exhibits the various scenes painted on the blocks.

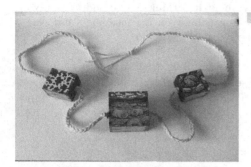 

**Fig. 1.** Augmented Block Jewellery physical artefacts

The flat surfaces of the block faces could easy be photographed and made into image targets for marker-based AR. Each block was painted to have two image targets to therefore display two pieces of virtual content. The virtual content on each block is as follows:

– *Block 1: Farm Animals* - One block face depicts a spotted cow pattern that produces an animated 3D model of a cow, and another block face depicts painting of a rooster that produces an animated 3D model of a chicken.
– *Block 2: Landscape Scenes* - One block face depicts hills and valleys, and another block face depicts a farm landscape with hay bails. Both block faces produce two audio recordings each that were recorded in the Cotswolds. The audio recordings start and stop according to the tracking state of the image targets.
– *Block 3: Cheese* - This block only has one augmented painting, which depicts cheeses. The cheese block face produces an over-layed recording of hand motions that were captured by the Leap Motion sensor. The recorded hand movements were produced during the actual painting process.

The virtual content for Block 1 and Block 2 were created using the Unity + Vuforia SDK. The virtual content for Block 3 was made and attached using Processing + NyARToolkit, as well as the recordings produced by Processing sketc.hes that captured the hand movements from a Leap Motion sensor. The virtual content for each block face is shown in Fig. 2.

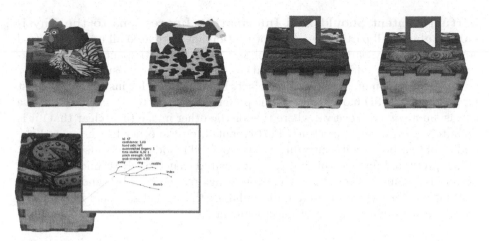

**Fig. 2.** AR craft prototype: virtual content per each block face

### 4.3   Methods

Each participant was provided a quick demo of the prototype including instructions on how to elicit the virtual content on a laptop screen. The participants were given free range to interact with the prototype to produce the AR virtual content themselves.

After this short demo the participants were asked several open-ended questions regarding aspects and opinions of the prototype to determine whether participants conceived the virtual content as contributing the same or differently to the AR craft prototype. These discussions were recorded and transcribed by the first author. The data was analysed using thematic analysis, which used to analyse the semi-structured interviews described in Sect. 3.1. The emergent themes are described below.

### 4.4   Results

**Disagreement as to Whether AR Craft Includes the AR Technical System:** Overall, all participants except participant 4 perceived the AR virtual content and technical system as an integral composite of the complete piece, rather than the physical artefact alone to be a complete or 'whole' object in itself. Participant 1 noted that the AR content appeared 'embedded' rather than 'added-on'. When asked specifically about the virtual content, participants 2, 3 and 4 sought clarification as to whether the virtual content referred to what was shown on the laptop screen, or the entire AR system. When prompted for their impressions, participants 2 and 3 claimed that the entire apparatus of the physical object, laptop with the camera and screen, and the virtual content on the screen were all part of one whole craft object. Participant 4 did not find the AR content to add anything interesting to the physical artefact.

**Virtual Content Should Add Information/Impressions to the Physical Artefact:** All participants (except participant 4) treated all types of virtual content as equally part of the whole object, however participants preferred virtual content that added new information or impressions to supplement their own interpretation of the physical artefact. Participant 1, for instance, was most interested in the 3D model content but preferred the cow to the chicken because it was "more of a discovery... where the for the other image (it is clear that) it is definitely going to be a chicken". Participant 3, on the other hand, favored the audio content because it contributed to the overall mood of the landscape image. When prompted further, this participant posited that the AR technologies were materials because material is "whatever brings out the mood", and audio was a supplementary and novel domain to enhance that mood as visual information was already available on the physical artefact.

**Virtual Content of Hand-Recordings Are Not Self-evident:** None of the participants endorsed the hand-recording virtual content as especially informative or interesting. Participant 2 posited that the recordings were not accessible to understand on their own, and required further information to know that these recordings were of the hand-making process. As a visual artist, participant 3 was not interested in an already known technique but preferred to see a novel technique displayed. Participant 1 suggested either showing more 'impressionistic' recordings that are open to interpretation, or more precise recordings that can be used as a teaching tool rather than evoke an aesthetic response.

## 5   Discussion and Conclusion

The purpose of this study was to explore craft makers reflections on their making processes and how AR elements comply with these processes and their associated values. We conducted semi-structured interviews to inquire directly with craft makers about their own making techniques, and supplemented this by having craft makers reflect on the role of different types of virtual content on an AR craft prototype. The primary finding was an elaboration of the concept of narrative as situated in a physical object. Overall, the narrative inherent in a craft object is tied to its physical properties. Since AR consists of virtual components and technological systems that do not have observable interactions with physical artefact, it is unclear to makers how virtual content can be meaningfully or purposefully incorporated as a craft material - one direction suggested in the second study, however, is using non-visual domains to supplement the visual properties of the physical artefact.

The results corroborate with articulated understandings of computation and craft described in Sect. 2. The participants substantiated the common attributes associated with craft making, including hand-making, purposeful and deliberate making processes. Traditional makers especially supported Pye's concept of 'the workmanship of risk' ([2], p. 342) through material interactions and evidence of trial and error in making. In addition, the participants maintained the notion of computation as 'immaterial', and necessarily a composite with a physical artefact [44, 45].

Overall, the findings suggest AR can be considered a medium rather than a material because of the priority of narrative communication, as well as taking into account the whole AR technology setup and physical artefact as part of the communicative system [20, 23]. Nevertheless, the rules and capabilities of the medium have yet to be defined. The virtual elements alone were described as being more or less amenable to craft practices depending on the maker's desire for a 'definitive' versus 'impressionistic' narrative. Traditional makers begin making through nascent, vague ideas that are filled in through feedback from material interactions, and subsequently deny virtual elements are able to contribute to this physical narrative. Other makers who highlighted social contexts that shape an object's narrative, however, open up room for virtual elements to contribute with supplementary information, which may also be more or less definitive/impressionistic. Only one participant fully endorsed personal recordings as narrative inherent, which suggests that while personal recordings may enhance an object (e.g. the Spyn toolkit [35, 36]), these additions have yet to be considered the narrative of the object itself.

## 6   Future Work

As the purpose of this research is to lay the groundwork for further work into AR craft in a critical making framework, there are many fruitful directions for future work. First, it would be beneficial to investigate making processes through collaborative workshops with craft makers and AR practitioners building (functioning or non-functioning) AR craft prototypes (e.g. [22]). A shortcoming of this study is the interviewees lack of making experience with AR technologies. While the reflections on their own making processes is useful as conceptual analysis, a next step is assessing how these conceptions can be expressed, or fall short of being expressed, in AR craft objects made by makers themselves.

Second, the second study can be expanded to include additional exploratory prototypes, as well as add participants from other craft making contexts. Additional prototypes could attempt to instantiate different levels of 'defined' narrative - for instance, personal audio recordings, impressionistic hand-recordings, virtual 'craft simulations' (e.g. [14]), and other methods of preserving cultural heritage [16, 21, 30] - and have makers further reflect on their role in contributing to the craft objects narrative.

## References

1. Adamson, G.: Thinking Through Craft. Bloomsbury Publishing (2018)
2. Adamson, G.: The Craft Reader. Berg, New York (2010)
3. Amin, D., Govilkar, S.: Comparative study of augmented reality SDKS. Int. J. Comput. Sci. Appl. **5**(1), 11–26 (2015)

4. Aukstakalnis, S.: Practical Augmented Reality: A Guide to the Technologies, Applications, and Human Factors for AR and VR. Addison-Wesley Professional (2016)
5. Azuma, R.T.: A survey of augmented reality. Presence Teleoper. Virt. Environ. **6**(4), 355–385 (1997)
6. Bardzell, J., Bardzell, S.: What is "critical" about critical design? In: SIGCHI Conference on Human Factors in Computing Systems, pp. 3297–3306 (2013)
7. Bardzell, S., Rosner, D., Bardzell, J.: Crafting quality in design: integrity, creativity, and public sensibility. In: Designing Interactive Systems Conference, pp. 11–20. ACM (2012)
8. Becker, H.S.: Arts and crafts. Am. J. Sociol. **83**(4), 862–889 (1978)
9. Bell, D., Jayne, M.: The creative countryside: policy and practice in the UK rural cultural economy. J. Rural Stud. **26**(3), 209–218 (2010)
10. Bell, E., et al.: The Organization of Craft Work: Identities, Meanings, and Materiality. Routledge, New York (2018)
11. Bogner, A., Littig, B., Menz, W.: Interviewing Experts. Springer, London (2009). https://doi.org/10.1057/9780230244276
12. Braun, V., Clarke, V.: Using thematic analysis in psychology. Qual. Res. Psychol. **3**(2), 77–101 (2006)
13. Braun, V., Clarke, V.: Thematic analysis. In: Cooper, H., Camic, P.M., Long, D.L., Panter, A.T., Rindskopf, D., Sher, K.J. (eds.) APA Handbooks in Psychology®. APA Handbook of Research Methods in Psychology, Research Designs: Quantitative, Qualitative, Neuropsychological, and Biological, vol. 2, pp. 57–71. American Psychological Association (2012). https://doi.org/10.1037/13620-004
14. Chotrov, D., Uzunova, Z., Maleshkov, S.: Real-time 3D model topology editing method in VR to simulate crafting with a wood turning lathe. Comput. Sci. Edu. Comput. Sci. **1**, 28–30 (2019)
15. Conefrey, C.M.: Creative rural places: A study of cultural industries in Stroud, UK. Ph.D. thesis, University of the West of England (2014)
16. Damala, A., et al.: Bridging the gap between the digital and the physical: design and evaluation of a mobile augmented reality guide for the museum visit. In: 3rd ACM International Conference on Digital Interactive Media in Entertainment and Arts, pp. 120–127 (2008)
17. Johnston, L.: Digital Handmade: Craftsmanship in the New Industrial Revolution. Thames & Hudson (2015)
18. Dunne, A., Raby, F. (2007). http://dunneandraby.co.uk/content/bydandr/13/0
19. Freeman, J., et al.: A concise taxonomy for describing data as an art material. Leonardo **51**(1), 75–79 (2018)
20. Fuchsberger, V., Murer, M., Tscheligi, M.: Materials, materiality, and media. In: SIGCHI Conference on Human Factors in Computing Systems, pp. 2853–2862 (2013)
21. Gheorghiu, D., Stefan, L.: Mobile technologies and the use of augmented reality for saving the immaterial heritage. In: VAST (Short and Project Papers) (2012)
22. Golsteijn, C., van den Hoven, E., Frohlich, D., Sellen, A.: Hybrid crafting: towards an integrated practice of crafting with physical and digital components. Pers. Ubiquit. Comput. **18**(3), 593–611 (2013). https://doi.org/10.1007/s00779-013-0684-9
23. Gross, S., Bardzell, J., Bardzell, S.: Structures, forms, and stuff: the materiality and medium of interaction. Pers. Ubiquit. Comput. **18**(3), 637–649 (2013). https://doi.org/10.1007/s00779-013-0689-4
24. Hsieh, H.F., Shannon, S.E.: Three approaches to qualitative content analysis. Qual. Health Res. **15**(9), 1277–1288 (2005)

25. Hung, S., Magliaro, J.: By Hand: The Use of Craft in Contemporary Art. Princeton Architectural Press, New York (2007)
26. Krugh, M.: Joy in labour: the politicization of craft from the arts and crafts movement to ETSY. Canadian Review of American Studies **44**(2), 281–301 (2014)
27. Lee, W.H., Lee, H.K.: The usability attributes and evaluation measurements of mobile media AR (augmented reality). Cogent Arts Hum. **3**(1), 1241171 (2016)
28. Manovich, L.: The Language of New Media. MIT Press, Cambridge (2001)
29. McCullough, M.: Abstracting Craft: The Practiced Digital Hand. MIT Press, Cambridge (1998)
30. Miyashita, T., et al.: An augmented reality museum guide. In: 7th IEEE/ACM International Symposium Mixed and Augmented Reality, pp. 103–106. IEEE Computer Society (2008)
31. Nitsche, M., et al.: Teaching digital craft. In: CHI 2014 Extended Abstracts on Human Factors in Computing Systems, pp. 719–730. ACM (2014)
32. Pöllänen, S.: Elements of crafts that enhance well-being: textile craft makers' descriptions of their leisure activity. J. Leisure Res. **47**(1), 58–78 (2015)
33. Ratto, M.: Critical making: conceptual and material studies in technology and social life. Inf. Soc. **27**(4), 252–260 (2011)
34. Risatti, H.: A Theory of Craft: Function and Aesthetic Expression. Univ of North Carolina Press, Chapel Hill (2009)
35. Rosner, D., Ryokai, K.: SPYN: augmenting knitting to support storytelling and reflection. In: 10th ACM International Conference on Ubiquitous Computing, pp. 340–349 (2008)
36. Rosner, D., Ryokai, K.: SPYN: augmenting the creative and communicative potential of craft. In: ACM SIGCHI Conf. Human Factors in Computing Systems, pp. 2407–2416 (2010)
37. Rosner, D.K., Ikemiya, M., Regan, T.: Resisting alignment: code and clay. In: Ninth International Conference on Tangible, Embedded, and Embodied Interaction, pp. 181–188. ACM (2015)
38. Sennett, R.: The Craftsman. Yale University Press, London (2008)
39. Shillito, A.M.: Digital Crafts: Industrial Technologies for Applied Artists and Designer Makers. Bloomsbury, London (2013)
40. Shiner, L.: "blurred boundaries"? rethinking the concept of craft and its relation to art and design. Philos. Compass **7**(4), 230–244 (2012)
41. Sundström, P., et al.: Inspirational bits: towards a shared understanding of the digital material. In: SIGCHI Conference on Human Factors in Computing Systems, pp. 1561–1570 (2011)
42. Tsaknaki, V., Fernaeus, Y., Schaub, M.: Leather as a material for crafting interactive and physical artifacts. In: Designing interactive systems Conference, pp. 5–14. ACM (2014)
43. Vaismoradi, M., Turunen, H., Bondas, T.: Content analysis and thematic analysis: implications for conducting a qualitative descriptive study. Nurs. Health Sci. **15**(3), 398–405 (2013)
44. Vallgårda, A., Redström, J.: Computational composites. In: SIGCHI Conference on Human Factors in Computing Systems, pp. 513–522 (2007)
45. Vallgårda, A., Sokoler, T.: A material strategy: exploring material properties of computers. Int. J. Des. **4**(3), 1–14 (2010)
46. Wiberg, M.: Methodology for materiality: interaction design research through a material lens. Pers. Ubiquit. Comput. **18**(3), 625–636 (2013). https://doi.org/10.1007/s00779-013-0686-7
47. Zoran, A., Buechley, L.: Hybrid reassemblage: an exploration of craft, digital fabrication and artifact uniqueness. Leonardo **46**(1), 4–10 (2013)

# The Application of Urban AR Technology in Cultural Communication and Innovation

Yueyun Fan[1(✉)] and Yaqi Zheng[2]

[1] Department of Design, Politecnico di Milano, Milan, Italy
sherryfan199592@gmail.com
[2] Department of Industry Design, Guangzhou Academy of Fine Art, Guangzhou, China
598859974@qq.com

**Abstract.** With the development of science and technology, today's society has gradually entered the Internet-led system, and based on the demand for creating a more perfect intelligent experience for users, augmented reality technology has become a hot spot for the development of the modern intelligent industry. AR applications such as AR games, AR smart furniture, AR navigation systems, etc. are constantly emerging. As technology evolves and matures, AR applications are beginning to shift from 2D data information (text or image-based descriptions) to 3D integration objects. The development of technology has provided more ways and possibilities for cultural communication and innovation. City around the globe is ready for the next step for their evolution to augmented cities. This paper aims to design an application of Location-based AR technology for exploring cities and discuss its important role in cultural communication and innovation.

**Keywords:** Augmented reality · Cultural communication · Cultural innovation

## 1 Introduction

The research on this project was originally based on the concept of "Italicity". The word "Italicity" was first proposed by Piero Bassetti, in order to describe a cultural value system that is not limited to the sense of the nationality, but take the way of thinking, the mode of behavior, and the living habits as the core for those people who have a sort of subtle connection with Italy to some extent.

This view breaks the usual understanding of cultural value systems. With the acceleration of globalization, it is clear that a cultural value system divided by the state will not be enough to support such a complex and diversified cultural system. However, despite the closer communication between regions, cultural exchanges are still in a subtle stage and lack of effective tools for cultural communication.

With the rapid development of the Internet and the emerging technologies such as AR (Augmented Reality) and VR (Vitual Reality), it has injected new vitality into the research of the cultural industry. VR technology is used to build past historical experiences, and AR technology is used to restore ancient cultural relics and ancient ruins. The emergence of these cultural innovation industries has greatly enhanced the intensity and depth of cultural dissemination.

© Springer Nature Switzerland AG 2020
C. Stephanidis et al. (Eds.): HCII 2020, LNCS 12428, pp. 70–79, 2020.
https://doi.org/10.1007/978-3-030-59990-4_6

At the same time, smart phones also provide users with a variety of information exchange systems. In today's society that 5G networks are beginning to prevail, the use of the Internet in smart phones can better support users to explore augmented reality environments. In an era of information explosion, people are becoming more accustomed to spending less time to obtain information, instant information becoming a way by users to get knowledges. This means that, the way of cultural dissemination from large amounts of textual information in the past must be appropriately adjusted and changed. With the continuous penetration of the Internet into human life, social platforms have gradually become one of the main ways for people to communicate, exchange and obtain information.

In 'the Smart City Expo World Congress 2017', Gregory Curtin pointed out: city around the globe is ready for the next step for their evolution to augmented cities. And by augmented cities, we mean using all of their various data, infrastructure, the connected things that now are coming online in smart cities and new urban design. And in the augmented city, really bring all together to change the city in reaction and engagement with citizens, residents, workers, businesses. It really change the entire city experience. So, an augmented city is using all the information, data, infrastructure cities have available now. It enables all that for augmented reality devices experiences in like. And it definitely will be one of next step in terms of urban design, city design, really creating a whole fabric for a city, for the citizens as users to engaging and navigating to really experience in city.

So, in terms of augmented reality, in using augmented reality is really a fancy term for new technology, is very usable, readily available, most of the young people are already behavior or change their required for augmented reality anyway. Playing games on their handheld, smartphones, navigating cities… younger people use smart phones for just about everything. So already that behavior changing is happening, or using smartphones for voice interaction, or using smartphones for direction. So there's no problem in transform using the smartphone to then moving to the visual. Augmented reality, is very visual navigation, visual interaction.

City image is the manifestation of city culture, and it is an urban form that can stimulate people's ideological and emotional activities. It is a cultural model with urban characteristics, created jointly by citizens in the long-term life process, and is the sum of urban living environment, lifestyle and living customs. It has the characteristics of complexity and diversification. We often visit museums in a certain country or city to learn about the local history, culture and customs. However, the city, as an all-encompassing huge museum, the cultural connotation hidden under the reinforced concrete constitutes the flesh and blood of it.

Public art design, as one of the cultural carriers for constructing today's urban appearance, bears the responsibility of cultural communication in a certain sense, which is mainly manifested in: the spread of historical culture, the spread of aesthetic culture, and regional characteristics. In today's information age, the scope of public art design has covered digital media and interactive application methods. This article discusses the way that uses art and design as media for cultural communication.

Artists and designers are the subject of this project. We will extract the cultural and design elements of artists and designer to build a sample library. Then AR technology is used to combine the virtual model constructed by the artist's design elements with the real city to create a new urban style and bring a new urban experience to users. The real city will become a huge canvas (background) in the future, providing a new way of expression for artist/designer style and ideas. For users, this is also a new cultural experience journey different from visiting a museum.

## 2 Literature Review

Many research papers have shown that AR technology has played a significant role in promoting cultural exchanges. In AR, one of the display areas with great potential is cultural and historical heritage, also known as virtual heritage. Reconstruct 3D visual representations of monuments, artifacts, buildings and other cultural and historical artifacts by using AR technology. This visual and interactive cultural experience is more realistic and detailed than traditional text or pictures to convey information. It is very useful to help users understand site information and history.

In addition to its role in the study of historical and cultural heritage, AR technology has also played an increasingly important role in the development of contemporary tourism which plays an important role in the cities. Tourists who come to the city are aiming to explore historical and touristic areas, social areas, entertainment and shopping centers. Many navigation systems based on augmented reality technology came into being, and many location-based AR city exploration apps evolved constantly.

**Go Find!** GoFind! can be used by historians and tourists and provides a virtual view into the past of a city. The system provides location-based querying in historic multimedia collections and adds an augmented reality-based user interface that enables the overlay of historic images and the current view.

**Nokia City Lens.** Nokia City Lens uses the device's camera to display nearby restaurants, stores, and other notable locations in augmented reality style, along with important information about each location — including reviews, directions, hours of operation, and more. And when you're not in the mood to have your reality augmented, you can also check out the same information through both a list and map view.

**AR Urban Canvas App for HKYAF.** HKYAF is a City Exploring Mobile App. Users can follow cultural tour routes, explore hidden neighborhood stories in Hong Kong, and view the hidden artworks with Augmented Reality technology.

Near the tail end of the 20th century, pseudonymous author and technologist Ben Russell released The Headmap Manifesto—a utopian vision of augmented reality referencing Australian aboriginal songlines and occult tomes, while pulling heavily from cybernetic theory and the Temporary Autonomous Zones of Hakim Bey. At turns both wildly hypothetical and eerily prescient, Headmap explores in-depth the implications of "location-aware" augmented reality as a kind of "parasitic architecture" affording ordinary people the chance to annotate and re-interpret their environment.

Around the same time, artist and scholar Teri Rueb began developing her pioneering, site-specific augmented "soundwalks," some of the earliest and most influential examples of GPS-based art practice. Influenced by land art practitioners such as Robert Smithson and Richard Serra, Rueb's work identified the critical potential of locative AR as a direct mediator of spatial experience, capable of revealing hidden layers of meaning within landscapes. Beyond land art, a number of early AR practitioners and theorists explicitly identified the Situationist International (and even Archigram) as conceptual touchstones for the kind of digital enhancement, and potential subversion, of space made possible through augmenting reality.

In 2019, Apple introduced Apple [AR]T Walk Project. Augmented Realities Co-Curated with New Museum. These experiential walks take participants through San Francisco, New York, London, Paris, Hong Kong and Tokyo as they encounter works by world-renowned artists, most of whom are working in AR for the first time. Works by Cave, Djurberg and Berg, Cao, Giorno, Höller and Rist connect participants to public spaces such as London's Trafalgar Square, San Francisco's Yerba Buena Gardens or New York's Grand Army Plaza in Central Park. Using AR, the artists have reimagined or invented new ways to express core themes of their art practice. Rist's "International Liquid Finger Prayer" bounces, taunts and sings as participants race to catch a shimmering form, while Giorno's "Now at the Dawn of My Life" is a rainbow journey of homespun wisdom, and "Through" by Höller takes viewers through a portal into a world with no perspective.

These studies show that there have been many attempts at location-based AR technology-based art exploration. Combining the inspiration obtained from these cases and the development trends and application areas of AR application software in the era, we have launched a project called City Art Project.

## 3  System Description

An AR mobile app called City Art through Vuforia using the SDK open source framework will be design in this project.

This is an urban exploration app that combines urban experience with cultural communication. City was chosen as the background for the exploration journey.

A city's common image is a mixture of many images. American urban research scholar Kevin Lynch, in his book "The Imageof The City" proposes the main elements of urban imagery: channels, edges, regions, nodes, landmarks these five aspects. Lynch mentioned that urban imagery is a basic part of our daily life. The city image is valuable, it is like having a map for people to taking directions, and taking action, and the scenery plays a social role. The place where all known causes a common feeling when mentioning about it. This symbolic sign will unite everyone and use it to create a common goal. Therefore, based on the analysis of urban characteristics and urban imagery, the city is divided into five parts: Popularity, Activity, Street view, Modern life, Transportation. These five parts will serve as a background for exploring city.

Milan was selected as a pilot city as a demonstration of the city's art journey. By decomposing the composition of the city, we identify the exploring objects of the city as the main buildings of the city——famous places of interest (such as Milan Duomo), historic buildings (such as the Monastery of Santa Maria della Grazie), and the Museum with Cultural Significance, etc.

## 3.1  Concept Interpretation

We have selected some artists and designers to build a sample library, analyze and decompose their works, and extract design elements and color features from them to establish independent material packages. These elements will be randomly combined to form a vitual 3D model cover the real scenery in the city in the app (Fig. 1).

Designer/Atist Work      >      Design element      >      Final result

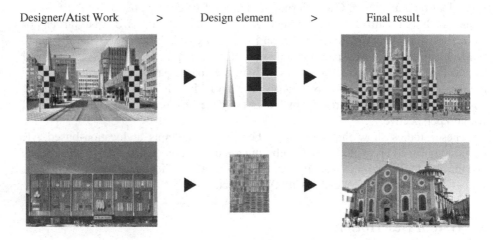

**Fig. 1.** Transformation process

## 3.2  System Overview

The application's system has three main components: User Block, Marker Tracking, and Augmented Reality Environment. The user block contains users and mobile devices. Marker Tracking with geographic recognition will obtain the user's location through GPS. At the same time, the corresponding 3D rendering model will be loaded in the Augmented Reality Environment. Finally, the output results are visualized to the user (Fig. 2).

After opening the app, users can choose to download their favorite designer material packages, then turn on the AR mode, call up the camera, and the images taken from the camera will be loaded onto the Vuforia SDK and a virtual 3D model with geo-tagging (that is, 3D model stitched by designer material package), and finally displayed to the user through a mobile display device.

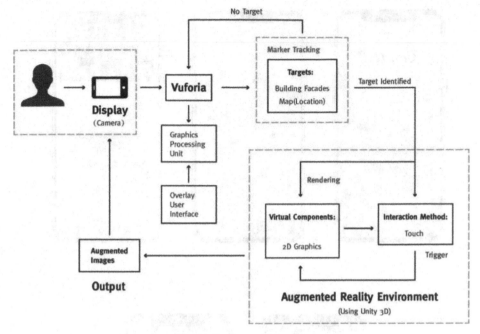

**Fig. 2.** System architecture

### 3.3  Application Design and Features

The design of the entire app mainly includes 5 major modules: Home, Map, Community, Favorite, and Me.

**Homepage.** The homepage is the portal for the user to start the city experience. First, by click the camera buttom, user will directly enter the download interface, where user can choose to download the artists/designers material pack that they prefered. The next step is to activate the camera to allow the user to experience the city in AR mode, and taking pictures or videos. The pictures and videos will be saving to 'Me' as user's creation. This is the main function of the app, for the display and recording of the combination of urban scenes and virtual 3D model (Fig. 3).

**Map.** The map plays the role as a guider. Users can find Milan's main city nodes and landmark buildings on the map. The map will highlight famous places of interest such, historic buildings, and culturally significant museums. Users can explore the city according to the planned routes.

**Community.** The design of the app includes community attributes. Users can record their city impressions through photos or videos, and share to the community or other social platforms. This constitutes a platform for cultural communication.

**Favorite.** After sharing to the community, users can look through the works in it, and collect their favorite works save them to Favorite.

**Fig. 3.** App interface design

**Me.** This part used to store user's own picture and video material that record during city tour. All the pictures and videos will automatically save to here.

### 3.4 Operating Process

After entering the homepage, click on the camera button to start downloading artist/designer material packages. After the authorization is completed, the camera is authorized to enter AR mode, and users can download multiple material packages (no more than 5). Then, authorize the camera to turn on the AR mode. After entering the AR experience mode, users can switch between different artist/designer roles to experience and observe different style features (Fig. 4).

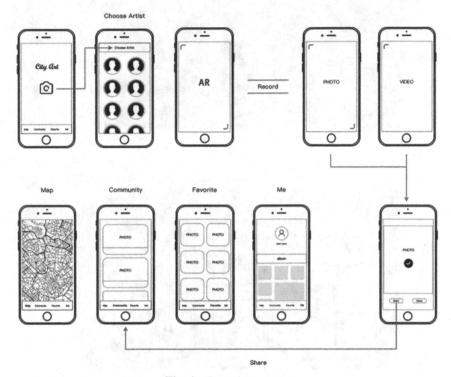

**Fig. 4.** Operating process

In simple terms, the solid scene covered by the 3D virtual model is like a layer of 'Filter'. After downloading a certain number of artist/designer material packages, users can switch between different artist/designer roles (such as When facing the Milan Cathedral, the user can choose the perspective of designer Alessandro Mendini or the perspective of Ettore Sottsass, which will produce different visual effects), or switch between different perspectives under the same character (such as MD01, MD02 in the material of Mendini Effects such as different design elements or design styles) (Fig. 5).

Of course, it is ideal first use map to determine the destination. Photographs and videos are all allowed recording. Recording results are automatically saved to 'Me' and allowed to be shared to 'Community' or other social media. After sharing, users can browse other people's records in 'Community' and add to 'Favorite'…

**Fig. 5.** Final effect

## 4 Discussion

This article actually uses Italian culture and representative city Milan in Italy as examples to explore the role of urban AR technology in cultural exchange and innovation. Italy, as the birthplace of the Renaissance, has deep historical and cultural heritage, especially in the field of art design, and has extremely rich resources and materials. The spiritual culture of Italian art design has been subtly influencing the development of contemporary design. However, this inflence has gradually been swallowed up by the cultural products of the great fusion in the torrent of cultural fusion, losing its independence and characteristics. By extracting distinctive element characteristics, that is, sorting out the characteristics of design elements of representative artists and designers, using innovative emerging AR technology and urban architecture, combined with the communication nature of communities and social media, this City Art Interactive urban

AR application, to some extent, is an effective means for quickly acquiring cultural information in accordance with the fast-moving modern society.

## References

1. Smith, T.F., Furht, B.: Handbook of Augmented Reality. Florida Atlantic University. Springer, New York (2011). https://doi.org/10.1007/978-1-4614-0064-6
2. Cirulis, A., Brigmanis, K.B.: 3D outdoor augmented reality for architecture and urban planning. In: 2013 International Conference on Virtual and Augmented Reality in Education. Procedia Comput. Sci. **25,** 71–79 (2013)
3. Faruk, Ö., Karaarslan, E.: Augmented reality application for smart tourism: GÖkovAR. In: 2018 IEEE
4. dela Cruz, D.R., et al.: Design and development of augmented reality (AR) mobile application for malolos, Kameztizuhan, Malolos Heritage Town, Philippines. In: 2018 IEEE Games, Entertainment, Media Conference (GEM) (2018)
5. Sauter, L., Rossetto, L., Schuldt, H.: Exploring cultural heritage in augmented reality with go find! In: 2018 IEEE International Conference on Artificial Intelligence and Virtual Reality (AIVR) (2018)
6. Marjury Díaz, H., Karen Barberán, C., Diana, M.-M., Gabriel López, F.: Offline mobile application for places identification with augmented reality. IEEE (2017)
7. Azevedo, J.N., Alturas, B.: The augmented reality in Lisbon tourism. In: 14th Iberian Conference on Information System and Tchnologies (CISTI) (2019)
8. Lee, H., Chung, N., Jung, T.: Examining the cultural differences in acceptance of mobile augmented reality: comparison of South Korea and Ireland. In: Tussyadiah, I., Inversini, A. (eds.) Information and Communication Technologies in Tourism 2015, pp. 477–491. Springer, Cham (2015). https://doi.org/10.1007/978-3-319-14343-9_35
9. Yung, R., Khoo-Lattimore, C.: New realisties: a systematic literature review on virtual reality and augmented reality in tourism research. Current Issue Tourism, 1–26 (2017)
10. Schechter, S.: What is markerless Augmented Reality| AR Bites (2014). https://www.marxen tlabs.com/what-is-markerless-augmented-reality-dead-reckoning/
11. McWhirter, J.: City skins: scenes from an augmented urban reality. FA Failed Architecture (2018). https://failedarchitecture.com/city-skins-scenes-from-an-augmented-urban-reality/
12. Curtin, G., SCEWC Urban Internet of Things: Augmented Reality and the Future of Urban Design. In: urbanNext (2017). https://urbannext.net/urban-internet-of-things/
13. Palos-Sanchez, P., Saura, J.R.: Ana Reyes-Menendez and Ivonne Vásquez Esquivel "USERS ACCEPTANCE OF LOCATION-BASED MARKETING APPS IN TOURISM SECTOR: AN EXPLORATORY ANALYSIS. J. Spat. Organ. Dyn. **6**(3) (2018). Marketing and Tourism
14. Li, R., Zhang, B., Shyam Sundar, S., Duh, H.B.-L.: Interacting with augmented reality: how does location-based AR enhance learning? In: Human-Computer Interaction – INTERACT 2013, p. 61 (2013)

# Reporting Strategy for VR Design Reviews

Martin Gebert⊙, Maximilian-Peter Dammann(⊠)⊙, Bernhard Saske,
Wolfgang Steger, and Ralph Stelzer

Technische Universität Dresden, 01062 Dresden, Germany
{martin.gebert,maximilian_peter.dammann,bernhard.saske,
wolfgang.steger,ralph.stelzer}@tu-dresden.de

**Abstract.** Design reviews are an established component of the product development process. Especially, virtual reality design reviews (VRDRs) can generate valuable feedback on the user's perception of a virtual product. However, the user's perception is subjective and contextual. Since there is a lack of strategies for recording, reproducing VRDRs is an intricate task. User feedback is therefore often fleeting. This makes deriving meaningful results from VRDRs a diffuse task.

In this paper, we present a strategy for recording VRDRs in structured and coherent reports. We suggest dividing all involved information into *structural* and *process information*. While the former describes the content and functionality of the VRDR, the latter represents occurring events.

The strategy provides means to store collections of events in the context they occurred in. Additional properties such as timestamps, involved users and tags are provided to create comprehensive VRDR reports. By storing the report in a database, sorting and filtering algorithms can be applied to support efficient data evaluation subsequent to the VRDR. Thus, such reports can be used as a basis for reproducible VRDRs.

**Keywords:** Design review · Virtual reality · Reporting strategy · Interaction logging · User perception · Reproduction · Product development

## 1 Introduction

Virtual Reality (VR) is an established tool for numerous visualization tasks that occur throughout the product development. VR creates high immersion and supports intuitive input devices, enabling a realistic impression of a virtual product [1]. Especially, virtual reality design reviews (VRDRs) take advantage of these characteristics and promise valuable feedback on a virtual product throughout the product development phase.

Gathering meaningful feedback from a VRDR is not a simple task. The feedback is based on the subjective user perception of a virtual product. User perception can only be understood by reproducing the relevant VRDR events in the context they occurred in. Unfortunately, comprehensive event logging in VR has not been thoroughly discussed in previous research [2]. The provided feedback is therefore mostly fleeting.

In this publication, we propose a report strategy that provides a basis for making VRDRs reproducible. For this purpose, we take a look at the information involved in a

© Springer Nature Switzerland AG 2020
C. Stephanidis et al. (Eds.): HCII 2020, LNCS 12428, pp. 80–90, 2020.
https://doi.org/10.1007/978-3-030-59990-4_7

VRDR. We collect the information and their semantic connection in a structured manner. The information is then stored in a database, making it accessible for sorting and filtering algorithms to enable targeted evaluation.

## 2  Literature Review

A systematic literature review on logging in VR was conducted in 2018 by Luoto [2]. The review concluded that publications about logging and report strategies in VR were relatively rare. In many cases, they did not discuss details about how data was gathered and stored. The most common reasons to log data in a VR application were to compare or verify different VR technologies. Others log data for the purpose of assessing the value of VR in design reviews [3].

The gathered data often comprised tracking data of the HMD or other tracking or input devices [4–6]. Other log data included specific interactions [7], text entries [8], user annotations [9], conversations [10] or videos [3]. Log data was used to automatically detect relevant operations or design tasks [3, 11] or to replay events in a VR application [12, 13].

Steptoe and Steed proposed a multimodal data collection pipeline that aimed at log standardisation [5]. They discussed the handling of tracking data with high refresh rates like eye tracking.

The data log of a VR application can be stored in different ways. While storage in a database was proposed by Koulieris et al. [14], other research suggests to save data streams of tracking data on a server [15] or to use log files [5, 13]. Ritchie et al. suggest the usage of an XML log file [12].

In most related research, logging is only performed to collect specific data required to carry out studies. Generally, the research focuses on using logs for the purpose of improving design reviews rather than improving the reviewed content.

## 3  Virtual Reality Design Review

A VRDR is a process intended for the evaluation of products. It is performed in a 3D scene that allows users to interact with future products in arbitrary settings. An information model is required to describe the 3D scene. The model contains the information from which the representation is generated. We propose differentiating between *structural information* and *process information*. While the former describes the appearance and behaviour of all scene content, the latter refers to all occurring events during the VRDR (see Fig. 1).

Structural information is managed within our *scene model* that integrates four main scene model components: a *scene component*, a *user component*, a *product component* and a *setting component*. The structural information of each component derives from the corresponding model description that is developed and managed in its domain-specific software applications. Further details on the components are provided in Sect. 4.

During a VRDR session, users interact with the virtual products. In this process, the components generate and receive process information on their behaviour. Our *recording model* describes how to log the information with contextual details of the scene model.

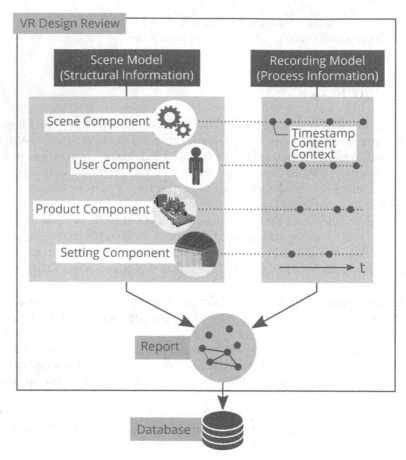

**Fig. 1.** The report consists of structural and process information

Furthermore, a timestamp is added to the log the moment it is created. The collected *report* is then stored in a *database*.

## 4 Scene Model

The content of a VRDR scene is derived from model descriptions that are developed in their specific domains. This content is a momentary image of selected data of constantly evolving model descriptions. It is of vital importance to the evaluation of a VRDR as it facilitates understanding user remarks and behaviour in their respective context. This context can only be provided by well-structured information. For this reason, we call it *structural information.*

Structural information must be organized in logical granules that constitute more complex, hierarchical structures such as the scene model components. These granules are often represented in 3D scenes by objects. Within the scope of the scene model, the logical granules are called *structural objects* (see Fig. 2).

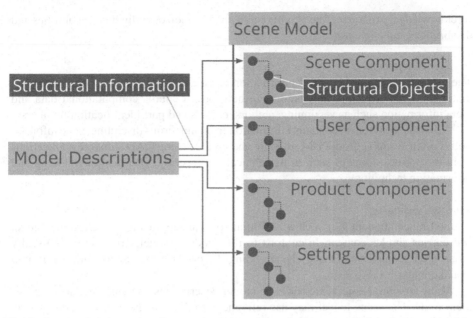

**Fig. 2.** Structural information is organized in structural objects within each scene model component

Each structural object implements a fraction of the appearance and behaviour of a scene model component. The relationship between structural objects is defined by the hierarchical structure of the scene model.

The content and structure of each component may vary in every particular VRDR scene. In Fig. 2, we suppose an exemplary set of structural objects within each component. In the following paragraphs, we explain the role of each component and suggest content and functionality that could be implemented by the contained objects. The contents are not intended to be exhaustive.

*Scene Component*
The scene component acts as the control centre of the VRDR scene, containing all information that enables interaction between the user, product and setting components. As it handles all process information that is generated during a VRDR, the scene component is essential to the VRDR report.

The scene component contains user interface functionality, integrates input and output hardware and implements further scene-specific functionality. Input and output hardware are means for users to interact with the scene. The hardware must be assigned to users when they enter the scene.

*User Component*
The user component characterises one or more users with a set of personal characteristics that are required to integrate them into the scene to the desired degree. These may include physical and cognitive parameters such as body height, age, pupillary distance,

foreknowledge, experience, etc. Additionally, information on individual capabilities and disabilities can be integrated.

*Product Component*
The product component contains information on one or multiple products' geometries, materials, structure and functionality. Furthermore, it entails computational data and meta information such as versioning, authors, editors and part identification numbers.

A product component contains information from numerous disciplines. The information is often available from PDM software and created using CAD software applications from the various domains such as mechanical engineering, electrical engineering and information technology.

*Setting Component*
The setting component is created with similar software applications as a product model and carries similar content. In contrast to the product model, this content is usually focused on geometry and materials, providing the product component with the desired visual context.

Many structural objects are based on model descriptions contained in data management tools. The scene model may use references to the management tools rather than containing the full description of structural objects, reducing data redundancy.

The described scene model can be represented by a scene graph. A previous publication proposed an enhanced scene graph to describe 3D scene content in a structured manner [16]. This approach can be applied to our scene model, allowing storage of its structural objects in a database and facilitating references from the recording model. For this purpose, all structural objects must be assigned a structural object identification (SOID) which is used to reference process information to the involved structural objects.

## 5   Recording Model

Considering the complexity of a VRDR, an incoherent, sequential recording of all events is not a sensible approach as searching such a recording for specific information is highly inefficient. However, many events have a semantic connection and create meaningful information when viewed as a collective.

For example, a motion profile of a user may consist of multiple tracking data streams. All recorded tracking events must indicate which data stream they belong to. Furthermore, each stream must indicate which user it belongs to. This way, the motion profile of a specific user can be accessed efficiently.

Our recording model reflects the semantic relationship between events. The core element of the recording model is a generic *process object*. It collects related events and offers additional properties that indicate semantic connections to other process objects.

A process object includes the following properties:

*Timestamp*
The point in time when the process object is created is important to relate it to other events. Therefore, every process object must contain a timestamp.

*Origin SOID*
A process object always originates from an event that was evoked by a structural object. This object could represent a user or an object that is controlled by a user. As this piece of information may be useful to the evaluation process, the *origin SOID* points to the structural object that created the process object. Using the SOID and the structural information attached, the context of the process object can be reproduced.

*POID*
Each process object is assigned a process object identification (POID) using a different nomenclature than applied for SOIDs. This facilitates pointing from one process object to another, e.g. when a process object builds on a previous one.

*Object Data*
Object data provides all the variable data of the process object. The data is of arbitrary type and complexity. While a button identifier and a Boolean can represent a simple button press event, a motion profile is a vast collection of tracking data. Object data may also contain file pointers to external data containers for enhanced functionality.

*Classification Tags*
Many process objects can be classified according to their type of content or must indicate their semantic connection with other process objects. For this purpose, an arbitrary number of classification tags can be added to the process object.

Our intention is not to define the specific contents of process objects that could be used to log all VRDR events. In fact, the content depends on the functionality of the VRDR and cannot be anticipated by the recording model. Instead, the functionality to create process objects and provide it with the required data should be implemented along with the scripts that implement the VRDR functionality.

In order to provide a better understanding of this implementation, we demonstrate the application of a process object for the purpose of storing user annotations in Sect. 7.

## 6   Report and Database

An evaluation of a VRDR is based on reproducing the events in the context of the scene they occurred in. For this reason, a VRDR report must include all process and structural objects.

Once the structural and process objects are stored, the database allows access to all information for evaluation purposes. The information can be filtered and sorted using the properties available from the objects. Additionally, the relationship of objects can be used to reproduce process information in a large context. This enables great opportunities for evaluation of a VRDR, e.g.

- reproducing the VRDR sequentially,
- searching for specific types of process information using keywords,
- gathering all process information related to a specific user or
- usability of (parts of) a product.

## 7  Application

A typical representation of process information in a VRDR are user annotations. Their purpose is to give users the means to provide intentional feedback on their individual observations. In earlier research, we proposed an annotation system based on predefined icons and short texts (see Fig. 3). These can be combined to represent complex information [9]. This system is especially suited to the needs of AR/VR-systems with limited interaction and input options.

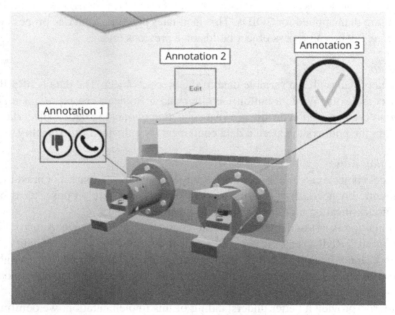

**Fig. 3.** Example annotations, generated in a VRDR

Each annotation in Fig. 4 is represented by a process object. These process objects contain properties that are common to any process object as described in Sect. 5. However, some properties are characteristic to this type of process object. They are provided in the object data property: priority, content, transformation matrix and reference objects (see Fig. 4).

The priority is selected by the user in the VRDR user interface. The content describes the actual user remark that can feature predefined icons, text snippets, sketches, audio files and video files. The transformation matrix defines the location, rotation and scale of the annotation in the VRDR scene. Finally, the reference object contains an SOID that points to a structural object which could represent a part of the product.

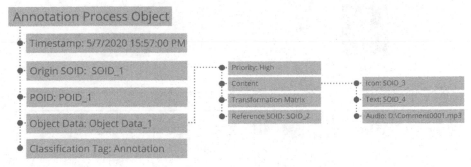

**Fig. 4.** Annotation process object

In order to create this type of process object, the relevant functionality needs to be implemented in the scene component. This requires only little effort since the functionality for the creation of a user annotation is already implemented. The predefined properties of the process object merely enforce a consistent description of the annotation.

The moment the user initializes the creation or change of an annotation, the scene component has to create a process object. The origin SOID points to the object that creates the annotation. Using the scene model, the user that is assigned to this object, can be identified. Finally, the process object is assigned the classification tag "Annotation".

The content of the annotation is a list of properties. Predefined content like icons and short texts are referenced with SOIDs. Data that is created during the VRDR such as video or audio files is referenced using a file path.

With the previously proposed methods, all changes of an annotation can be logged and analysed. Every change of the annotation results in the creation of a new process object (see Fig. 5). Therefore, single states of the annotation are not lost and each change can be traced. Process objects belonging to an annotation keep the same POID throughout the VRDR. The timeline for all process objects with the same POID is mapped by the timestamps.

The provided example demonstrates a typical use case of a process object for engineering scenarios. The annotation may contain valuable information to an engineer that aims to improve a virtual product subsequent to the VRDR. This use case is anticipated and the functionality to create and log the annotation is deliberately implemented.

In a different scenario, the actual content of the annotation may not be of interest. Instead, a user interface designer may want to assess the functionality that lets the user create the annotation in the VRDR. The designer could be interested in the course of action in order to evaluate the performance and error rate of the user when creating the annotation.

In contrast to the annotation, the course of action that leads to the annotation's creation is a complex combination of diverse user actions. This semantic connection of the involved events can hardly be anticipated. For this reason, we suggest to store all VRDR events in process objects. These process objects organise the events in a fashion that may not be ideal for the discussed use case, but relevant events can still be reproduced. It is the responsibility of the creator of the scene component to decide on the definition and content of the process objects.

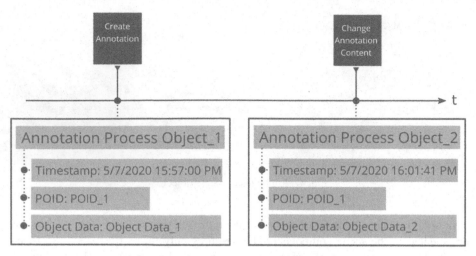

**Fig. 5.** Example for process object timeline

# 8   Conclusion

In this paper, we propose a strategy for VRDR reports. The scene model defines a hierarchical structure of the structural objects involved in a VRDR. This provides each structural object with identification and semantic context. The recording model is based on process objects. They collect and store related types of events that occur throughout a VRDR. Furthermore, their properties contain details on the time of creation, involved structural objects, and classification tags.

By combining the two models, we are able to create a comprehensive report of a VRDR. The report is capable of reproducing user actions in their performed context using the properties provided in the process objects. With access to the report from a database, the VRDR events can be sorted and filtered according to the needs of evaluation.

In our exemplary application, we illustrate how to store user annotations in a process object. This process object is intentionally implemented for the purpose of evaluation as it may provide engineers with useful feedback from the VRDR. However, the events of interest for evaluation cannot always be anticipated. Therefore, we propose to store all occurring events as process objects to enable reproduction of the VRDR for arbitrary evaluation purposes.

In conclusion, the report strategy provides the means to collect meaningful feedback from a VRDR. It is a basis for reproducing subjective user perception and gathering input for the product development process.

# 9   Outlook

As presented in this publication, our report strategy has been successfully applied to describe and store user annotations created in a VRDR. Consequently, a fully functional prototype should be developed to apply the strategy to a wider range of process information. The prototype should include storage of the report in a database that allows us to experiment with the recorded VRDR data. It is important to derive rules and recommendations to those who implement the report strategy in the future.

Furthermore, the generated process objects of our approach are capable of storing the course of actions of a VRDR. This can be utilized during a review to undo changes on objects or to record and replay courses of movement.

Until now, we have aimed our report strategy to be applicable for VRDRs in the area of mechanical engineering, more specifically as a strategy to improve the product development process. As our approach is very generic, it could certainly be applied to other areas of VR applications. It could be used in industrial VR applications of architectural and civil engineering but also in research areas where VR is used to conduct studies, e.g. in architecture or psychology.

As the approach promises to assist in understanding user perception by making their history of actions accessible from a database, insights could be gained on human-machine-interaction. A specific task for the near future is to investigate adaptive user interfaces while logging the user actions. The goal is to provide a suitable model for adaptive user interfaces, enabling fast implementation and flexible experiments to conduct studies on user perception.

**Acknowledgements.** The research presented in this publication is funded by
- the Deutsche Forschungsgemeinschaft (DFG, German Research Foundation) – 319919706/RTG2323
- the AiF Projekt GmbH, a branch of the AiF (Arbeitsgemeinschaft industrieller Forschungsvereinigungen "Otto von Guericke" e.V.). The AiF Projekt GmbH is a loaned project manager for ZIM (Zentrales Innovationsprogramm Mittelstand) cooperation projects. The ZIM is a funding program by the German Federal Ministry for Economic Affairs and Energy - focussed on the cooperation of medium-sized companies and research institutes. - ZF4123401LF5 and ZF4123404SS7.

# References

1. Steger, W., Kim, T.-S., Gebert, M., Stelzer, R.: Improved user experience in a VR based design review. In: Volume 1B: 36th Computers and Information in Engineering Conference, ASME 2016 International Design Engineering Technical Conferences and Computers and Information in Engineering Conference, Charlotte, North Carolina, USA, 21–24 August 2016. American Society of Mechanical Engineers (2016). https://doi.org/10.1115/detc2016-59840
2. Luoto, A.: Systematic literature review on user logging in virtual reality. In: Proceedings of the 22nd International Academic Mindtrek Conference, the 22nd International Academic Mindtrek Conference, Tampere, Finland, 10 October 2018–10 November 2018, pp. 110–117. ACM, New York (2018). https://doi.org/10.1145/3275116.3275123
3. Liu, Y., Castronovo, F., Messner, J., Leicht, R.: Evaluating the impact of virtual reality on design review meetings. J. Comput. Civ. Eng. **34**, 4019045 (2020)
4. Williams, B., McCaleb, M., Strachan, C., Zheng, Y.: Torso versus gaze direction to navigate a VE by walking in place. In: Hoyet, L., Sanders, B.W., Geigel, J., Stefanucci, J. (eds.) Proceedings of the ACM Symposium on Applied Perception, the ACM Symposium, Dublin, Ireland, 22–23 August 2013, p. 67. ACM, New York (2013). https://doi.org/10.1145/2492494.2492512

5. Steptoe, W., Steed, A.: Multimodal data capture and analysis of interaction in immersive collaborative virtual environments. Presence Teleoperators Virtual Environ. (2012). https://doi.org/10.1162/PRES_a_00123
6. Piumsomboon, T., Day, A., Ens, B., Lee, Y., Lee, G., Billinghurst, M.: Exploring enhancements for remote mixed reality collaboration. In: Billinghurst, M., Rungjiratananon, W. (eds.) SIGGRAPH Asia 2017 Mobile Graphics & Interactive Applications, SIGGRAPH Asia 2017 Mobile Graphics & Interactive Applications, Bangkok, Thailand, 27–30 November 2017, pp. 1–5. ACM, New York (2017). https://doi.org/10.1145/3132787.3139200
7. Bani-Salameh, H., Jeffery, C.: Evaluating the effect of 3D world integration within a social software environment. In: Latifi, S. (ed.) 2015 12th International Conference on Information Technology - New Generations (ITNG), Las Vegas, Nevada, USA, 13–15 April 2015, pp. 255–260. IEEE, Piscataway (2015). https://doi.org/10.1109/ITNG.2015.47
8. McCall, R., Martin, B., Popleteev, A., Louveton, N., Engel, T.: Text entry on smart glasses. In: 2015 8th International Conference on Human System Interactions (HSI), Warsaw, Poland, 25–27 June 2015, pp. 195–200. IEEE, Piscataway (2015). https://doi.org/10.1109/HSI.2015.7170665
9. Dammann, M., Steger, W., Stelzer, R., Bertelmann, K.: Aspekte der Interaktionsgestaltung in mobilen AR/VR Anwendungen im Engineering. In: Putz, M., Klimant, P., Klimant, F. (eds.) VAR$^2$ 2019 – Realität erweitern 5. Fachkonferenz zu VR/AR-Technologien in Anwendung und Forschung an der Professur Werkzeugmaschinenkonstruktion und Umformtechnik. VAR$^2$ 2019 - Realität erweitern., Chemnitz, 04–05 December 2019, pp. 151–161 (2019)
10. Sunayama, W., Shibata, Y., Nishihara, Y.: Continuation support of conversation by recommending next topics relating to a present topic. In: Matsuo, T., Kanzaki, A., Komoda, N., Hiramatsu, A. (eds.) 2016 5th IIAI International Congress on Advanced Applied Informatics - IIAI-AAI 2016, Kumamoto, Japan, 10–14 July 2016, pp. 168–172. IEEE, Piscataway (2016). https://doi.org/10.1109/IIAI-AAI.2016.90
11. Sung, R.C.W., Ritchie, J.M., Robinson, G., Day, P.N., Corney, J.R., Lim, T.: Automated design process modelling and analysis using immersive virtual reality. Comput. Aided Des. (2009). https://doi.org/10.1016/j.cad.2009.09.006
12. Ritchie, J.M., Sung, R.C.W., Rea, H., Lim, T., Corney, J.R., Howley, I.: The use of nonintrusive user logging to capture engineering rationale, knowledge and intent during the product life cycle. In: Kocaoglu, D.F. (ed.) Portland International Conference on Management of Engineering & Technology 2008, PICMET 2008, Cape Town, South Africa, 27–31 July 2008, pp. 981–989. IEEE Service Center, Piscataway (2008). https://doi.org/10.1109/PICMET.2008.4599707
13. Stelzer, R., Steindecker, E., Arndt, S., Steger, W.: Expanding VRPN to tasks in virtual engineering. In: Volume 1B: 34th Computers and Information in Engineering Conference, ASME 2014 International Design Engineering Technical Conferences and Computers and Information in Engineering Conference, Buffalo, New York, USA, 17–20 August 2014. American Society of Mechanical Engineers (2014). https://doi.org/10.1115/DETC2014-34277
14. Koulieris, G.-A., Bui, B., Banks, M.S., Drettakis, G.: Accommodation and comfort in headmounted displays. ACM Trans. Graph. (2017). https://doi.org/10.1145/3072959.3073622
15. Kobayashi, K., Nishiwaki, K., Uchiyama, S., Yamamoto, H., Kagami, S.: Viewing and reviewing how humanoids sensed, planned and behaved with mixed reality technology. In: 2007 7th IEEE-RAS International Conference on Humanoid Robots (Humanoids 2007), Pittsburgh, PA, USA, 29 November 2007-1 December 2007, pp. 130–135. IEEE, Piscataway (2007). https://doi.org/10.1109/ICHR.2007.4813859
16. Gebert, M., Steger, W., Stelzer, R.: Fast and flexible visualization using an enhanced scene graph. In: Proceedings of the ASME International Design Engineering Technical Conferences and Computers and Information in Engineering Conference – 2018, Quebec City, Canada, 26–29 August 2018, V01BT02A025. The American Society of Mechanical Engineers, New York (2018). https://doi.org/10.1115/DETC2018-85750

# Video Player Architecture for Virtual Reality on Mobile Devices

Adriano M. Gil, Afonso R. Costa Jr., Atacilio C. Cunha$^{(\boxtimes)}$,
Thiago S. Figueira, and Antonio A. Silva

SIDIA Instituto de Ciência e Tecnologia (SIDIA), Manaus, Brazil
{adriano.gil,afonso.costa,atacilio.cunha,thiago.figueira,
antonio.arquelau}@sidia.com

**Abstract.** A virtual reality video player creates a different way to play videos: the user is surrounded by the virtual environment, and aspects such as visualization, audio, and 3D become more relevant. This paper proposes a video player architecture for virtual reality environments. To assess this architecture, tests involved comparisons between an SXR video player application, a fully featured 3d application, and a video player implemented in Unity. Tests generated performance reports that measured each scenario using frames per second.

**Keywords:** Virtual reality · Mobile · Video player · Architecture · Unity · SXR

## 1 Introduction

Virtual Reality (VR) applications provide great interaction with multimedia content. This content becomes a unique experience for the end-user as applications work as a media gallery inside the VR environment, such as the Samsung VR [14].

Those applications, which are usually built using 3D engines or frameworks, target mobile devices like smartphones. In the Android platform, there are two main 3D engines for creating such applications for mobile devices: Unity [16] and the Samsung XR (SXR) [15]. Those engines are a good choice, but it is important to mention that these applications will run in a mobile platform that has limited resources like memory, battery, and computational power. In this scenario, this paper defines a software architecture to organize the communication between the render and the platform layers, therefore providing better performance results.

Mobile platforms like Android already have their native media player. This kind of application has its architecture and components well defined and designed to manage performance and resource consumption, which means that the media codecs were chosen/defined and are organized to support the most common media formats and data sources.

Supported by Sidia.

A video player application is built using two main parts: a media player (that takes digital media in and renders it as video) and a media controller (an User Interface - UI, with transport controls to run the player and optionally display the player's state). Video player applications that are being built for VR environments need to have this kind of organization as well, once they are using 3D engines to render digital media.

This work proposes a high-performance architecture that can be implemented for video players on mobile platforms to run videos in VR environments. This architecture is compared to the Unity and SXR engines in the Android platform.

The structure of this paper is as follows. In Sect. 2, we provide an overview of the video players usage in VR context. Then, in Sect. 3, the methodology used in this work will be detailed. In Sect. 5, the proposed architecture will be shown and explained. Finally in Sect. 6, we will provide some final remarks.

## 2   Related Work

Media consumption is a relevant activity for users in the digital world, and it has been growing according to [10]. In this work, this consumption is shown as a way to deal with tedious situations, sharing experience and sharing knowledge.

Applications can use media as a way to create an specific interactions with users. In [4] is cited an example of applications to language learning using a video player focusing on language manipulation inside the media.

In [13] are listed different codecs that can by used by stereoscopic videos. For such kind of media the compressed data streaming have a depth component together with RGB channels. So a VR video player architecture should take account of different video formats that can be decoded and rendered in specific ways, as it is the case of stereoscopic videos.

The work [18] mentions Video Players should be being accessible to people with special needs. According to the article, it is necessary that the video player has subtitles support, audio description, media transcriptions, support for change volume and color contrast.

MR360 defined in [11] proposes a composition of 3D virtual objects during a 360 video execution. The panoramic video is used both as a light source to illuminate virtual objects how much to compose the backdrop to be rendered. That's an interesting example of a different way to render videos in VR environments. How to design an architecture that make possible a high range of different ways to show videos in VR?

## 3   Methodology

The following methodology is being used to evaluate the proposed architecture:

1. Definition of the video format.
2. Definition of ways to visualize the video in a virtual environment.

3. Comparison between two different render implementations (Unity and SXR) in the Android platform (using two different devices).
4. Comparison between two different native video players (ExoPlayer [2] and Android Player [1]).

The chosen videos are 360-degree videos in the *.mp4* format since it is widely used in Android devices. The performance evaluation tool is the OVR Metrics Tool [7], and the metric is frame-rate (FPS), which is most significant for the user experience.

## 4  VR Video Player

A video player is a powerful feature to implement in VR applications, however, they require attention to key dependencies. One of the major ones is the rendering quality. [12] analyzes some the video players available and compares them in terms of visual quality.

Besides good rendering, VR applications should explore other aspects such as the user immersion. In a virtual reality environment, the user does not have a mouse to click or a keyboard to type, so other ways to allow user interaction with the video player should be created. The article by [9] presents solutions for user interactions with video players and a comparison between them.

The visualization mode is the main feature users interact with and requires attention. It is applicable in different forms: 2D is the most common one; in the 180° mode, a semi-sphere renders the video; in the 360° mode, a full-sphere presents the video around the user, as shown in Fig. 1.

Performance is a critical non-functional feature in VR video players because any frame drop is noticed by the user, causing nausea and discomfort. Some studies measure the experience in VR video players such as [19], where the author proposes a model to build a good experience for video players.

## 5  Architecture

In the search for an optimized way to use a video player in mobile virtual reality platforms, the architecture is divided in two main layers:

1. Platform layer: native implementation to handle I/O operations and file system.
2. Rendering layer: use of a render framework to make the 3D virtual universe visible.

This architecture (Fig. 2) aims to (1) organize the communication between all modules present in the layers; (2) organize the code to be used by different render engines or different native players; and (3) provide good performance in all media codecs and texture renders.

The platform layer is responsible for media consumption, file system, allocation, and memory management. Some multimedia applications need not only to render digital media but also allow the user to interact with this player.

**Fig. 1.** A video running in a sphere surface

## 5.1 Rendering Layer

The rendering layer is responsible for getting the digital media content and make it visible on the display once it is processed by the codec/native player and provided to the memory buffer. Therefore, it is necessary to use a rendering engine that can receive the memory buffer and render the content on a surface.

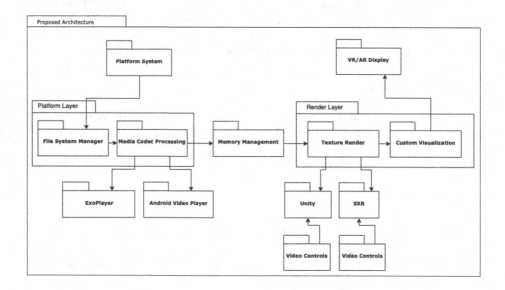

**Fig. 2.** Proposed architecture for Video Player

Both rendering engines (Unity and SXR) used in this work are capable of applying this rendering process in a good and efficient way. Both engines can use low level graphics API, such as Vulkan [17] and OpenGL ES [8], which provides a flexible and powerful interface between software and graphics acceleration hardware for embedded devices.

Rendering in textures is not so discussed and is one of the least documented parts of literature. The use of textures allows the update and continuous rendering of dynamic components such as websites and videos. For static objects, the best approach is to work with bitmap, for example.

In this context, it is known that applications that use the graphics API commonly create several textures during its execution time. In this case, the higher the resolution textures, the better the graphics are. But with the increased resolution, more memory is required for management, and more time is spent during texture loading. One way to improve the application's performance, battery consumption, device heating and even decrease the size of the generated application binary is to make use of hardware accelerated formats.

Another interesting aspect of textures usage is that distinct ways of showing a video is possible. From the received texture can be created materials for rendering in a curved plane like a fancy TV in a virtual environment, as implemented by Oculus version of Netflix app. It also possible to apply video texture in a sphere in order to visualize panoramic videos, or even a more complex visualization as a volumetric video. So, a VR Video player can contain different rendering implementations, therefore it is advisable to have a abstract layer to isolate rendering from the remain aspects of a video player implementation.

## 5.2   Platform Layer

The Platform layer is composed of two parts that provide the baseline for the Video Player Architecture for Virtual Reality on Mobile Devices. They are:

- *File System Manager*: in charge of the media file usage and organization;
- *Media Codec Processing*: process media files that are loaded for Render Layer.

These parts will be explored in the following sections.

**File System Manager.** The main objective of the File System Manager is to provide a centralized and organized data layer composed of multimedia files. This layer must provide all files transparently so that the player has an unique way of working with any type of media.

A robust file system manager must implement an interface to output data in the order the video player handles it, even if it comes from different sources (Fig. 3). In this case, the file system manager needs to parse all data received from any cloud server, from its own mobile storage or even from a DLNA (*Digital Living Network Alliance*) server and use the defined interface to provide a unique data type that will be processed by the Media Codec.

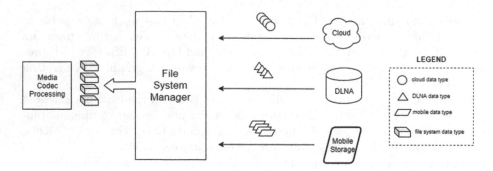

**Fig. 3.** File System Manager environment

**Media Codec Processing.** Any multimedia application that plays video must not only render digital media but also needs to provide the status of the currently playing file and controls that enable interaction. These requirements are important, considering they define the possibilities each media player provides for the end-user.

In the Android platform, it is possible to create a media player using one of the following technologies:

– *Media Player class* [6]: provides basic controls to reproduce audio and video files;
– *Exo Player library*: the ExoPlayer [2,3] library defines standards for audio and video reproduction. It was built using the MediaCodec API with features such as DASH (*Dynamic Adaptive Streaming over HTTP*) and HLS (*HTTP Live Streaming*), which are not available in the MediaPlayer class.

The MediaCodec class offers several possibilities for ExoPlayer by giving access to media codecs at a low level of implementation. This is possible because codecs convert input data into output data through multiple I/O buffers processed asynchronously (see Fig. 4).

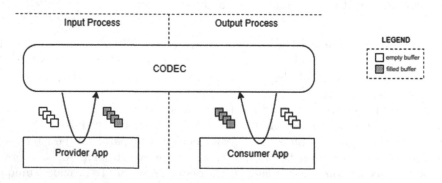

**Fig. 4.** Codec process

MediaCodec has the following types of data [5]: compressed, raw audio, and raw video. Any of these types can be processed using objects of the ByteBuffer class.

The codec performance is improved when Surface is used for raw video data. This is due to the use of native buffers where it is not necessary to map or copy data for ByteBuffers objects. It is worth mentioning that it is impossible to access raw data when using a Surface. In this case, an object of the ImageReader class is necessary to obtain video frames.

The mimetype of a file defines the compressed data for the input and output buffers. For videos, it is usually just a single frame of compressed video. Regarding video buffers in ByteBuffer mode, they are defined according to the color format that can be *native raw video format, flexible YUV buffers* or *others* - generally supported by the ByteBuffer mode through specific format.

As presented on [5], every codec has states that map actions during its life cycle. They can assume the *Stopped, Executing* and *Released* states. The Stopped state maps a set of other sub-states that are *Uninitialized, Configured* and *Error*. The Executing state builds on other sub-states that are *Flushed, Running* and *End-of-Stream*. It's important to know its behavior in order to plan how codec will be used in way that it does not impact other applications.

## 5.3   Experiments and Results

The experiments were made using three different applications: VR Gallery, SXR Video player, and a demo app. The first application is a fully-featured Unity application supporting many other parallel features besides the video player; the second one is a video player application on the SXR platform, and the last one is a Unity application that implements the architecture presented in this paper.

The graphs below shows the FPS registered in the VR Gallery application.

According to Fig. 5 and Fig. 6, the SXR framework maintains 60 FPS in all test cases. However, the initial experiments have shown that the video player of the VR Gallery (application developed in Unity) does not perform well. While Samsung Galaxy S8 Gallery has an fps variation between 55 and 60 FPS, in Samsung Galaxy S6, it stays between 40 and 50 FPS. The reason for these results is the heavy environment VR Gallery has, which is not seen in the other applications.

The graph below shows the FPS registered by SXR application:

Even with the mentioned remarks, the user experience was not affected in any of the tests, the video player performed well. The user cannot perceive the difference between both applications and the frame-rate difference is unnoticeable.

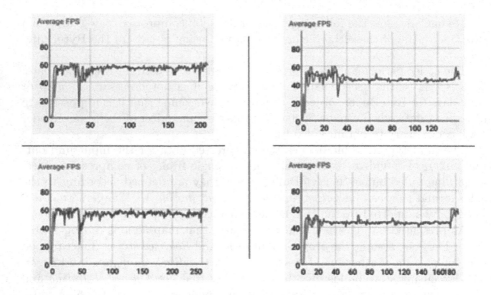

**Fig. 5.** FPS in VR Gallery

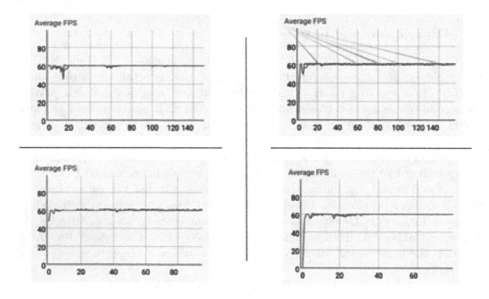

**Fig. 6.** FPS on SXR

The difference between both applications is visible. The SXR application has better results, considering its FPS count is around 60 with some small variations. Even though this is a good result, a better behavior can be observed in the last application where the FPS only varies during the application loading and is stable during video execution.

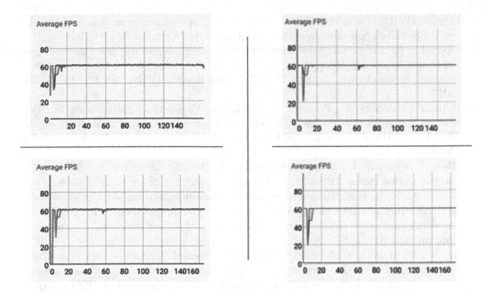

**Fig. 7.** FPS in the isolated application.

## 6  Conclusion

According to the tests, the VR Gallery video player does not have good performance when compared to SXR, however the difference between then can be explained by the fact that the VR Gallery is a complete and robust application, with providers and 3d effects. Even so, the video player of Gallery gives users a complete experience of a good performance video player in a VR environment.

Besides that the video player implemented in unity, that was presented, shows that the proposed architecture delivers a 60 FPS application, without frame drops and lagging, and proved to be very stable even in a full 3d environment application.

## References

1. Android video player (2019). https://developer.android.com/guide/topics/media-apps/video-app/building-a-video-app
2. Exoplayer (2019). https://developer.android.com/guide/topics/media/exoplayer
3. Exoplayer - hello world (2019). https://exoplayer.dev/hello-world.html
4. Hu, S.H., Willett, W.J.: Kalgan: video player for casual language learning. In: Extended Abstracts of the 2018 CHI Conference on Human Factors in Computing Systems, p. LBW053. ACM (2018)
5. Mediacodec    (2020).    https://developer.android.com/reference/android/media/MediaCodec
6. Mediaplayer    (2020).    https://developer.android.com/reference/android/media/MediaPlayer

7. Oculus: OVR Metrics Tool (2018). https://developer.oculus.com/documentation/mobilesdk/latest/concepts/mobile-ovrmetricstool/. Accessed 09 Oct 2018
8. Opengl es (2020). https://developer.android.com/guide/topics/graphics/opengl
9. Pakkanen, T., et al.: Interaction with WebVR 360 video player: comparing three interaction paradigms. In: 2017 IEEE Virtual Reality (VR), pp. 279–280. IEEE (2017)
10. Repo, P., Hyvonen, K., Pantzar, M., Timonen, P.: Users inventing ways to enjoy new mobile services-the case of watching mobile videos. In: Proceedings of the 37th Annual Hawaii International Conference on System Sciences, 2004, p. 8. IEEE (2004)
11. Rhee, T., Petikam, L., Allen, B., Chalmers, A.: Mr360: mixed reality rendering for 360 panoramic videos. IEEE Trans. Visual Comput. Graphics **4**, 1379–1388 (2017)
12. Sari, C.A., Mala, H.N., Rachmawanto, E.H., Kusumaningrum, D.P., et al.: A comparative study of some video players based on visual output quality. In: 2018 International Seminar on Application for Technology of Information and Communication, pp. 313–317. IEEE (2018)
13. Smolic, A., Mueller, K., Merkle, P., Kauff, P., Wiegand, T.: An overview of available and emerging 3D video formats and depth enhanced stereo as efficient generic solution. In: Picture Coding Symposium, 2009. PCS 2009, pp. 1–4. IEEE (2009)
14. Samsung vr - content portal (2019). https://samsungvr.com/portal/about
15. Samsung xr sdk (2018). http://www.samsungxr.com/
16. Unity for all (2019). https://unity.com/
17. Vulkan (2020). https://www.khronos.org/vulkan/
18. Wild, G.: The inaccessibility of video players. In: Miesenberger, K., Kouroupetroglou, G. (eds.) ICCHP 2018. LNCS, vol. 10896, pp. 47–51. Springer, Cham (2018). https://doi.org/10.1007/978-3-319-94277-3_9
19. Yao, S.H., Fan, C.L., Hsu, C.H.: Towards quality-of-experience models for watching 360 videos in head-mounted virtual reality. In: 2019 Eleventh International Conference on Quality of Multimedia Experience (QoMEX), pp. 1–3. IEEE (2019)

# A Shader-Based Architecture for Virtual Reality Applications on Mobile Devices

Adriano M. Gil$^{(\boxtimes)}$ and Thiago S. Figueira$^{(\boxtimes)}$

SIDIA Instituto de Ciência e Tecnologia, Manaus, Brazil
{adriano.gil,thiago.figueira}@sidia.com

**Abstract.** As new technologies for CPUs and GPUs are released, games showcase improved graphics, physics simulations, and responsiveness. For limited form-factors such as virtual reality head-mounted displays though, it is possible to explore alternatives components to harness additional performance such as the GPU. This paper introduces a shader-based architecture for developing games using shared resources between the CPU and the GPU.

## 1 Introduction

Virtual reality (VR) brings the promise of a revolution in the way entertainment is consumed in present times. The user is placed at the center of the action and perceives content from every direction.

Games, for their part, transport players to a world envisioned by game designers and developers. As the technology for other form factors such as PC and console advances, VR players want life-like graphics and improved responsiveness.

Modern mainstream game consoles and PC sets allow parallel computing to be performed. In order to harness this extra computing power, it is necessary to move tasks from the single threaded game loop and place the ones that can run in parallel in different processors.

VR devices, on the other hand, have a limited form factor. Issues like the heat generated by the processing components have to be taken into account which means adding more computing power is not possible without having side effects for the final user through the current form factor.

In VR games development, the most used game engine is *Unity* [12] and even though it is optimized, we believe there is an opportunity in exploring graphics cards for additional performance.

In this work, we propose an implementation of the classic game Snake using a shader-based architecture. By using a logic based of parallel execution we achieved a performatic virtual reality application in which every visual element is defined and rendered by the shader in a unique mesh.

Related research is explored in Sect. 2. An explanation about the game is provided in Sect. 3. Important concepts for this topic, Game Loop and Game Architecture are explored in Sects. 4 and 5. The proposed architecture is explained in

---

Supported by SIDIA.

Sect. 5.1. Section 6 describes how the game logic is managed. Section 7 analyzes how shaders are used for rendering the game. At the end, results are in Sect. 8.

## 2    Related Work

The literature is scarce about approaches using GPU development for Virtual Reality though there are some initiatives in mobile and computer environments.

The two-dimensional game *GPGPUWars* [5] has its code structure based on *shaders*, similar to the architecture presented here where the GPU performs all the processing of the game. The mobile game MobileWars is a massive 2D shooter with top-down perspective [6] which uses the GPU to process the game logic and the CPU for the data acquisition step. However, applications in virtual reality differ from other applications because there is the need to fill the three-dimensional space to provide content for 3 degrees of freedom (3DoF-*3 Degrees of Freedom*). For example, the application described in [15] uses computational vision to generate a panoramic view of an 8-bit console game.

Some works explore different game architectures: AlienQuiz Invaders [4], for example, is an augmented reality game that implements cloud services to improve overall game quality. [5–7] explore the GPU for performance in mobile and PC environments.

## 3    The Game - VRSnake

VRSnake is the virtual reality version of the classic 2D game Snake. In the original game, the player moves the snake to collect the elements that appear randomly during gameplay. In its virtual reality adaptation, VRSnake, the player decides where to position the collectable items rather than directly controlling the snake. In other words, the player stands in the center of the game world surrounded by the inverted sphere where the snake moves. For this new gameplay design, we defined the following rules:

1. The player sets the collectible object position;
2. The snake continuously and automatically seeks the collectable object placed by the player and grows in a unit when it reaches this object;
3. The victory condition is making the snake hit itself during the pursuit for the object.

## 4    Game Loops

The game loop is the foundation which upon games are built. Games are considered real-time applications because their tasks rely on time constraints. According to [7], these tasks can be arranged into three steps: data acquisition, data processing and presentation. The first step is about collecting input data from the input devices (for VR devices, it is the Head-Mounted Display (HMD) and the joysticks); the second step is applying player input into the game as well as game rules and other simulation tasks; the last step is providing to the user the current game state through visual and audio updates.

There are two main groups of loop models proposed by [13]: the coupled model and the uncoupled model. In the simplest approach, all steps are executed sequentially and it runs as fast as the machine is capable of. The uncoupled model separates rendering and update steps in different threads, but this may cause the same unpredictable scenario of the Coupled Model when executed in different machines. The Multi-thread Uncoupled Model feeds the update stage with a time parameter to adjust its execution with time and allow the game to behave in the same way in different machines.

Due to its interactive nature, these steps should be performed as fast as possible in games or performance may jeopardize user experience. For VR applications, this constraint is even heavier as VR games should run at 60 frames per second [11] to avoid nausea and other negative user effects, this is why Virtual Reality software requires powerful CPU and GPU hardware [2].

Modern mainstream game consoles and PC sets allow parallel computing to be performed. In order to harness this extra computing power, it is necessary to move tasks from the single threaded game loop and place the ones that can run in parallel in different processors (Figs. 1 and 2).

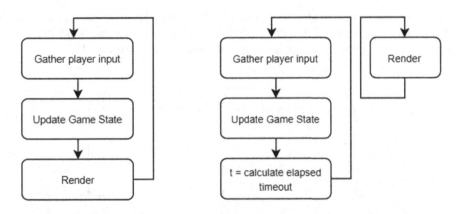

**Fig. 1.** Coupled model          **Fig. 2.** Multi-thread uncoupled model [13]

## 5   Game Architectures

Games are a software product, therefore the architecture of a game is comparable to that of software and defines how the game is built. Usually, it is not apparent to the player, except for performance [1].

According to [8], it is possible to classify any sub-system in a game in one of the three categories (see Fig. 3): the application layer, the logic layer, and the game view layer. The first layer deals with the operating system and hardware. The second layer manages the game state and how it changes over time. The game view layer presents the current game state with the graphics and sound outputs.

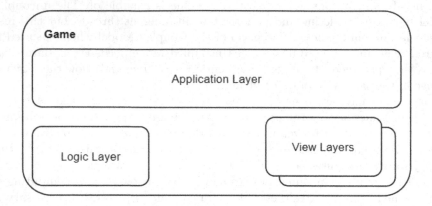

**Fig. 3.** High-level game architecture

Game engines provide *"software that is extensible and can be used as the foundation for many different games without major modification"* [2], and as such, they encapsulate the general game architecture (see Fig. 4).

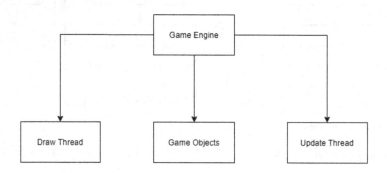

**Fig. 4.** Simplified view of a game engine architecture

The game engine is responsible for the Update Thread, the Draw Thread and the Game Objects. Each correspond to a general step defined in the general game architecture. The Game Objects are the items the player may or may not interact with during game-play. The Update Thread is responsible for updating game objects as far as it can whereas the Draw Thread updates all elements visually on the output device [10].

Game architectures can be updated to better reflect the needs of the game. A well-defined architecture enables better performance and overall results with the final game. In fact, most game engines (as well as the architectures behind it) are crafted for specific platforms and, in some cases, for particular games [2].

## 5.1 Proposed Architecture

We propose a game architecture based on two layers: one layer handles the game logic whereas the other manages rendering. The logical layer is CPU-bound and has tasks such as the search for the collectible objects and snake movement.

The visualization layer involves a shader, a piece of code that runs in the Graphics Processing Unit (GPU), that renders all game objects on the output device, which includes the collectible objects as well as the snake.

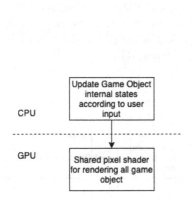

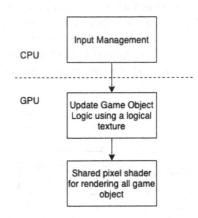

**Fig. 5.** Logic layer runs on CPU

**Fig. 6.** Both logic and visualization layers runs on GPU

In other words, the logical layer manages the collision and movement of the snake and selects the most promising path given a randomization factor; The visualization layer renders all the elements arranged in the output device, the CPU does not influence these objects.

Our implementation is based on pixel shaders since we want to guarantee support for Android-based VR platform that don't have support for more specialized shader types, like computer shaders for instance.

Figures 5 and 6 show diagrams of how our proposed architecture can be devised considering CPU and GPU-bounded scenarios. For mitigating heavy computation on the GPU, a logic layer on CPU is recommended, in the other hand the intensive usage of multiple elements on the logic layer can indicate the GPU as the best solution.

# 6    Logic Layer

The logic layer is responsible for handling user input and its correspondent actions inside the game virtual word. Thus it is necessary to model how the game works. In VRSnake, the snake modeling includes how it moves and how it interacts with each game element.

In VRSnake, the user input is based on an external joystick. As a sensor, however, these controllers are subject to noisy interference during the virtualization of the event represented in the real world by the user's movement. In order to improve the signal capture and translate the player's intentions more faithfully, the Kalman filter [14] is applied to the readings of the joystick events.

## 6.1    Managing Game Objects

Our framework is based on managing objects data that are rendered by pixel shaders. Thus we made use of textures to store logic data, that we named as logical textures. The visualization layer should localize the right position of the data inside the logical texture to render the content. One of the disadvantages of such method is more expensive to spread big changes (Fig. 7).

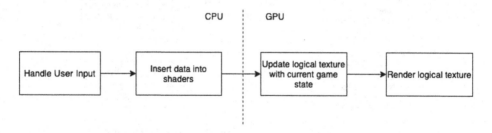

**Fig. 7.** Our proposed data management approach for game elements

The logical texture is generated by blit operations, i.e., copying the output of a pixel shader to a texture buffer. Its final size is defined by overall size of the game world. As we designed a 50 × 50 grid, it's required a 50 × 50 texture to store all the information regarding the game elements. Thus an identification value for each game element is stored in the right position of the grid. After each game update, the next blit call should update the current snake position. By looking ahead of the movement direction in a 3 × 3 neighborhood, it's possible to know whether a snake block will move.

## 6.2    Snake Movement Agent

The movement of the snake is managed by a state evaluation function that analyzes each possible action at any given time. In essence, the serpent is always seeking the collectible objects, so it evaluates the shortest distance course in the

X and Y axes in UV space and, provided that there is no possibility of hitting itself, proceeds through this path and repeats the process. The function below illustrates this procedure:

$$F(A) = R * (D + O) \tag{1}$$

Where $R$ is a randomization factor; $D$ represents the *Manhattan* distance between the current position and the collectable object; And $O$ is a value attributed to the existence or not of obstacles in this path.

## 7    Visualization Layer

The visualization layer has the code necessary to make game elements visible to the user. In the proposed architecture, the visualization layer is based entirely on a pixel shader running in a single mesh.

Given a logical texture, i.e. a texture representing the current game state, the centralized pixel shader must render the position of each game element: snakes and food. The role of the visualization layer is to represent the game element according to a style: 2D or 3D for instance. In this work, we implemented a 2D representation on a sphere and a raymarched visualization on a cube. However many others are possible, like using particles for example.

The following subsections explain how snakes in VRSnake are drawn. Subsect. 7.1 explains the general method, whereas the others explore the specifics such as inverted spheres (Subsect. 7.2), 2D shader snakes (Subsect. 7.3) and raymarched snakes (Subsect. 7.4).

### 7.1    Rendering VR Snakes

Pixel shaders are scripts that carry the mathematical calculations and algorithms to compute the color of each pixel rendered on the output device. For 2D and 3D snake rendering, the shader in the visualization layer requires an external texture that contains the logical information needed to draw the final image.

### 7.2    Virtual Reality in a Inverted Sphere

The immersion sense that comes with virtual worlds requires visual information available from all angles. Given that our proposition considers a 2D game, the challenge is displaying two-dimensional content in a 3D scenario with the user at the center.

An inverted sphere, a sphere that has only its inner side rendered, makes it possible to fill the entire field of view, it also is the endorsed solution to display equirectangular images in 360°. The procedural generation of a sphere can follow one of the two approaches below:

1. An icosphere, i.e., a sphere which vertices are evenly distributed;
2. Generation of vertices based on longitude/latitude coordinates.

For this work, the second approach was adopted due to the possibility of using longitude/latitude as a way to map the UV coordinates through the equation below:

$$R^2 \leftarrow R^3 : (\lambda, \theta) \rightarrow (x, y, z) \tag{2}$$

To calculate the positions of the sphere vertices, given $N_{latitude}$ latitude values and $N_{longitude}$ longitude values, the value $R_{longitude}$ is defined as the angular longitude size of a cross-section of the sphere, as seen in the Eq. 3.

$$R_{longitude} = \frac{2\pi}{N_{longitude}} \tag{3}$$

The total angular size of an amount of $i$ of longitude values can be given by the Eq. 4.

$$\alpha_i = i * R_{longitude} \tag{4}$$

Equations 5 and 6 define the X and Z positions of sphere points belonging to a cross section of the sphere that has radius $D$.

$$x_i = D * \sin(\alpha_i) \tag{5}$$

$$z_i = D * \cos(\alpha_i) \tag{6}$$

In a longitudinal cut, it is possible to notice that the radius $D$ of the cross section is variable along the height of the sphere. It is then determined a value $R$ as the angular size of a latitude value of the sphere, as seen in the Eq. 7.

$$R_{latitude} = \frac{\pi}{N_{latitude}} \tag{7}$$

The total angular size of an amount of $i$ of latitude values can be given by the Eq. 8.

$$\alpha_{latitude} = i * R_{latitude} \tag{8}$$

The Y position of the sphere points, considering unit radius, can be given by the Eq. 9.

$$y_i = \cos(\alpha_{latitude}) \tag{9}$$

The radius $D_{yi}$ obtained in a cross section at latitude $i$ is defined in the Eq. 10 as:

$$D_{yi} = 2 * \sin(\alpha_{yi}) \tag{10}$$

By applying the Eq. 10 in the Eqs. 5 and refequation4 the X and Z positions of the sphere vertices are obtained according to their longitude and latitude coordinates.

$$x_i = 2 * \sin(\alpha_{latitude}) * \sin(\alpha_{longitude}) \tag{11}$$

$$z_i = 2 * \sin(\alpha_{latitude}) * \cos(\alpha_{longitude}) \tag{12}$$

Equations 10, 11 and 12 allow to define a vector $P_i$ with components $(x_i, y_i, z_i)$ for each uv position. By using all those equations we can find all the vertice positions of a procedural sphere.

## 7.3   Simple 2D Snakes

In VRSnake, the shader responsible to render the snake requires an external texture that defines the current game state. This texture as well as other game elements such as snake color and game background can be defined in the user inspector in Unity (see Fig. 8).

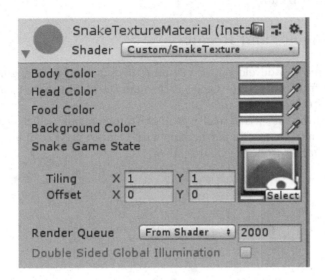

**Fig. 8.** Snake rendering shader in Unity Editor

The external texture is managed by the logic layer, which updates the snake during the update loop of the game. The shader layer sweeps this texture and gathers the red parameter of the RGB color system. The shader reads those values and converts each pixel to screen information according to arbitrary value ranges as seen below:

– Between 0–0.25 - Draws the head of the snake
– Between 0.26–0.5 - Draws the body of the snake
– Between 0.51–0.75 - Draws the collectible object
– Between 0.76–1 - Draws the background

## 7.4   Raymarching Cube-Based 3D Snakes

Raymarching is an iterative approach for rendering a scene. In each iteration, a ray moves through a line towards a fixed direction and only stops when it reaches an object. Differently from raytracing, it is not necessary to calculate the intersection point between a line and a geometric model. It should be calculated only if the current point is in the geometric model.

Using signed distance function (SDF) techniques [3], it is possible to find the distance from the current point position until the target geometry. By means of SDFs, it is possible to accelerate the iterations of raymarching by jumping straight to right position of the intersected geometry.

Our proposal can also be implemented as fake 3D objects rendered by pixel shaders in a plane mesh. We employed raymarching to render cube-based 3D snakes inside a real cube.

## 8    Experiments and Results

The Snake game was developed for virtual reality in the Unity engine. Figure 9 illustrates the frame rate in the GearVR through the Oculus performance assessment tool, the *OVR Metrics Tool* [9]:

The GearVR is the virtual reality headset from Samsung which is designed to interact with smartphones. In partnership with Oculus, it is possible to download VR apps and games from the Oculus store.

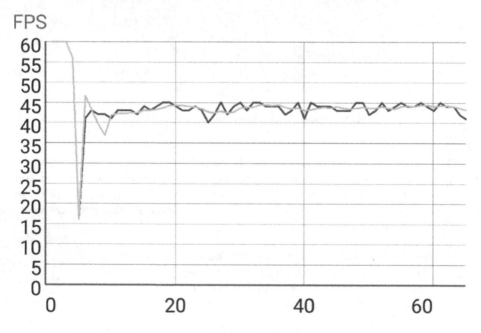

**Fig. 9.** Frames-per-second recored by the OVR Metrics Tool

The application presented an average frame-rate of 43.96 fps (frames-per-second), with a minimum 16 fps and maximum of 60 fps. As represented by the Fig. 9, the application does not 60 fps and this is most likely due to the kalman filter implementation in the data collection step of the game loop and the garbage collector.

# 9    Conclusions

We believe it is possible to explore the GPU for additional performance of virtual reality games in mobile devices. The architecture defined here can be explored as an additional tool for developers to complement the way the game looks and manage resources of VR applications. For future work, we see an opportunity to bring the steps currently present in the logic layer to the visualization layer, further exploring the GPU. We also envision bringing machine learning techniques to the visualization layer to allow multiple snakes to share the same space thus increasing game difficulty. There is the possibility of further improving performance through optimizations to the input gathering implementation as well.

# References

1. Croft, D.W.: Advanced Java Game Programming. Apress, New York (2004)
2. Gregory, J.: Game Engine Architecture, 3rd edn. CRC Press, Taylor & Francis Group, Boca Raton (2019)
3. Hart, J.C.: Sphere tracing: a geometric method for the antialiased ray tracing of implicit surfaces. Vis. Comput. **12**(10), 527–545 (1996)
4. Joselli, M., et al.: An architecture for mobile games with cloud computing module. In: XI Brazilian Symposium on Computer Games and Digital Entertainment. SBGames 2012 - Computing Track (2012)
5. Joselli, M., Clua, E.: GpuWars: design and implementation of a GPGPU game. In: 2009 VIII Brazilian Symposium on Games and Digital Entertainment, pp. 132–140. IEEE (2009)
6. Joselli, M., Silva, J.R., Clua, E., Soluri, E.: MobileWars: a mobile GPGPU game. In: Anacleto, J.C., Clua, E.W.G., da Silva, F.S.C., Fels, S., Yang, H.S. (eds.) ICEC 2013. LNCS, vol. 8215, pp. 75–86. Springer, Heidelberg (2013). https://doi.org/10.1007/978-3-642-41106-9_9
7. Joselli, M., et al.: A game loop architecture with automatic distribution of tasks and load balancing between processors. In: Proceedings of the VIII Brazilian Symposium on Computer Games and Digital Entertainment, January 2009
8. McShaffry, M.: Game Coding Complete, 3rd edn. Course Technology PTR, Boston (2009)
9. OVR Metrics Tool. https://developer.oculus.com/downloads/package/ovr-metrics-tool/
10. Portales, R.: Mastering android game development. https://www.packtpub.com/game-development/mastering-android-game-development
11. Pruett, C.: Squeezing performance out of your unity gear VR game, May 2015. https://developer.oculus.com/blog/squeezing-performance-out-of-your-unity-gear-vr-game/
12. Unity for all (2019). https://unity.com/
13. Valente, L., Conci, A., Feijo, B.: Real time game loop models for single-player computer games (2005)
14. Welch, G., Bishop, G.: An Introduction to the Kalman Filter. University of North Carolina, Department of Computer Science (1995)
15. Zünd, F., et al.: Unfolding the 8-bit era. In: Proceedings of the 12th European Conference on Visual Media Production, p. 9. ACM (2015)

# Emotions Synthesis Using Spatio-Temporal Geometric Mesh

Diego Addan Gonçalves(✉) and Eduardo Todt

Universidade Federal do Paraná, Curitiba, PR, Brazil
{dagoncalves,etodt}@inf.ufpr.br

**Abstract.** Emotions can be synthesized in virtual environments through spatial calculations that define regions and intensities of displacement, using landmark controllers that consider spatio-temporal variations. This work presents a proposal for calculating spatio-temporal mesh for 3D objects that can be used in the realistic synthesis of emotions in virtual environments. his technique is based on calculating centroids by facial region and uses classification by machine learning to define the positions of geometric controllers making the animations realistic.

**Keywords:** Emotion · Facial expression · 3D avatar · Spatio-temporal data

**Fig. 1.** Emotion classification based on 2D datasets used for training 3D landmark controllers.

## 1 Introduction

Emotions can include meaning in virtual performers like 3D avatars, adding context to messages like irony, sadness, fear, etc. In computer graphics, realistic animations are a challenge since they require calculations and objects with high computational cost. Emotions in this context are even more complex since details such as the corners of the mouth, micro expressions, or even the head motion can impact realism [1]. These details have an effect on perceived emotional intensity, and how it affects the affinity for facial expressions (Fig. 1).

© Springer Nature Switzerland AG 2020
C. Stephanidis et al. (Eds.): HCII 2020, LNCS 12428, pp. 112–120, 2020.
https://doi.org/10.1007/978-3-030-59990-4_10

Techniques that allow classifying facial expressions using artificial intelligence are increasingly optimized but are seldom used in the synthesis process. Most approaches use Spatio-temporal information to classify facial data [2, 3, 12]. Visualization techniques of four-dimensional geometrical shapes rely on limited projections into lower dimensions, often hindering the viewer's ability to grasp the complete structure, or to access the spatial structure with a natural 3D perspective [4].

Different mathematical models for temporal data representation are found in the literature [5, 6–9]. The general definition commonly used for 4D data is for an object observed temporarily in a defined space. The spatial data type, as well as the temporal environment, define the specificity of the models developed.

If a 3D object is the increment of a depth dimension in relation to a two-dimensional object, we can mechanically observe that an object in n-dimensions would be the increment of n new spatial axes, being 'x' for width, 'y' for height, 'z' for depth and 'n' for temporality. This temporality refers to observations of the same 3D object geometrically linked at the end vertices.

The most traditional concept to illustrate this type of observation is a tesseract, or a cube within a cube, connected by the four vertices at its end creating facets that prevent the rendered observation of the total object.

Other concepts of a fourth dimension, widely used in the field of philosophy, include sensory layers such as the addition of feeling, sound, or another unobservable channel of perception. Even in mathematical modeling, the fourth dimension is not necessarily linked to a physical dimension but an extension of the three-dimensional object.

In this work, the 4D object concept will be used to represent an observable extension of a geometric region and the visualization of this point will be calculated in temporal slices.

With this it will be possible to calculate displacement of specific vertices linked to virtual controllers that will act as a guide for facial regions from centroids. In the process of synthesis of facial animation, understanding the minutiae of sectorized behavior of the face is a fundamental element to create realistic emotions represented in a virtual environment.

This paper proposes a method of extracting spatio-temporal meshes based on spatio-temporal trajectories for the synthesis of emotions. The techniques presented can be applied to 3D avatars and virtual environments and can assist in the realism of facial animations based on classification with machine learning.

## 2   Spatio-Temporal Data for Geometric Emotion Synthesis

This work proposes a method for the extraction of the displacement geometric meshes of each controller regarding the centroids of the main facial regions. These results allow us to define the behavior of each controller and generate trajectories that, observed as curves, allow the complete analysis of the synthesis process of emotions and the generation of complex facial expression interpolations.

For this process, extracted landmarks for a 3D avatar based on the model proposed on [10] were transformed into function curve structures. In order to represent a function curve, the values of the points represented by t are a part of the sequence called

Knot Vector and determine the base function that influences the shape of the B-Spline trajectory.

Knot vector is represented by $t = (t0, t1, ...tn)$ (e.01) in range $t[t0, tn]$ [6]. Each centroid is a spatial point in a trajectory in the synthesis process. In this way, for a trajectory referring to the synthesis of an expression $E$ of degree $n$ and Controller Points represented by the centroid of the region $\alpha$ observed as a time point $t(\alpha)$ as follows:

$$t = p_0, p_1, \ldots, p_n$$

$$t = \forall p(\alpha) \in [0, 1] = (p_0, p_1, \ldots, p_n) p_i \geq p_i - 1$$

$$t(C) = \frac{\sum_{Nir \in \Re_r} \frac{|dn_{v_{ir}} - de_{v_{ir}}|}{dn_{v_{ir}}}}{NV_r}$$

$$B(t) = \sum_{i=0}^{n} N_{i.k}(t) C(t) + W_{e(t)}$$

For all control points, pi as centroid $t(C)$ and the knots in B-Spline consider the parameter $W$ based on the value of influence extracted from Principal Component Analysis (PCA) and Euclidean Distance Variance Analysis (EDVA) algorithms applied to a 3D avatar. These algorithms were applied to the facial points of the 3D avatar in order to extract the Centroids from facial regions, reducing the dimensionality of the geometric mesh [10].

The influence parameter $W$ is considered in the animation generation process, where a lower value of $W$ corresponds to less important regions that can be ignored in the expression synthesis, reducing the computational load.

Elabessi et al. calculated features over detected Spatio-Temporal Interest Points (STIP's) using Histograms of Optical Flow (HOF) and Histograms of Oriented gradient (HOG) besides Histograms of Oriented 3D spatio-temporal Gradients (HOG3D) [14]. The Spatio-temporal points of interest used in this work follow a model similar to the one presented in [14] where two HOF/HOG histograms are joined together to form a single vector, which describes the object appearance and motion space-time neighborhoods of detected interest points. The method presented by the authors defines the size of the descriptor as $\Delta_x(\sigma), \Delta_y(\sigma), \Delta_i(t)$ where $\sigma$ is spatial scale value and $t$ is temporal scale value [14].

$$\Delta_x(\sigma) = \Delta_y(\sigma) = 18\sigma$$

$\Delta_i(t) = 8t$ [14]

To identify actions, machine learning techniques can be used, such as K-Nearest Neighbors (KNN) algorithms from these tracked landmarks, or STIP's (Fig. 2). These actions are generally defined by patterns of spatial position and displacement. If we observe the points of interest as facial points and the action displacements (eg movement of arms or legs) as the local facial displacements (eg corners of the mouth, frowning), we can use the same technique for the extraction of Spatio-Temporal trajectories.

In Fig. 2. we can observe that from sequential images of a common action, in the case of a person running, it is possible to track points of interest and automatically classify an action based on the patterns of displacement and positions of landmarks [14].

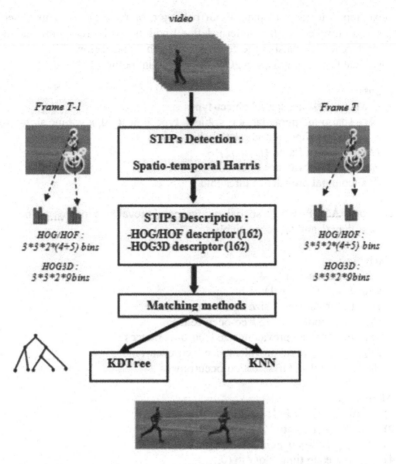

**Fig. 2.** Human action recognition using machine learning through spatio-temporal interest points [14].

For the identification of facial regions and their centroids, facial Action Units (AU) can be used [11], which defines facial sub-areas and classify behaviors based on spatial displacements in a temporal sequence of readings. Furthermore, Deep Learning has been widely used for the classification of facial emotions [13] using random forest algorithm to extract local binary features (LBF) for each face landmark, $\phi^1 = \{\phi_i^1\}$ indicates the random forest corresponding to the l-th personal face landmark in the t-th phase. According to the authors, a random forest algorithm is used to establish the tree corresponding to the face landmarks, and the global linear regression, is trained by using the two-coordinate descent method to minimize the result following:

$$min_w \sum_{i-1}^{n} \left|\left| \Delta S_i^t - W^t \phi^t \left( I_i, S_i^{t-1} \right) \right|\right|_2^2 + \lambda \left|\left| W^t \right|\right|_2^2$$

where each face landmark is trained with the same LBF [13].

The next step is to identify patterns of behavior, or facial actions, in general by observing the occurrences of the pattern defined within Spatio-Temporal readings of facial landmarks calculated using the centroids of each facial region.

For this, Mete Celtik proposes in 2011 the algorithm below [15]:

**Inputs:**
   $E$: a set of distinct spatial object-types
   $ST$: a spatio-temporal dataset <object_type, object_id, x,y, time slot>
   $R$: spatial neighbourhood relationship
   $TF$: a time slot frame $\{t_0,...,t_{n-1}\}$
   $O_s$: a spatial prevalence threshold
   $O_t$: a temporal prevalence threshold

**Output:** PACOPs whose spatial and temporal prevalence indices are no less than $O_s$ and $O_t$, respectively.

**Variables:**
   $k$: co-occurrence size
   $t$: time slots $(0,....,n-1)$
   $T_k$: set of instances of size $k$ co-occurrences
   $C_k$: set of candidate size $k$ co-occurrences
   $SP_k$: set of spatial prevalent size k co-occurrences
   $TP_k$: set of temporal prevalent size k co-occurrences
   $PAP_k$: set of PACOP size $k$ co-occurrences

**Algorithm**
1)   initialization : $k=1$, $C_k=E$, $PAP_k(0)=ST$
2)   **while** (not empty $PAP_k$)   {
3)       $C_{k+1}(0)=gen\_candidates(C_k, PAP_k)$
4)       **for each** time_slot $t$ in $(0,...,n-1)$
5)           $T_{k+1}(t)=gen\_instances(C_{k+1}(t), T_k(t),R)$
6)           $SP_{k+1}(t)=discover\_spatial\_prev\_patterns\ (T_{k+1}(t), C_{k+1}(t), O_s)$
7)       }
8)       $TP_{k+1}=calculate\_temporal\_prev\_indices\ (SP_{k+1})$
9)       $PAP_{k+1}=discover\_PACOP\_prev\_patterns\ (TP_{k+1}, O_t)$   }
10)  $k=k+1$
11) }
12)  return union $\{ PAP_2, ..., PAP_{k+1}\}$

Called the PACOP-Miner algorithm [15] this technique is used to identify the k-spatial prevalent Partial Spatio-temporal Co-Occurrence Pattern (PACOPs) for all time slots and discover prevalent patterns of these size k spatial prevalent patterns. Using size k PACOP prevalent patterns size k + 1 candidate PACOPs will be generated. The Algorithm follows the steps below [15]:

– **Step 3:** Generating candidate co-occurrence patterns;
– **Step 5:** Generating spatial co-occurrence instances;
– **Step 6:** Discovering spatial prevalent co-occurrence patterns;

- **Step 8 and 12:** Calculate temporal prevalence indices;
- **Step 9 to step 12:** Discovering PACOP prevalent patterns;

The trajectories can be extracted from these behaviors, classifying from the machine learning application that can be trained using classes of examples extracted from bases of 2D images of emotions and facial expressions.

The next section presents the synthesis process using 4D meshes extracted from training with the machine learning algorithm Support Vector Machine (SVM) and K-Nearest Neighbor (KNN).

### 2.1   Facial Expression Classification and Synthesis

The Spatio-temporal region model was presented by Schneider et al. [5] where the trajectory of a spatial point, based on temporal readings can be observed as a region if its spatial displacement is considered. When the shape of the curve changes, the region displacements are expanded or retracted.

The same concept is used in this work, once the centroids' trajectories along the generated facial expression animation can be represented by dynamic curves, controlling the edges that connect the vertexes landmarks, producing the movement or deformation of facial regions. The relevant coordinates occur in the transition between the neutral expression to the synthesis of one of the six base emotions of the Ekman model since with these landmarks it is possible to observe the specific impact of these expressions on the 3D mesh.

The trajectories of the Facial Expression Landmarks (FEL) of the facial model were extracted using intervals of 60 frames for each Expression by:

$$Traj = \sum_{i=1}^{n} Cent_{R1}, [E_0, E_1]$$

When in a range between the neutral expression and the base expression *[E0, E1]*, the coordinates of the FEL are extracted by the region centroid $Cent_{R1}$, and their displacement. The centroids displacements of the region define their behavior and influence for each expression. 4D meshes were extracted based on the geometric displacement calculated by the Three-dimensional Euclidean Distance (3DED) of the Centroid coordinates observed at 50 $t$ ($t$ relating to an element of the Keyframes vector for the emotion synthesis).

Each 4D mesh defines the Centroid trajectory by each base expression, the displacement projection can be considered the 4D data as a trajectory mesh [2], once represent a geometric controller in a time slice. The 3D Euclidean Distance values of each Centroid reading are assigned to the matrix sequenced by the Keyframe of the observed animation. Based on 3D Euclidean Distances.

Figure 3 presents an example of classification of emotions where both classification and training using SVM and KNN and the extraction of 4D meshes are used. The training and classification by machine learning uses 2D bases of emotions for the training of the Centroids and then the calculations are adjusted based on the calculation of the three-dimensional Euclidean distance.

The extraction of 4D meshes, made from the temporal trajectories of facial centroids, helps in adjusting the position of geometric controllers that point to the patterns defined in the ML training phase.

A sequence of Multi-points Warping for 3D facial landmark [12] works in a similar way, considering the 4D trajectory extracted using the technique of this work as the sequence of facial landmarks. Even the accuracy of classification of emotions, when extracted using the constructed algorithm, reinforces the general accuracy of more than 90% using machine learning, the expressions of fear and surprise being those that presented more false positives, as well as in the technique presented in [11].

The results presented, although preliminary, introduce the use of three-dimensional geometric meshes, observed over time, as an instrument in the process of classification and synthesis of complex facial expressions, being an interesting resource in the process of understanding basic emotions.

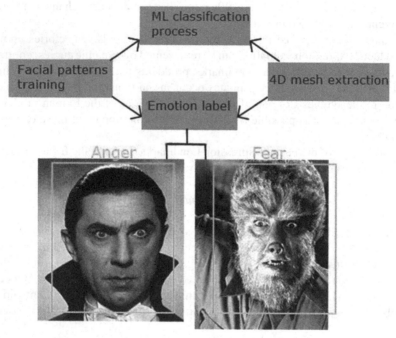

**Fig. 3.** Emotion classification using extracted 4D mesh.

## 3 Conclusion

The addition of emotions in 3D avatars is an important tool for communication in virtual environments. Realistic facial expressions, which efficiently express an emotion, are complex animations with high computational cost. This work studied the relationship between facial regions and their deformations in the synthesis of facial expressions using a 3D avatar. The behavior of facial landmarks was analyzed through Spatio-temporal modeling and the extraction of 4D mesh. which defines the correlation between facial regions and the main expressions.

The results are preliminary and consist of calculations that can be used in 3D avatars to generate facial animations and complex emotions. Its important to understand how to computationally generate complex facial expression interpolations, and optimize the synthesis process. Future works can define relations between base expressions and secondary interpolations. With Spatio-temporal meshes, new studies are possible focusing on optimizing 3D controllers and relating interpolations with base expressions creating intermediate labels of secondary expressions that can be integrated in the emotion and facial expression datasets.

Future work should focus on the integration of the 4D geometric mesh extraction process in the synthesis process in a functional avatar. In addition, calculations of spatial and temporal trajectories can have an impact on the analysis of interpolations of facial expressions and justify further investigation in this regard.

**Acknowledgement.** This work was financially supported by the São Paulo Research Foundation (FAPESP) (grants #2015/16528-0, #2015/24300-9 and Number (2019/12225-3), and CNPq (grant #306272/2017-2). We thank the Universidade Federal do Paraná (UFPR) for making this research possible.

# References

1. Cao, Q., Yu, H., Nduka, C.: Perception of head motion effect on emotional facial expression in virtual reality. In: 2020 IEEE Conference on Virtual Reality and 3D User Interfaces Abstracts and Workshops (VRW), Atlanta, GA, USA, pp. 750–751 (2020). https://doi.org/10.1109/vrw 50115.2020.00226
2. Le, V., Tang, H., Huang, T.: Expression recognition from 3D dynamic faces using robust spatio-temporal shape features. In: 2011 IEEE International Conference on Automatic Face Gesture Recognition and Workshops (FG 2011), pp. 414–421 (2011)
3. Yan, P., Khan, S., Shah, M.: Learning 4D action feature models for arbitrary view action recognition. In: IEEE Conference on Computer Vision and Pattern Recognition, CVPR 2008, pp. 1–7 (2008)
4. Li, N., Rea, D.J., Young, J.E., Sharlin, E., Sousa, M.C.: And he built a crooked camera: a mobile visualization tool to view four-dimensional geometric objects. In: SIGGRAPH Asia 2015 Mobile Graphics and Interactive Applications, SA 2015, pp. 23:1–23:5. ACM, New York (2015)
5. Erwig, M., Schneider, M., Güting, R.H.: Temporal objects for spatio-temporal data models and a comparison of their representations. In: Kambayashi, Y., Lee, D.-L., Lim, E.-p., Mohania, M., Masunaga, Y. (eds.) ER 1998. LNCS, vol. 1552, pp. 454–465. Springer, Heidelberg (1999). https://doi.org/10.1007/978-3-540-49121-7_40
6. Oh, E., Lee, M., Lee, S.: How 4D effects cause different types of presence experience? In: Proceedings of the 10th International Conference on Virtual Reality Continuum and Its Applications in Industry, VRCAI 2011, pp. 375–378. ACM, New York (2011)
7. Sikdar, B.: Spatio-temporal correlations in cyberphysical systems: a defense against data availability attacks. In: Proceedings of the 3rd ACM Workshop on Cyber-Physical System Security, CPSS 2017, pp. 103–110. ACM, New York (2017)

8. Yu, S., Poger, S.: Using a temporal weighted data model to maximize influence in mobile messaging apps for computer science education. J. Comput. Sci. Coll. **32**(6), 210–211 (2017)

9. Suheryadi, A., Nugroho, H.: Spatio-temporal analysis for moving object detection under complex environment. In: 2016 International Conference on Advanced Computer Science and Information Systems (ICACSIS), pp. 498–505 (2016)

10. Gonçalves, D., Baranauskas, M., Reis, J., Todt, E.: Facial expressions animation in sign language based on spatio-temporal centroid. In: Proceedings of the 22nd International Conference on Enterprise Information Systems - Volume 2: ICEIS, pp. 463–475 (2020). ISBN 978-989-758-423-7. https://doi.org/10.5220/0009344404630475

11. Lou, J., et al.: Realistic facial expression reconstruction for VR HMD users. IEEE Trans. Multimedia **22**(3), 730–743 (2020). https://doi.org/10.1109/TMM.2019.2933338

12. Telrandhe, S., Daigavane, P.: Automatic fetal facial expression recognition by hybridizing saliency maps with recurrent neural network. In: 2019 IEEE Bombay Section Signature Conference (IBSSC), Mumbai, India, pp. 1–6 (2019). https://doi.org/10.1109/IBSSC47189.2019.8973018

13. Deng, L., Wang, Q., Yuan, D.: Dynamic facial expression recognition based on deep learning. In: 2019 14th International Conference on Computer Science & Education (ICCSE), Toronto, ON, Canada, pp. 32–37 (2019). https://doi.org/10.1109/iccse.2019.8845493

14. Elabbessi, S., Abdellaoui, M., Douik, A.I.: Spatio-temporal interest points matching in video. In: 2015 Global Summit on Computer & Information Technology (GSCIT), Sousse, pp. 1–4 (2015). https://doi.org/10.1109/GSCIT.2015.7353335

15. Celtic, M.: Discovering partial spatio-temporal co-occurrence patterns. In: Proceedings 2011 IEEE International Conference on Spatial Data Mining and Geographical Knowledge Services, Fuzhou, pp. 116–120 (2011). https://doi.org/10.1109/icsdm.2011.5969016

# An Augmented Reality Approach to 3D Solid Modeling and Demonstration

Shu Han, Shuxia Wang$^{(\boxtimes)}$, and Peng Wang

Cyber-Physical Interaction Lab, Northwestern Polytechnical University, Xi'an 710072, China
hhanshu@163.com

**Abstract.** This paper presents an intuitive and natural gesture-based methodology for solid modelling in the Augmented Reality (AR) environment. The framework of Client/Server (C/S) is adopted to design the AR-based computer aided design (CAD) system. The method of creating random or constraints- based points using gesture recognition is developed to support modelling. In addition, a prototype system of product 3D solid modelling has been successfully developed, we have compared it with traditional CAD systems through several basic design modeling tasks. Finally, analysis of questionnaire feedback survey shows the intuitiveness and effectiveness of the system, and user studies demonstrate the advantage of helping accomplish the product early design and creating and manipulating the 3D model in the AR environment.

**Keywords:** Augmented Reality (AR) · Virtual Reality (VR) · Gesture recognition · Computer Aided Design (CAD) · 3D solid modeling

## 1 Introduction

Augmented reality (AR) can augment the designers' visual perception by computer-generated information [1], and it is a novel human computer interaction (HCI) technique [2, 3]. With the rapid development of computer graphics technology, AR technology provides innovative and effective ways to help design the 3D solid modeling of products, and AR technology is being used as an interface in CAD tools extending the designer to perceive 3D design models over a real environment [4]. Obviously, in the AR design environment, the designer can simultaneously see both physical space surroundings and the virtual world through head mounted displays (HMD). With the help of AR technology, the information about the surrounding real world of the designer becomes interactive and manipulable, designers can design the product of 3D solid models taking full advantage of the real objects within the AR workspace as the interaction tools [5].

In manufacturing, the product development is a component of the engineering process, specifically the conceptual design is a creative and iterative process [4]. Computer-aided design (CAD) contains a wide field of academic research supporting design processes, which will renovate even as the fundamental technologies are evolving [6]. In view of something more promising of the next generation of CAD system, we would

C. Stephanidis et al. (Eds.): HCII 2020, LNCS 12428, pp. 121–140, 2020.
https://doi.org/10.1007/978-3-030-59990-4_11

prefer to concentrate upon AR-based CAD, i.e., CAD develops and deploys AR technology. The AR-based CAD system has advantages of AR technology and the conventional CAD systems. Therefore, it is a great potential and promising tool in the conceptualization design stage. Conceptual design is not only an early stage of the design process, but also the main process in which the designer in an inspirational mode attempts to articulate the broad outlines of decomposed functions and overall structure in a flexible way. On the one hand, common means of conceptual design are concept sketches and virtual 3D models. With the emergence of new technology like VR/AR as a tool for product design and development, it can help promote the perfection of two common means of conceptual design. On the other hand, products' shape primarily depend on the spatial relationships of environments, but there is a lack of bridge between the conceptual design process and the real environment in the current situation [4]. At the same time, it still remains unexplored that the influence of conceptual design of products whose configuration, shape depends mainly on the context using AR technology [4]. As a result, the AR-based CAD system can provide a perfect platform to create and show the blueprint of virtual conceptual products in a natural and intuitive manner. That is the designer can immerse in the virtual-real fusion environment to design product model, which lets the designers real-timely consider the spatial restrictions and innovate the virtual concepts on the basis of the requirements of design context.

In this paper, we tackle the problem of 3D solid modeling via gesture recognition in the AR environment. The main contributions of our work are threefold:

(a) Present a framework of AR-based CAD system for 3D solid modeling.
(b) Propose an intuitive and natural gesture-based method for solid modelling in the AR environment, and a method generating a set of modeling points and adding constraints to them via gestures.
(c) Put forward a 3D solid model rendering mechanism in the AR environment.

## 2  Related Works

In the past several decades, AR has got an increasing amount of attention by researchers and practitioners in many domains such as the manufacturing technology community (product design, maintenance, operation guidance and assembly) [1–5, 8, 9], medical fields [10, 11], games [12] and education [13, 14] etc. Particularly in CAD, evidently AR technology has a wider space for lots of advantages. At first, in the virtual-real fusion context, the designers can interact with real objects around them. During the design process, they can inspect 3D CAD models of products and walk around them [4, 5, 15], which allows them to focus on the product itself and its relation with the context [4]. Nevertheless, for typical modern commercial CAD systems, the 3D CAD models of the part to be designed are displayed on computer screens to assist the designers in deciding part's geometrical and functional characteristics as well as features for a new product. It is clear that they are principally computer-centric systems where the design models' spatial relationships are vague, since the display of the blueprint often takes a projection onto a 2D monitor [5]. Secondly, HCI will be more natural and intuitive. 2D WIMP (Windows-Icon-Menu-Pointer) interface is still the predominated way by which the designers interact with conventional CAD systems, and it is typically controlled by the right hand [16]. When manipulating a 3D model and locating modifying features, they lack direct interaction between users and the components not by hands naturally, but 2D interfaces such as a monitor, a keyboard and mouse. Meanwhile, designers must decompose 3D design tasks into 2D or 1D detailed and specific modeling operations with 2D interface, which makes the design process/task lengthy and tedious [17]. However, the designers can manipulate the virtual or real objects in a natural and intuitive way by two hands in the AR design context. Additionally, compared with virtual reality(VR) where designers are entirely immersed in a virtual environment, AR enriches the way that designers experience the real world by embedding virtual world to be compatible and interactive with physical world [18]. In such case, the AR-based CAD system will provide more safe and real feeling, and inspire the designers' inspiration, insights, and awaken innovative subconscious by the pretty good immersion [19].

Bergig et al. [20] presented an AR-based in-place 3D sketching system of authoring and augmenting mechanical systems. The system can convert 2D sketches based on hand sketching or already in a book into 3D virtual models and augment them on top of the real drawing, and the user also can interact with the application via modifying sketches and manipulating models in a natural and intuitive way. However, its interaction is limited and vague 2D. Haller et al. [21] developed a collaborative tabletop AR environment for discussion meetings with 2D drawing, Xin et al. [22] proposed a tool using a tablet as the sketching platform for creating 3D sketches of artistic expression on a physical napkin using mixed reality (MR) techniques, Prieto et al. [23] presented a novel 3D interface to help carry out structural design in 3D MR environment. Although these proposed AR/MR CAD systems can help the designers realize their blueprint at early stages of new product design, they support for creating wireframes model, which increases the difficulty of comprehending some complex models and designers' mental as well as physical fatigue to some extent.

Other researchers have also demonstrated the ability to AR helps modelling. Kim et al. [24] introduced the development of AR-based co-design applications, where the designers can interact with the virtual/real objects using the tangible interfaces. Similarly, Shen et al. [25, 26] have reported an AR-based design system in which the designers can manipulate the models and the related design information can be augmented in the designer's view. In the above mentioned AR-based CAD systems, the designers were incapable of directly creating or modifying the models. Ong et al. [5, 7, 27, 28] proposed the AR-based co-design system based on the Client/Server(C/S) architecture, which supports for creating and modifying the 3D solid model. However, the designers interact with the virtual model via a virtual stylus rendered on a marker not by hands or natural and intuitive gestures. Piekarsk et al. [29] presented a creating city models system with AR wearable computer—Tinmith-Metro, which can achieve outdoor building design while the way of the HCI is traditional 2D interface making the manipulations complex. Santiago et al. [4] investigated Air-Modelling—a tool for gesture-based sold modelling in the AR environment, which allows to create virtual conceptual models by gestures and make the conceptual design process more efficient.

From the related researches we know AR technology certainly can help the earlier stages of product design like conceptual design process in which the designers want to real-timely express their inspirations and visualize it flexibly. Meanwhile in the AR environment, the interaction techniques would largely influence the effectiveness and intuitiveness of the visualizing and manipulating activities [5, 30]. In such case, the AR-based CAD system can provide an enhanced experience based on the innovative ways of natural and intuitive interaction such as gestures, voice, etc. Furthermore, it can reduce cost and time of showing the blueprint of the early design stage and also improve the quality of design evaluations integrating the context. Therefore, it is a great promising and potential CAD tool.

## 3   The System Framework and Overview of the Approach

### 3.1   The Framework of AR-Based CAD System

Figure 1 shows the general framework of AR-based CAD system. To support the product of 3D solid modeling and its information visualization, the system framework is based on the client/server (C/S) structure. This framework involves five parts: HCI module, 3D CAD model elements rendering module, Open Cascade (OCC) module, sever info database module, and AR main loop module. Here we give a brief description of each part respectively:

(1) HCI module. This step realizes flexible and dexterous modeling interactive operations between designers and AR-based CAD system. With interaction function, designers can create the modeling points, add constraints to the modeling points, and trigger the modeling commands by gestures recognition using the related algorithm, which will be elaborated in Sect. 4.
(2) 3D CAD model elements rendering module. In other words, this module can be described as OpenGL module for it will implement the main functions of the part. The points, planes and 3D model of a virtual product will be rendered by OpenGL in the designer's workspace.
(3) OCC module. This is the modeling and transformation module. It is in charge of constructing the 3D CAD model and transforming the model into the STL model.
(4) Sever info database module. This module is responsible for managing design and modeling information, such as the modeling operations record, the connection status between client and server, and the 3D solid model topological info, etc. Besides, it will record some clients' info e.g. clients' name, type, IP, etc.
(5) AR main loop module. It is composed of following six elements: camera capture, registration and tracking, image analysis and processing, interaction, modeling/design info management, and rendering. This module will integrate the four parts above mentioned into a system for paving the way to realize the capability of 3D CAD modeling in the AR environment.

Furthermore, the application of AR-based CAD design system adopts the multiple-threaded method that communicates with clients over networks based on the standard TCP/IP socket [31]. During the modeling process, the server always listens for incoming connections. Certainly, it will create a new pair of threads for each received client of successful connection. The server main thread will be created for each client to deal with all of the communication between client and server during the modeling process.

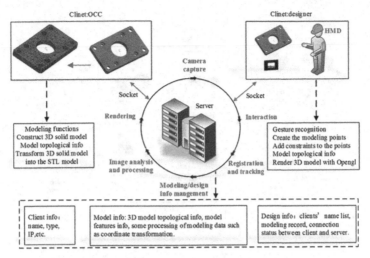

**Fig. 1.** The framework of AR-based CAD system

## 3.2 Overview of the Approach

In this section, we introduce the intuitive and natural gesture-based method for 3D solid modelling in the AR environment. Figure 2 shows the general framework of our approach to designing 3D model in the AR environment. There is the process of implementation.

Firstly, fix the design coordinate. A marker-based tracking method is used to superimpose a 3D CAD model onto a real design workspace. We use the ARToolkit library in the system. The designer's local coordinate will be fixed by the ARToolkit marker "Hiro".

Secondly, create the modeling points and add constraints to the points using gesture recognition.

Thirdly, the points and planes of the 3D model will be rendered by OpenGL in the designer view.

Fourthly, render 3D solid model and transform the CAD model format. The server sends model data created by the designer's modeling operations to the OCC module. Then a 3D CAD model will be modeled with OCC that is one of the most used kernels for 3D surface and solid modeling, visualization, data exchange, and rapid application development. To realize the 3D CAD model that can be rendered by OpenGL in the designer workspace, the OCC module transform the solid model into STL model, and the related STL model data (e.g. the numbers of triangular and vertex, vertexes coordinate, and triangular surface normal vectors, etc.) are transmitted via the server to the other client.

After that the 3D CAD model is rendered by OpenGL in the designer' AR view.

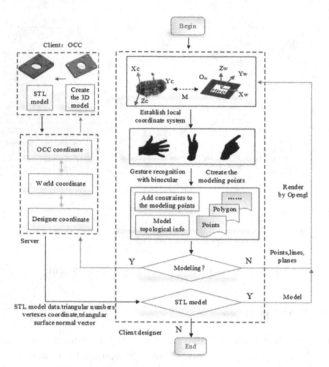

**Fig. 2.** General framework of our approach

## 4 Key Technologies

### 4.1 Gesture Recognition of 3D Solid Modeling

The AR-based CAD system compared with the traditional CAD systems, there is an obvious difference, that is, it enables the user to design the virtual product model in the virtual-real fusion surroundings. In such a design environment, the designer can simultaneously see both physical surroundings and the virtual space through HMD. That is, the AR technology can make the designer' real objects interactive during the 3D modeling process. Apparently, it is inconvenient to interactively design by traditional input devices such as keyboards and mouse. However, the interaction techniques determine the effectiveness and intuitiveness of AR-based CAD system [5]. Owing to their convenience and naturalness, vision-based hand gesture recognition means are gaining remarkable attention [12]. A lot of researches have been proved that hand pose estimation has potential to enable more natural interaction, and it is beneficial to many higher lever tasks such as HIC, AR applications [4, 33–37]. Therefore, it is safe to conclude that recognition of hand gesture is an intuitive natural and desirable way of interacting with virtual/real objects to design 3D CAD model in the AR environment.

The problem related to recognizing gestures falls into three steps: hand detection, pose estimation and gesture classification [32]. Generally, gesture recognition can be divided into two groups which are static and dynamic. In our proposed the AR-based

CAD system, we put to use the static gesture recognition which focuses on techniques to detect hand in the frame from a regular RGB binocular camera.

**The Outline of Gesture Recognition Algorithm**

Figure 3 shows the algorithm flowchart of gesture recognition. The interactive semantic model of gesture recognition consists of three layers: physical layer, lexical layer and syntax layer. Here we give a description of each layer respectively:

i)   Physical layer. This layer is the interface with which the interactive semantic model connects the input devices. It will get the primitive data from the RGB binocular camera in the field of view.
ii)  Lexical layer. It is the critical and abstract layer in the interactive semantic model of gesture recognition. The original data from the physical layer are analyzed and processed such as converting the color space from BGR to HSV, thresholding and filtering (e.g. morphology), etc.
iii) Syntax layer. The layer can identify the specific meaning of gesture by the defined rules and logic reasoning. Therefore, gestures can be recognized by hand information such as contours with convex polygons, convexity defects, and the angle between the lines going from the defect to the convex polygon vertices, etc. Finally get the operational intentions which is resulting in the corresponding modeling command.

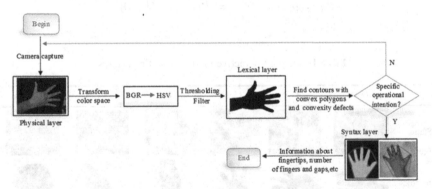

**Fig. 3.** The algorithm flowchart of gesture recognition

**Define the Basic Static Gestures**

To support the 3D modeling operations, we define 5 basic static gestures (see Table 1). They can be denoted as follows:

$$G ::= \{g_n(i, m)\}\, 0 \leq n \leq 5,\ 0 \leq i \leq 5,\ 0 \leq m \leq 4\quad n, i, m \in N \qquad (1)$$

Where $G$ is the set of static gesture, $n$ is the gestures' order, $i$ is the number of extended fingers, $m$ is the number of finger gaps.

## Define the Rules of Gesture Recognition

There are five operational intentions we defined as shown in Table 1. Each has two parts: command semantics (cse) and operational objects (opo). Then they can be denoted as follows:

$$H ::= \{h_n(cse\,,\ opo)\,\}\ \ 0\ \le n\ \le 5\ \ n \in N \tag{2}$$

where $n$ has the same meaning as in formula (1), $g_n$ & $h_n$ are one-to-one correspondence. i.e. $g_n \to h_n$ is a bounded linear surjection, cse is the modeling command as shown in Table 1, opo are points, lines, planes, polygons, 3D solid model, and the STL model, etc.

In the AR-based CAD system, the algorithm adopts vision-based method which requires binocular camera to capture the image for the natural interaction. Therefore, the rules of gesture recognition can be denoted as follows:

$$\begin{cases} R_R ::= \{g_{R,n}(i\,,\ m)\,\} \\ R_L ::= \{g_{L,n}(i\,,\ m)\,\} \end{cases}\ \ 0\ \le n\ \le 5,\ 0 \le i \le 5,\ \ 0\le\ m\ \le 4\ \ \ n, i, m \in N \tag{3}$$

Only when $R_R = R_L$, the rules are perfect. here the subscript: $L$ and $R$ stand for left camera and right camera of the binocular camera, $n$, $i$, $m$ have the same meaning as in formula (1).

Without loss of generality, the rule of operational intention $h_1$ is as follows:

$$R_R = g_{R,1}(1\,,\ 0)\ \ \ = R_L = g_{L,1}(1\,,\ 0) \tag{4}$$

**Table 1.** Define interactive modeling static gestures

| Sign | $g_5(5,4)$ | $g_4(4,3)$ | $g_3(3,2)$ | $g_2(2,1)$ | $g_1(1,0)$ | $g_0(0,0)$ |
|---|---|---|---|---|---|---|
| Static gestures | | | | | | |
| Threshold & filtering | | | | | | |
| Command semantics | Next | Undo the last gesture command | Transform the points' coordinate | Add constraints | Create the modeling point | Clear redundant points |
| Intention | $h_5$ | $h_4$ | $h_3$ | $h_2$ | $h_1$ | $h_0$ |

## 4.2  Add Constraints to the Modeling Points

The modeling points can be randomly created by gesture when there is no special requirement of point sets in the AR design workspace. We should add constraints to the modeling pints, when there are some special requirements such as the length between two points and the angle between two vectors, etc. Figure 4 presents a flowchart outlining our two-step approach to creating points and adding constraints. When creating the second point, the designer should pay attention to the length between two points as show in Fig. 7. However, to create the next modeling point, when there are more than two points, the length and angle should be considered (see Fig. 7 and text for details).

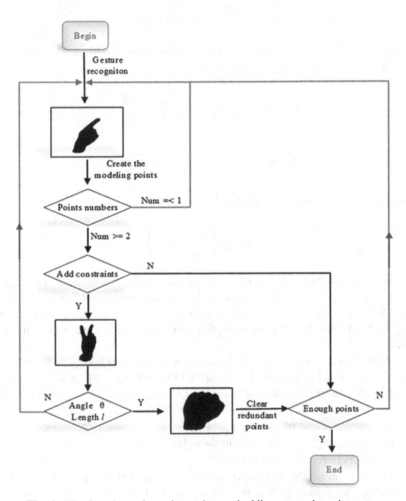

**Fig. 4.** The flowchart of creating points and adding constraints via gesture

**Create Region**

Without loss of generality, supposing the coordinate of three points A, B, C created by gesture are $A(a_x, a_y, 0)$, $B(b_x, b_y, 0)$, $C(c_x, c_y, 0)$ as show in Fig. 5. When creating the point A, we can real-timely get the coordinate and the length of AB:

$$l_{AB} = \sqrt{(a_x - b_x, a_y - b_y)} \tag{5}$$

when the point A is created, the next point C will be uniquely determined by two factors: $l_{AC}$ and $\theta$.

$$l_{AC} = \sqrt{(a_x - c_x, a_y - c_y)} \tag{6}$$

while creating the point C, the point A and B have determined. Thus we can easily have vectors $\overrightarrow{AB}$ & $\overrightarrow{AC}$ and their unit vectors $\vec{k}$ & $\vec{j}$ .

$$\overrightarrow{AB} = (b_x - a_x, b_y - a_y), \ \vec{k} = \frac{\overrightarrow{AB}}{l_{AB}} \tag{7}$$

$$\overrightarrow{AC} = (c_x - a_x, c_y - a_y), \ \vec{j} = \frac{\overrightarrow{AC}}{l_{AC}} \tag{8}$$

Therefore, the angle between $\overrightarrow{AC}$ and $\overrightarrow{AB}$ is $\theta$.

$$\theta = \arccos\left(\frac{\overrightarrow{AB}}{l_{AB}} * \frac{\overrightarrow{AC}}{l_{AC}}\right) \tag{9}$$

In order to ensure the efficiency of creating points and the robustness of the system, the points are created by the static gesture $g_1(1, 0)$ when the finger touches the design plane and keeps still for several seconds, which the designer has haptic force feedback avoiding the hand jitter. As a consequence, this way can reduce fatigue and facilitate the operations.

The angle $\theta$ can be determined in a common way that is to first quantize the orientations and to use local histograms like in SIFT [38] or HOG [39]. In addition, we are motivated by the capture function of traditional CAD software. Figure 6 and Table 2 show how to determine the angle $\theta$. In other words, the practical angle is $\alpha$ which is always the several multiple of five when adding the constraints to the modeling points.

Take the vertical angle as an example to show the idea. An assumption in this approach is that the angle between $\overrightarrow{AC}$ and $\overrightarrow{AD}$ is $\delta$ ($0 \leq \delta \leq 5^0$, see Fig. 7), when creating the point C. Thus we can get the point D(dx, dy) by formula (10), and the redundant points can be clear by the gesture $g_0(0, 0)$ as show Table 1. Finally, the region is created (Fig. 10(a)).

$$\left|\overrightarrow{AC}\right| = \left|\overrightarrow{AD}\right|, \ \overrightarrow{AD} \times \overrightarrow{AB} = 0 \tag{10}$$

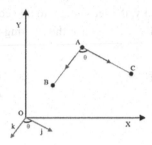

**Fig. 5.** Create points

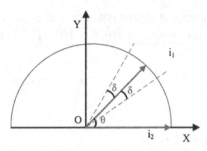

**Fig. 6.** Voted an angle

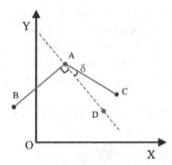

**Fig. 7.** Create the special point

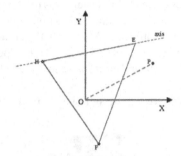

**Fig. 8.** Create the rotating axis

**Table 2.** Voted angles

| Real-time $\theta$ /$^0$ | [0, 2.5) | [2.5, 7.5) | [7.5, 12.5) | [$\alpha$ -2.5, $\alpha$ +2.5) | [172.5, 177.5) | [177.5, 180) |
|---|---|---|---|---|---|---|
| Voted $\alpha$ /$^0$ | 0 | 5 | 10 | $\alpha$ | 175 | 180 |

## Create Path

After that we created the region, the path has to be created by static gestures according to the specific operation such as sweep and revolve, etc.

For sweep, the path stands for the height and direction. To ensure that the normal vector of region is parallel with the design plane in which the path is located, the coordinates of region must be transformed by the gesture $g_3(3, 2)$ before creating points of path as show Fig. 11(b) & (c).

For revolve, the path is a rotating axis. That is, the region and path are coplanar (see Fig. 8 and Fig. 12(a)). Supposed the first point of path is the coordinate origin, and the second point is P(px, py) as show Fig. 8. Then we can get the rotating axis by calculating the cosine value of between vectors. Supposing the region is EHF, and their coordinates are E($e_x$, $e_y$), H($h_x$, $h_y$), F($f_x$, $f_y$). Then we can easily get their vectors:

$$\vec{HE} = (e_x - h_x, e_y - h_y), \vec{EH} = (h_x - e_x, h_y - e_y), \vec{OP} = (p_x, p_y) \qquad (11)$$

and others can also be got in the same way. For simplicity, the set λ refers to all vectors in the region such as $\overrightarrow{HE}$, $\overrightarrow{EH}$, $\overrightarrow{HF}$, $\overrightarrow{FH}$, $\overrightarrow{FE}$, $\overrightarrow{EF}$. . Finally we can have the rotating axis as follows:

$$\lambda^* = \arg \max_{\lambda} \cos < \vec{\lambda}, \overrightarrow{OP} > \tag{12}$$

## 5   User Study

### 5.1   Apparatus and Experimental Setup

The AR-based CAD system was implemented on a 3.8 GHz, 16 GB RAM laptop, a set of HTC Vive devices, and a RGB binocular camera. Moreover it has been established on the software platform of Microsoft Visual udio 2013(C++), Unity3D5.3.4[1], and some open source libraries: OCC, Opencv3.0.0, OpenGL, ARToolkit5.3.2[2], Steam VR Plugin[3]. The system setup and working scenario of 3D modeling in the AR environment can be seen in Fig. 9.

Here we will design two 3D CAD models to demonstrate our approach.

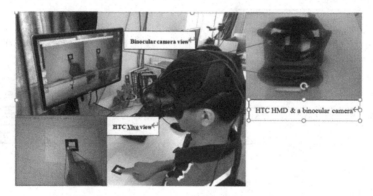

**Fig. 9.**   The system setup and working scenario

### 5.2   Task

**Task1: Sweep a Model**
The modeling process can be divided into three parts: creating a region, creating a path and rendering a model.

*Create a Region*
Firstly, the sketch plane(the black grid region) was determined by the ARToolkit marker

---

[1] Unity 3d. https://unity3d.com/.
[2] https://artoolkit.org/.
[3] https://developer.viveport.com/cn/develop_portal/.

'Hiro', and the origin of local coordinate system coincides with the marker' center as shown in Fig. 10(a).

Secondly, create the modeling points. The first two points randomly create with the gesture, and the next point is confirmed by constraints of length and angle, which real-timely can be seen to locate where to put the finger (Fig. 10(b)). The order was triggered by the static gesture $g_2(2, 1)$ (Fig. 10(c)). Create the last point without any constraints, so we can have four points as shown in Fig. 10(d).

Finally, the region of a trapezoid was created by the gesture $g_5(5, 4)$ (see Fig. 10(e) and (f)).

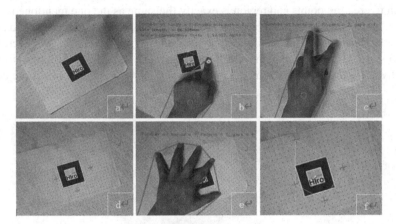

**Fig. 10.** Create a region

## Create a Path

On the one hand, transform the coordinates of region. To sweep, the region's coordinates have to be transformed by the gesture (see Fig. 11(a)).

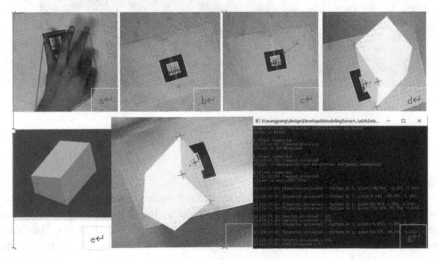

**Fig. 11.** Sweep

On the other hand, create the two points of path as the same operations of creating the region's points, and the path, the red arrow line, is rendered as shown in Fig. 11(c).

*Render a Model*

After specific operations, the model is rendered in the AR environment. Figure 11(d) and (f) are two different designer perspectives. Figure 11(e) is the model rendering by OCC, and all design history can be seen in Fig. 11(g).

## Task 2: Revolve a Model

The modeling process can also be divided into three parts like the sweep modeling. The approach to creating points and region is exactly the same as the one mentioned above. However, there is different when creating the path. The right vector can be found by the formula (10), and it confirms the rotating axis as shown in Fig. 12(a). Moreover, other operations are the same as the above. Figure 12(b) and (c) are two different designer perspectives. Figure 12(d) is the model rendering by OCC, and Fig. 12(e) is the design history.

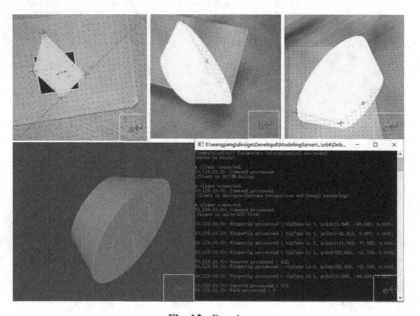

**Fig. 12.** Revolve

### 5.3  Participants and Procedure

User studies have been performed to evaluate the AR-based CAD system and the interfaces. A total of 28 participants took part in the study, 16 males and 12 females. Participants' ages ranged from 21 to 29 years (mean 23.82, SD 1.90, SE 0.36). They were primarily graduate students in our university. The focus of user studies are to measure the usefulness and flexibility of the prototype. 22 participants of them have experienced VR/AR applications, and all participants were familiar with commonly used traditional CAD systems, e.g., AutoCAD, Catia, Pro/E, UG NX or SolidWorks. In this comparative experiment task, we chose Pro/E as a representative of traditional CAD software systems. Before the trials, each participant was asked to learn about the systems. The experiment conductor explained the processes and provided a complete demonstration. Each participant completed the tasks in about 4 min. After each trial we collected quantitative feedback about the AR-based CAD prototype system from the participants by the questions shown in Table 3. Answers where captured on a Likert scale of 1 to 5 in which 1 is "strongly disagree" and 5 is "very agree". Condition ARCAD was using the AR-based CAD system, and Condition traditional CAD was using Pro/E. After finishing all two conditions, participants answered a post experiment questionnaire. Finally, the experiment ended with a debriefing and the opportunity for participants to offer open-ended comments.

**Table 3.** Survey questions

| Q1 | It is easy to use |
|----|-------------------|
| Q2 | It inspires innovation and new ideas |
| Q3 | It is physical or mental challenging to use |
| Q4 | It is natural and intuitive to create a 3D model |
| Q5 | It helps comprehend to the structure of the 3D model |
| Q6 | It is natural and intuitive to manipulate the model, e.g., translation and rotation, etc |

### 5.4 Hypotheses

We designed a set of comparative experiments to compare the application of ARCAD system and traditional CAD system in conceptual design, and then studied the comprehensive performance of common modeling operations such as sketch rendering, migration and rotation in the two systems. The focus of our research is to evaluate inspiration creating, reachability evaluating and amenity of the systems when designers express their ideas, and the direction of future conceptual design systems. We hypothesized that AR-based CAD system provides a more efficient and more inspiring environment for model designer. Based on this, we make the hypothesis that:

H1: Performance. There will be a significant difference in performance time between different conditions, and the Gesture-based AR design system provides faster task performance.
H2: Subjective experience. There will be a significant difference in subjective feeling and creative experience between two systems. It is natural and inspiring in the context of augmented reality.
H3: Amenity. The 3D AR-based system interface is more intuitive and natural and easy to accessible when designers expressing their originality.

# 6 Results

## 6.1 Performance Time

Our initial desire is to explore whether participants could adapt to the AR gesture-based design system quickly and complete the transformation from creativity to 3D digital prototyping naturally. We measured and compared the time required to complete the basic tasks of design modeling in ARCAD and traditional CAD systems.

Statistical analysis was conducted to find out if there were significant differences between the AR CAD and Pro/E. As can be seen in Fig. 13, descriptive statistics showed that on average, the users took about 38.84% more time to complete design modeling tasks when using AR CAD design interactive interface (M = 183.71, SD = 16.61, SE = 5.03) than Pro/E (M = 132.32, SD = 18.35, SE = 5.35). The Cronbach's alpha indicated good internal consistency among items ($\alpha = 0.757$). Furthermore, the paired t test ($\alpha = 0.05$) revealed that there was the statistically significant difference between two conditions on the average time (t(27) = 16.375, p < 0.001).

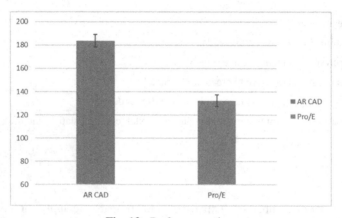

**Fig. 13.** Performance time

## 6.2 Questionnaire Feedback Evaluation

To explore the differences between 3D AR-based system and traditional CAD system in design experience, we analyzed the participants' answers to the Likert scale questionnaires on user evaluation about the AR-based 3D system and Pro/E assisted 3D modeling design. To compare the Likert scale ratings between the two conditions, we have conducted the Wilcoxon Signed-Rank Test ($\alpha = 0.05$). The results of the Likert scale questionnaires are presented in Figs. 14.

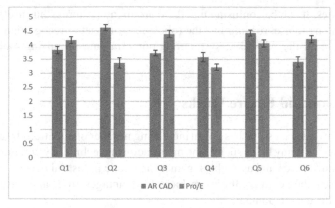

**Fig. 14.** Average results of survey questions

For Q1, we find participants felt that it is easy to use the traditional CAD systems than the AR-based CAD system for several factors, e.g. the robustness of gesture recognition algorithm, etc. Pro/E interface is easier to use, which can be reflected in the comparison of performance time. For Q2, the answers show the system can inspire designer's innovation, the participants said that they could immerse in the virtual-real fusion design environment which provides good conditions for inspiring new ideas and blueprint. For Q3, it is more physical or mental challenging to use ARCAD than Pro/E for the jitter of view. Obviously, the participants hold that creating the 3D CAD model by gestures in the AR environment is natural and intuitive for Q4, and it helps comprehend to the 3D shapes and spatial relationships of the model for Q5. Finally, similarly we find that it is more natural and intuitive to manipulate the model, e.g., translation, and rotation, etc. Overall, these results show that our approach is better than the traditional one to some extent.

## 7  Discussion

Most participants felt that the AR-based system is potential for natural and intuitive interaction. About the system usability, it is reasonable to conclude that the usability needs to be improved. During the modeling process, the robustness of gesture recognition algorithm has a critical impact on the usefulness and effectiveness of the system. Although our vision-based approach is simple and easy to realize, lots of condition challenges have to be faced such as the complex background, other skin-color objects, and lighting variation. Obviously, it largely influences the user experience. For manipulations, it is more natural and intuitive to use gestures in the AR environment which leads the user easily understand the 3D shapes and spatial relationships of the model. Moreover, flexibility was an advantage when using AR-based CAD compared to conventional CAD systems as the users can inspect 3D CAD models of products and move around it during product modelling just like they are used to looking at real objects around them, which will inspire innovative ideas. Finally, participants also provided some ideas for improvements: (a) reduce the numbers of gestures and define the comfortable, memorable and efficient gestures. It can be confusing that there are many gestures, thus users could not be sure which one to use. (b) improve the robustness of gesture recognition algorithm. Certainly, all in all the AR-based system is a potential and promising tool to design product.

## 8   Conclusion and Future Works

For more than two decades, AR has got an increasing number of attentions by researchers and practitioners in many fields such as the manufacturing technology community (product design, maintenance and assembly), games, aesthetic industrial design, etc. A large number of researches confirm there are lots of advantages from applications of AR techniques in the design domain.

In this paper, we present an AR-based CAD system in which designer can create 3D models of products by the static gesture. The presented system adopts the C/S architecture and its communication is based on the standard TCP/IP socket between server and clients. In addition, we propose the rendering mechanism of a 3D CAD model and the algorithm of gesture recognition for modeling. In the end, the usefulness and effectiveness of our approach is demonstrated by case studies and user studies. From the testing, it was observed that our approach can potentially support a new more natural and intuitive design method with direct 3D interaction and fully perceived for design evaluation. Although the approach is still an exploratory stage for product design, we have shown its usability and effectiveness for 3D solid modelling. Thus we conclude that the AR-based CAD system is potential and promising.

The study has some limitations that we will have to address to improve the practicability of our approach in the future as follows:

(1)   Improve the robustness of gesture recognition algorithm. We will explore more ways to improve the algorithm performance such as depth images [34] got by RGB-Depth camera or the latent regression forest [33] to recognize gestures, etc.
(2)   Augment the system's modeling function. We plan to augment the modeling function such as supporting the creation and edit of more complex models, etc.

**Acknowledgements.**  We would especially like to thank PhD student Peng Wang for his special contribution to this research. We would also like to appreciate the anonymous reviewers for their invaluable criticism and advice for further improving this paper.

**Funding Information.**  This research is partly sponsored by the civil aircraft special project (MJZ-2017-G73), National Key R&D Program of China (Grant No. 2019YFB1703800), Natural Science Basic Research Plan in Shaanxi Province of China (Grant No. 2016JM6054), the Programme of Introducing Talents of Discipline to Universities (111 Project), China (Grant No. B13044).

# References

1. Wang, Y., Zhang, S., Yang, S., He, W., Bai, X., Zeng, Y.: A LINE-MOD-based markerless tracking approachfor AR applications. Int. J. Adv. Manuf. Technol. **89**(5), 1699–1707 (2016). https://doi.org/10.1007/s00170-016-9180-5
2. Nee, A.Y.C., Ong, S.K., Chryssolouris, G., Mourtzis, D.: Augmented reality applications in design and manufacturing. CIRP Ann. – Manuf. Technol. **61**, 657 (2012)
3. Wang, X., Ong, S.K., Nee, A.Y.C.: A comprehensive survey of augmented reality assembly research. Adv. Manuf. **4**(1), 1–22 (2016). https://doi.org/10.1007/s40436-015-0131-4
4. Arroyave-Tobón, S., Osorio-Gómez, G., Cardona-Mccormick, J.F.: AIR-MODELLING: a tool for gesture-based solid modelling in context during early design stages in AR environments. Comput. Ind. **66**, 73–81 (2015)
5. Shen, Y., Ong, S.K., Nee, A.Y.C.: Augmented reality for collaborative product design and development. Des. Stud. **31**(2), 118–145 (2010)
6. Goel, A.K., Vattam, S., Wiltgen, B., et al.: Cognitive, collaborative, conceptual and creative — four characteristics of the next generation of knowledge-based CAD systems: a study in biologically inspired design. Comput. Aided Des. **44**(10), 879–900 (2012)
7. Ong, S.K., Shen, Y.: A mixed reality environment for collaborative product design and development. CIRP Ann. – Manuf. Technol. **58**(1), 139–142 (2009)
8. Lima, J.P., Roberto, R., Simões, F., et al.: Markerless tracking system for augmented reality in the automotive industry. Expert Syst. Appl. **82**, 100–114 (2017)
9. Henderson, S., Feiner, S.: Exploring the benefits of augmented reality documentation for maintenance and repair. IEEE Trans. Visual. Comput. Graphics **17**(10), 1355 (2011)
10. See, Z.S., Billinghurst, M., Rengganaten, V., Soo, S.: Medical learning murmurs simulation with mobile audible augmented reality. In: SIGGRAPH ASIA 2016 Mobile Graphics and Interactive Applications, p. 4. ACM (2016)
11. Zou, Y., Chen, Y., Gao, M., et al.: Coronary heart disease preoperative gesture interactive diagnostic system based on augmented reality. J. Med. Syst. **41**(8), 126 (2017)
12. Lin, S.Y., Lai, Y.C., Chan, L.W., et al.: Real-time 3D model-based gesture tracking for multimedia control, pp. 3822–3825 (2010)
13. Hsu, T.C.: Learning English with augmented reality: do learning styles matter? Comput. Educ. **106**, 137–149 (2017)
14. Wang, Y.H.: Exploring the effectiveness of integrating augmented reality-based materials to support writing activities. Comput. Educ. **113**, 162–176 (2017)
15. Januszka, M., Moczulski, W.: Augmented reality for machinery systems design and development. In: Pokojski, J., Fukuda, S., Salwiński, J. (eds.) New World Situation: New Directions in Concurrent Engineering. Advanced Concurrent Engineering, pp. 91–99. Springer, London (2010). https://doi.org/10.1007/978-0-85729-024-3_10
16. Wang, R., Paris, S.: 6D hands: markerless hand-tracking for computer aided design. In: ACM Symposium on User Interface Software and Technology, Santa Barbara, CA, USA, October 2011, pp. 549–558. DBLP (2011)
17. Gao, S., Wan, H., Peng, Q.: An approach to solid modeling in a semi-immersive virtual environment. Comput. Graph. **24**(2), 191–202 (2000)
18. Zhou, F., Duh, B.L., Billinghurst, M.: Trends in augmented reality tracking, interaction and display: a review of ten years of ISMAR. In: IEEE/ACM International Symposium on Mixed and Augmented Reality, pp. 193–202. IEEE (2008)
19. Schnabel, A.M., Thomas, K.: Design, communication & collaboration in immersive virtual environments. Int. J. Des. Comput. **4** (2002)
20. Bergig, O., Hagbi, N., Elsana, J., et al.: In-place 3D sketching for authoring and augmenting mechanical systems. In: IEEE International Symposium on Mixed and Augmented Reality, pp. 87–94. IEEE (2009)

21. Haller, M., Brandl, P., Leithinger, D., et al.: Shared design space: sketching ideas using digital pens and a large augmented tabletop setup. In: Advances in Artificial Reality and Tele-Existence, International Conference on Artificial Reality and Telexistence, ICAT 2006, Hangzhou, China, 29 November–1 December 2006, pp. 185–196, Proceedings. DBLP (2006)
22. Xin, M., Sharlin, E., Sousa, M.C.: Napkin sketch: handheld mixed reality 3D sketching. In: ACM Symposium on Virtual Reality Software and Technology, VRST 2008, Bordeaux, France, October 2008, pp. 223–226. DBLP (2008)
23. Prieto, P.A., Soto, F.D., Zuniga, M.D., et al.: Three-dimensional immersive mixed reality interface for structural design. Proc. Inst. Mech. Eng. Part B J. Eng. Manuf. **226**(B5), 955–958 (2012)
24. Kim, M.J., Maher, M.L.: The impact of tangible user interfaces on designers' spatial cognition. Hum.-Comput. Interact. **23**(2), 101–137 (2008)
25. Shen, Y., Ong, S.K., Nee, A.Y.C.: AR-assisted product information visualization in collaborative design. Comput. Aided Des. **40**(9), 963–974 (2008)
26. Shen, Y., Ong, S.K., Nee, A.Y.C.: A framework for multiple-view product representation using augmented reality. In: International Conference on Cyberworlds, pp. 157–164. IEEE Computer Society (2006)
27. Shen, Y., Ong, S.K., Nee, A.Y.C.: Product information visualization and augmentation in collaborative design. Comput. Aided Des. **40**(9), 963–974 (2008)
28. Shen, Y., Ong, S.K., Nee, A.Y.C.: Collaborative design in 3D space. In: ACM SIGGRAPH International Conference on Virtual-Reality Continuum and Its Applications in Industry, p. 29. ACM (2008)
29. Piekarski, W., Thomas, B.H.: Tinmith-metro: new outdoor techniques for creating city models with an augmented reality wearable computer. In: IEEE International Symposium on Wearable Computers, p. 31. IEEE Computer Society (2001)
30. Shen, Y., Ong, S.K., Nee, A.Y.C.: Vision-based hand interaction in augmented reality environment. Int. J. Hum.-Comput. Interact. **27**(6), 523–544 (2011)
31. Dunston, P.S., Billinghurst, M., Luo, Y., et al.: Virtual visualization for the mechanical trade. Taiwan University (2000)
32. Sohn, M.K., Lee, S.H., Kim, H., et al.: Enhanced hand part classification from a single depth image using random decision forests. IET Comput. Vision **10**(8), 861–867 (2016)
33. Tang, D., Chang, H.J., Tejani, A., et al.: Latent regression forest: structured estimation of 3D articulated hand posture. In: IEEE Conference on Computer Vision and Pattern Recognition, pp. 3786–3793. IEEE Computer Society (2014)
34. Xu, C., Nanjappa, A., Zhang, X., et al.: Estimate hand poses efficiently from single depth images. Int. J. Comput. Vision **116**(1), 21–45 (2016)
35. Jang, Y., Jeon, I., Kim, T.K., et al.: Metaphoric hand gestures for orientation-aware VR object manipulation with an egocentric viewpoint. IEEE Trans. Hum.-Mach. Syst. **47**, 113–127 (2017)
36. Cheng, H., Yang, L., Liu, Z.: Survey on 3D hand gesture recognition. IEEE Trans. Circuits Syst. Video Technol. **26**(9), 1659–1673 (2016)
37. Chavan, V.B., Mhala, N.N.: A review on hand gesture recognition framework
38. Lowe, D.G.: Distinctive image features from scale-invariant keypoints. Int. J. Comput. Vision **60**(2), 91–110 (2004)
39. Dalal, N., Triggs, B.: Histograms of oriented gradients for human detection. In: IEEE Computer Society Conference on Computer Vision and Pattern Recognition 2005, CVPR 2005, pp. 886–893. IEEE (2005)

# Quick Projection Mapping on Moving Object in the Manual Assembly Guidance

Weiping He, Bokai Zheng, Shuxia Wang$^{(\boxtimes)}$, and Shouxia Wang

Northwestern Polytechnical University, Xi'an, People's Republic of China
colinzhengbokai@126.com, Shuxiaw@nwpu.edu.cn

**Abstract.** In the modern assembly, manual assembly is one of the essential works in the factory. By the help of the robots and any other automatic equipments, modern assembly is quite important to improve the efficiency of the necessary manual assembly. Projection is one of the most significant assisted method to solve the assemble problem. This paper presents a quick projection method for the slightly moving object in the manual assembly guidance. By using the closed-loop alignment approach, the proposed method is low-latency and auto tracking to the target object. We designed a tracking and matching system to realize the projecting on the target object. The result showed that the precious of the projecting information can satisfy the needs which the workers followed the projecting guidance to do the assembly. This system also released workers pressure because of the wrong operation and made them comfortable to do their works in the factory.

**Keywords:** Projector · Guidance · Quick

## 1 Introduction

In the daily, the projector is used to present information in a public environment. Projector is a kind of interactive equipment using the simple and cheap tools to show the excellent guide. Using the projector to show the information is more effective than the ways using the paper and pen. During the assembling, Projector-camera systems can release the pressure of the assemble workers and improve the assemble efficiency for the whole workshop. In the MOON Boeing company, it shows that using the projector help assemble the pipes and the wires in the airplane can reduce 50% time compared with using the traditional instructions to do the assembly. The systems can present the information directly on the assembling part expressing the natural word to help the workers do the right things.

It is known that the HTC vive or HoloLens can express the information in the virtual scene or the augment reality using the head mounted devices (HMD). Henderson [1] used the HMDs to help the mechanical engineer repair the tank. He relied on the 3D arrows and some word directions to help the army's engineers maintain the tank. But, different with the HTC vive or HoloLens, the major of people can see the projection without any extra hardware and feel more comfortable than the HMDs. Schwerdtfeger [2, 3] compared two different methods in detecting the welding spot of the car. He

C. Stephanidis et al. (Eds.): HCII 2020, LNCS 12428, pp. 141–151, 2020.
https://doi.org/10.1007/978-3-030-59990-4_12

used two ways to project the same welding spot's information on the part of the car. It showed that the laser projector is more stable when projecting the information than the HMD. It also showed that users felt more comfortable when using the projector than the HMD. Rodriguez [4] used the mixed reality based on the projecting reflection helped the operators do the assembly. It showed that the assisted assemble technology can guide the workers without any prepared knowledge to finish the special assembly successfully.

In the study of the projector, distortion and misleading are the important problems. Only when the projector project the image in the vertical direction, the image may be the normal without any distortion. But, when the object have the slightly movement, the information will project to the wrong place and mislead the operators to do the wrong assembly. The wrong job will decrease the efficiency and cause the big problem in the product. Yang [5] used the calibrating image to solve the tracking problems. They added the calibrating image into the projecting image. When the camera captures the images contained the calibrating code, the computer can recognize the code information and calculate the code. Then the computer compares the original image to the real image and output the correct images.

There are some researches to solve the dynamic projecting problems. Narita [6] used the high-speed camera and high-speed projector to complete the instant tracking and compensating the moving object. They used the invisible infrared ink to print the code on the target object. The high-speed camera captured images contained the code in 30 fps. Then the computer recognized and matched these code points, reconstructing the same code gridding and changing the image fast. Kagami [7] made a high-speed camera-projector system as a basement on the fast plane equipment. The camera recognizes the AR markers. Using the hardware accelerating methods, the system can detect the plane's pose and project the image quickly. They got a excellent work in the tracking and projecting, but they used the equipment that is not usually used in the factory. We try to use the normal equipment to solve the problem.

In this work, we present a novel and automated algorithm for the situation that when the object are slightly moved, the projector-camera systems can track the object easily and quickly. This new algorithm is based on the consumer-grade camera and projector to make sure that most factories can use it easily. Also, we provide a simple system to check whether the algorithm can use in the assembly.

## 2 Projection-Camera System

This section shows the hardware platform which we set up to do the research and introduce the basic pipeline to show how our system works.

Our system uses the consumer-grade projector and normal industrial camera. All the hardware equipments are normal and can easily bought in the society. As the Fig. 1 shows, we used a Epson laser projector and a Basler industrial camera.

The Basler camera is 1.3 million resolutions using the Charge-coupled Device as its own optical sensor. The camera's lens is equipped with a zoom lens. The focal length is from the 8 mm to 12 mm. The camera changed its focal length by hand instead of automatic. The auto zoom length has a bad disadvantage that it will change the camera's

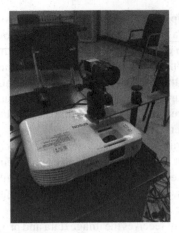

**Fig. 1.** The camera-projector system.

intrinsic parameter when the worker adjusts the focal length. The camera uses the web wire to make the connection with the computer that can ensure the transfer speed of the video stream.

The projector is a 1080P resolution projector. In our system, we set up the resolution in 800 × 600. In that resolution, the system can be calibrated easily. The projector is the laser projector that has lighter and more clear screen. The projector uses the HDMI to receive the information from the computer.

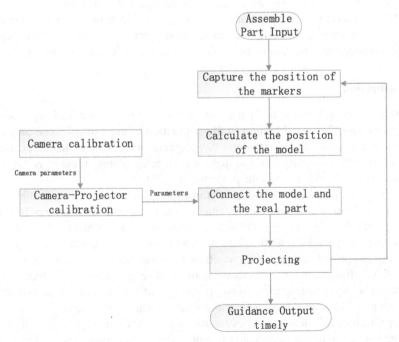

**Fig. 2.** The camera-projector system's pipeline.

The Fig. 2 shows the pipeline of the projector-camera system. We set up a client-server system. All the camera and projector are clients. The computer is the server. In this system, clients only need to collect information and show the information. The server needs to collect the data and then uses the algorithm to calculate the data which we need. In the end, the server sends these calculated data to the client.

Firstly, the camera captures the interesting area. It will capture the image in 24 fps to make sure the server can receive the image in time. Then the camera client transforms the observed image to our server.

After the server get the raw image, our system will detect the simple marker which are marked on the goal object. The marker, the high contrast circle, can be easily detected by the algorithm. Then the server uses the algorithm to calculate the matrix and get the new matrix to calculate new points. Then the server transfers the new images to the projector client.

Finally, the projector client receives the image data and project on the interesting part. The new image will be captured again and recalculated again to ensure the information our system providing are correct.

## 3    Tracking Projection Algorithm

### 3.1    Initialization

For initialization, a circle marker and a chessboard marker are prepared. The chessboard calibration marker is given in the projector image. The asymmetric circle calibrating marker is pasted on a black board. The circle marker is also prepared to guide the algorithm recognizing the interesting area. Once detected, the four corners of the target area will be labeled by four circle marker. In our system, the four circles are used as the control points to specify the area. The projector only needs to show the information in this area. Outside the edge, that is the invalid area or we called not interesting area.

### 3.2    Calibration

Our system firstly calibrates the camera and the projector. As the tracking procedure described in Sect. 2, the calibration in this algorithm is used to calculate an exactly matrix connecting the camera axes and the projector axes. Our system uses the two different kinds of calibrating board to make a connection between the physical world, the virtual world, the camera and the projector (Fig. 3).

We set a size of 4 rows and 11 lines asymmetric circle image as a calibrating board. Our algorithm firstly detects the whole center of the circle and then captures 20 images which the circle board is placed in the different area or set in different angles to the camera. Our algorithm used the least square method to solve the contradictory equations. Twenty images provide twenty different equations, and we just try to solve the equations with only five unknowns. The least square method can solve the problem quickly and preciously. After getting the intrinsic parameter of the camera, the algorithm will calculate a nine unknowns' error matrix and then present an average error.

The projector can't detect the chessboard calibration by itself. Our algorithm uses the camera which has been calculated recently to help the projector calibrate itself.

**Fig. 3.** The calibration board.

The projector will present a 6 rows and 9 lines chessboard on its projecting image. We also use the asymmetric circle board to assist calibrate the projector. When the camera captured the two different kind of calibrating boards in the same image, the algorithm will work automatically. Our algorithm analyzes the two different boards and divided into different groups. Then it matches the corners' position in the projector's pixel-coordinates to the position in the camera's pixel-coordinates. After the matching it will transfer a matrix. Our algorithm solves the same contradictory equations like the camera calibration. Finally, we combine two matrixs and solve the intrinsic parameters of the projector.

The intrinsic parameters are the character of the camera and the projector. After calibration, our algorithm will not change the intrinsic parameters. And these parameters are the base to solve the matching and tracking problems.

### 3.3  Tracking

Getting the intrinsic parameters can correct the distortion of the camera and projector because of the physical equipment's optic error. In this section, the algorithm will set up a connection to support the tracking. The image is transferred from camera client to computer server. Our algorithm changed the colored image into gray-scale image firstly. Then we use the Canny and Gauss operators to find the edge of the object in the image. Labeling all the edge and sending to the filters. Our algorithm set the boundary of the ellipses' perimeter and area to screen out the other edges that is not the circles which we needs. If the edges are the ellipses or the circle, it will be marked into a matrix (Fig. 4).

After the edges detected, the algorithm extracts the central point of each circles and recognizes the basic point. In all the circles, we set a circle with different radius. When the algorithm finds the basic point, its pixel coordinates will be connected to the virtual object coordinates. When the edges transfer to the central points, we can solve

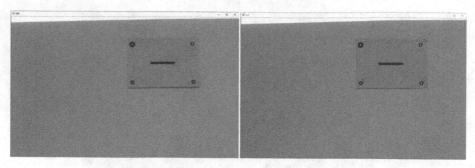

**Fig. 4.** The tracking results.

the problem as how to match all the point correctly. Then our algorithm uses the convex hull method to connect all the central points into one of the whole shape. The algorithm will find the points according to the clockwise order and label all the points by the order of the convex hull.

For tracking the real object correctly, each point corresponds to an equation. Combining all the equations, the algorithm solves the contradictory equations and gets a transform matrix. The matrix contains all the translation and rotation.

In the end of the tracking, the virtual object will be combined the transform matrix and the parameters of the hardware equipments and then output a new vector and rotation to change the projector image.

### 3.4  Projecting

When the algorithm calculates the new matching solutions, the computer server transforms the matrix to the virtual reality. In the virtual reality, we build a same size object and set a virtual camera and virtual projector. The virtual projector changed its position according to the transform matrix. And the virtual object also changed its position and rotation (Fig. 5).

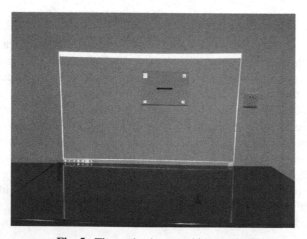

**Fig. 5.** The projecting matching results.

Then the server transfers the image which captured by the virtual projector to the real projector client. And the algorithm finishes all the tracking and matching works, then it goes to do another transform from the camera image to the projector image.

## 4   Evaluation Method

Our goal is to achieve projecting quickly on the target assemble part area. We designed an experiment to simulate the real scene in the laboratory. We simulate the real assembling circumstance: the projector and the camera are placed 2 m far from the target object. We present a paper with some circles and lines to simulate the real assemble holes and wires. Then, we print the circle label on the corner of the paper. During the experiment, we measure the average error of the projecting objects to the real objects. We also asked some people to do the assembly and let the simulate paper move slightly. After the experiment, we asked them how they felt about the guidance. To know the accuracy of the tracking, we also measure the error of the calibration.

We use the camera to capture the image and measure the error during the paper moving slightly. For each image, we exchanged the pixel to the real distance for knowing the error to detect the latency of the projecting. Figure 6 shows how the real experiment set.

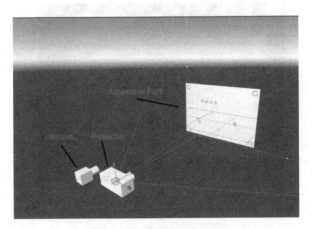

**Fig. 6.**  The experiment set.

In the virtual reality, the system uses the Unity to build a simulative scene. Unity is a strong game engine to develop an excellent scene with many interactions between the people and the game. In the Unity, the virtual target object is set the same size as the real physical world. For importing the assembly in the industrial field, all the objects are made by the 3D modeling software like CATIA, ProE, UG NX and so on. Using the 3DMax, we can transfer the model to the Unity. It will make the factory feel easy when a new model need to be assembled (Fig. 7).

The algorithm are written by the C++ and packed in the dynamic link library (dll). Using the dll, the Unity can make a strong interaction with the algorithm. We also use

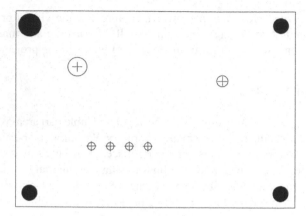

**Fig. 7.** The experiment assemble model.

the C# in the Unity to write some scripts to connect the algorithm to the virtual scene. Figure 8 shows how the virtual scene is set. When the operator clicks the "start" button, the system works automatically.

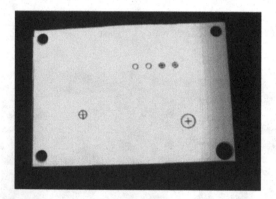

**Fig. 8.** The projecting results.

We asked six volunteers to do this experiment. As a opposite, we set a control group that all the assembly needs to finish with the help of the written instructions rather than the projecting guidance directly. All the job is about simulate the assembly of installing the fastener and putting the bolt model into the holes which they should be put in. After they finish the job we asked how they are feeling as a feedback.

## 5  Results

We measured one of the experiments' data. Figure 9 shows the average error when the object has slightly movement. And we use the imagej software to transform the pixel distance to the physical distance.

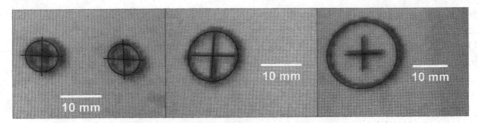

**Fig. 9.** The aeesmble character's projecting results.

The result shows that the error between the projecting position and the real position is near 0.8 mm. The circle edges' error is larger than the lines. Comparing the error of the automatic robot assembly, the projecting is a little bad to the robot in the precision. We also notice that the circles are more fuzzy than the lines because of the projector (Fig. 10).

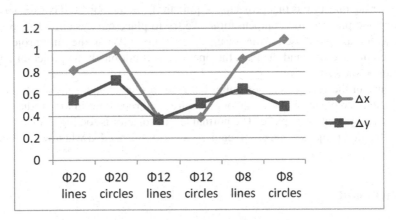

**Fig. 10.** The projecting error results.

Then we got the feedbacks from the volunteers. In the manual assembly, the volunteers spent more time when operated by the instruction of the assemble order than by the projecting the information directly on the target part. When the object slightly moved, most volunteers didn't feel uncomfortable to operate. Some volunteers said that the system improves their working efficiency. Some volunteers complained that in some place, they must use an uncomfortable post to do the assembly to avoid blocking the projecting guidance.

We test the latency of the projector. It showed that the average latency of the tracking is less than 0.5 s. Although the volunteers could see a little delay of the image. But it mostly didn't influence the volunteers operating and could guide the process in time.

## 6   Limitations and Discussion

Our work uses the simple tracking and matching algorithm to solve the dynamic projecting in the assembly. In the assemble factory, the projector is set in the back of the assemble parts. The new methods give the operators a quite direct instruction to do their jobs. Our laboratory also uses the HTC Vive, the HoloLens to do some assisted assemble guidance. The projecting uses the traditional methods and normal hardware equipments to finish the job. It's quite important to increase efficiency of the manual assembly.

However, the algorithm still has some latency to be cover. Because of the screen door effect of the projection, in the next step, we try to use the high distinguish camera and projector to upgrade the hardware. The system needs to prepare lots of things before doing the guidance of the manual assembly, so we try to use the faster algorithm to calculate the matrix and consider reduce some steps that may be not necessary in the preparation. We try to find a new method to track the natural labels or markers instead of the manual labels. We can track the edge of the assemble part, also we can track the holes or the lines on the target part. That will be more convenient for all the factories.

Another big question is that projector is quite far from the object. It caused so much occlusions and photometric compensation. We try to place our system into a portable mini projector and put these in front of the assemble part. When the mini projector set up between the operator and the part, the operator will not block the projector and the screen can be showed entirely.

Not only in the real factory assembly but also in our laboratory, we establish the system in an open place. When we layout the wires and pipe lines in the product, it will be operate in a very narrow place. The normal laser projector is too big to place in such a room. The mini projector and micro camera seems to be needed when we provide the guidance.

## 7   Conclusion

This paper described an approach to achieve a quick projecting guidance in the manual assembly. A closed-loop system has been achieved by the tracking markers and matching the labels using the normal hardware equipment. It was found that the tracking algorithm can project the right information in the appropriate position. In a future study, we aim to use the mini equipment to solve the operators blocking the view of the projector. Also, we try to use the nature characters in the assemble part to be the marker, that will be convenient to prepare the guidance.

**Acknowledgments.** Part of this work was supported by 111 Project. (B12024).

## References

1. Henderson, S.J., Feiner, S.: Evaluating the benefits of augmented reality for task localization in maintenance of an armored personnel carrier turret. In: IEEE International Symposium on Mixed and Augmented Reality, pp. 135–144 (2009)

2. Schwerdtfeger, B., Pustka, D., Hofhauser, A., et al.: Using laser projectors for augmented reality. In: Proceedings of the 2008 ACM Symposium on Virtual Reality Software and Technology, pp. 134–137 (2008)
3. Schwerdtfeger, B., Hofhauser, A., Klinker, G.: An augmented reality laser projector using marker-less tracking. In: Demonstration at 15th ACM Symposium on Virtual Reality Software and Technology (VRST 2008) (2008)
4. Rodriguez, L., Quint, F., Gorecky, D., et al.: Developing a mixed reality assistance system based on projection mapping technology for manual operations at assembly workstations. Procedia Comput. Sci. **75**, 327–333 (2015)
5. Yang, T.J., Tsai, Y.M., Chen, LG.: Smart display: a mobile self-adaptive projector-camera system. In: 2011 12th IEEE International Conference on Multimedia and Expo (2011)
6. Narita, G., Watanabe, Y., Ishikawa, M.: Dynamic projection mapping onto deforming non-rigid surface using deformable dot cluster marker. IEEE Trans. Vis. Comput. Graph. **23**(3), 1235–1248 (2017)
7. Kagami, S., Hashimoto, K.: Animated Stickies: fast video projection mapping onto a markerless plane through a direct closed-loop alignment. IEEE Trans. Vis. Comput. Graph. **PP**(99), 1 (2019)

# Design and Implementation of a Virtual Workstation for a Remote AFISO

Thomas Hofmann[1]($\boxtimes$), Jörn Jakobi[2], Marcus Biella[2], Christian Blessmann[1],
Fabian Reuschling[2], and Tom Kamender[1]

[1] University of Applied Sciences Osnabrueck, Osnabrueck, Germany
t.hofmann@hs-osnabrueck.de
[2] DLR - Deutsches Zentrum für Luft- und Raumfahrt, Brunswick, Germany

**Abstract.** On the basis of given use cases for the professional group of flight controllers (AFISO) and air traffic controllers (ATCO), a human machine interface with two different interaction concepts for virtual reality was developed.

The aim was to facilitate the cooperation between air traffic controllers and air traffic and to enable the remote monitoring of AFIS airfields with the help of a VR headset.

ATCOs have more tasks and a higher degree of authorizations, but also perform the same activities as AFISOs and can use the virtual workstation in the same way.

Software and hardware solutions were identified, the usage context of the AFISO was recorded and usage and system requirements for the user interface were formulated. Based on this, an overall concept was developed that includes the virtual work environment, the head-up display and interactive objects. A user interface and two different VR prototypes for interaction were developed. The focus lied on the implementation of the AFISOs tasks and the design of his virtual workplace. Ergonomic and usability relevant aspects as well as the physical environment of the user were considered. In a series of tests with seven ATCOs and two AFISOs, the VR prototypes were tested at the DLR research center in Braunschweig and evaluated through user interviews and questionnaires.

**Keywords:** Industrial design · Virtual reality · User interaction · Ergonomics · Prototyping · Leap Motion

## 1 Introduction

The daily work of Aerodrome Flight Information Service Officer (AFISO) at hub airports or larger airports is increasingly characterized by the use of highly technical, networked systems, which increase safety by reducing the workload. The integration and homogenization of the individual systems in terms of presentation and interaction is simplifying the human-computer interaction (HCI) - the systems are becoming increasingly ergonomic and easier to use.

While research focuses heavily on major airports, changes in HCI and ergonomics at regional airports or aerodromes and their AFISOs are few and far between. Especially here, the pent-up demand in favour of a user-centered and tasks appropriate design of

© Springer Nature Switzerland AG 2020
C. Stephanidis et al. (Eds.): HCII 2020, LNCS 12428, pp. 152–163, 2020.
https://doi.org/10.1007/978-3-030-59990-4_13

the technical equipment is very large. In many cases, many non-integrated technical systems are being used, which make efficient and safe interaction difficult. In addition, many HCI systems hardly represent the state of interaction and visualization familiar to users from the private sector. The overall situation is made more difficult by the fact that the coordinating staff at small airfields are not original AFISOs.

## 2  Project Focus

As part of the project presented here, a system has been developed which should enable Aerodrome Flight Information Service Officers (AFISO) at regional airports or local airfields, via Augmented Reality in combination with familiar basal interfaces, to control the air traffic quickly and intuitively. The aim of the - still ongoing - project is to build the control and management of air traffic and taxiing at smaller airports on the use of a HeadSet VR system. In contrast to previous VR systems, however, the VR headset has a different meaning: While previous approaches usually assume that a VR headset is a new system that users must first deal with, the VR headset represents the concept described here a further development of the AFISOs known binoculars.

Binoculars still represent the prioritized hardware interface for users at smaller airports. The approach based on research and observation, assumes that moving many HMIs into binoculars, as well as enriching them with different forms of interaction, can be an innovative and intuitive concept for the future.

## 3  Main Emphasis

The overall system should have numerous technical features that make the interaction intuitive and ergonomic, as well as enabling a completely new user experience. The system should have been equipped with:

- a real-time external view visualization
- supplementary, high-resolution external view display of defined zoom ranges (binocular function)
- presentation of relevant additional information
- Leap-motion gesture control for intuitive operation of different GUI
- Use of additional HMI on a standard VR headset

## 4  Concept

The overall concept of the HMI consists of three basic elements:

- the virtual environment of the user
- the Head-up display and
- various interactive objects

The virtual environment is presented to the user by a spherical panoramic image displayed. It is always located in the center of a sphere, which is the video stream is recorded. The sphere is rigid and is only used as a projection surface. A complete 360° live environment offers the user the possibility to look around free of camera delay. By putting on a VR headset he is automatically in a three-dimensional environment. Content that was previously displayed and perceived in 2D can now be used to fill the room. The sphere viewed from the inside simulates a natural environment and stimulates the user to use the full potential of the panoramic view. The PTZ camera is used to scan all areas of the panorama and to increase its size for the user. The head-up display is always within the user's field of view. It is arranged between the VR headset and panoramic environment. The position of the display is fixed in the field of view of the headset and moves along continuously. Here the user is provided with all relevant information for the implementation of his activity. In addition, an interactive compass is embedded in the upper field of view, which is Movement readjusted in real time. In the center of the field of view there is a special window for the insertion of the PTZ image marked stage. The additional camera image is displayed in this area. In addition, process sequences are visualized to the user in the head-up display.

Various interactive objects enable the user to make entries in the system and navigate within the application. There are interaction concepts which offer the user the highest possible degree of flexibility with the lowest possible intellectual and physical effort. Furthermore, reliability and accuracy in operation are decisive. The interaction must not impair the user's situational awareness through complex and fine-motor Restrict entries or time-consuming processes.

The interaction concepts are presented in the next steps and then evaluated according to ergonomic and usability relevant criteria. The design of the virtual workplace and of the functional models continues to be based on the criteria mentioned above, the influence of which has been is visualized.

## 4.1  Hardware Setup

As basis for the work presented here, a camera setup consisting of a panoramic camera and a pan-tilt-zoom camera (PTZ-camera) has been erected at the Braunschweig-Wolfsburg aerodrome, as depicted in Fig. 1. The setup is placed at approximately the same height and distance from the runway as the aerodromes tower is.

The panoramic camera is an AXIS P3807-PVE covering a field-of-view of 180° horizontally and 90° vertically with a resolution of 3648 × 2052 pixels. As PTZ-camera a Pelco Esprit Enhanced ES6230-12 is used, capturing a video stream with 1920 × 1080 pixels. The camera pans 360° horizontally and tilts 130° vertically. These cameras provide the required real-time external panoramic view and supplementary view with zoom functionality, respectively (Fig. 2).

- PTZ Camera
- Samsung Gear 360 Panorama Camera/180° Panorama Camera
- HTC - Vive Pro System (incl. controller & lighthouses)
- VR Computer
- Leap Motion IFR-Tracker

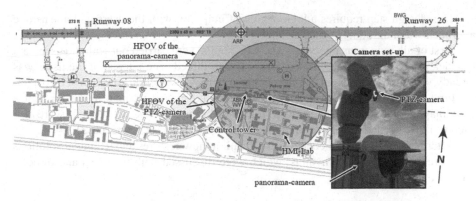

**Fig. 1.** Camera setup with 180° Panorama at the Braunschweig-Wolfsburg aerodrome

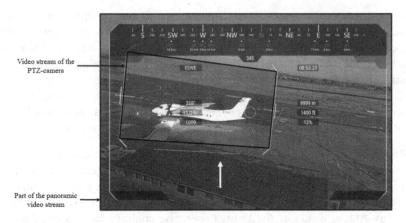

**Fig. 2.** The video streams are displayed with an HTC Vive Pro VR-Headset.

## 4.2 Methodical Approach

The experience from past research projects in the ATC environment has shown that a massive involvement of ATCOs is an essential aspect of a user-friendly design process. This aspect became all the more important as the project introduced a technique that was mostly unknown to users. Work was carried out on a massively iterative user-centered development model developed in the Industrial Design Lab of the UAS Osnabrueck.

This means that the users were continuously involved in the development from the requirements analysis to the prototype - from the first interviews to the support at work to the design of individual GUI elements and interaction gestures. Only then is there a high level of identification with a research project.

Especially when a new artifact is presented as an interaction element, the integration of the users is essential. In this case it was about the replacement of a traditional binocular, computer mouse and keyboard. The mouse and keyboard should be completely eliminated, the binoculars should get completely new functions.

This design step requires a high degree of empathy with the users and a sensitive approach when introducing such a concept [1].

First of all, numerous interviews and workshops were carried out with the users and the idea of augmented reality binoculars was introduced. The interest of the users was already very high at this point and the openness to new interaction concepts helped a lot.

First the concept was formulated as an idea and played through in the form of a gameplay [2]. Ideas and visualizations from well-known science fiction films were also used to illustrate the innovative strength and possibilities of the concept [3].

After basic mockups for concept visualization, new features and technical solutions were presented in weekly rhythms, which were discussed with the users and directly modified in workshops.

## 4.3 Design and Implementation

The implementation of the user interfaces and interaction concepts were carried out in constant exchange with the remote AFIS team of DLR (Braunschweig) and were developed in the HMI laboratory of the Osnabrueck University of Applied Sciences. The Osnabrueck University of Applied Sciences was able to provide valuable input regarding the usability and design development of the HMIs. DLR has numerous relationships with the core users of the system who have been involved in this process from the beginning. In addition, DLR has many years of experience in the development of ATM systems. These prerequisites are essential for a user-centric design of the system.

**Head-Up Display.** Via the Head-Up-Display (HUD) the user takes information visually. It is fixed in front of the user's headset in the virtual environment in the middle of his field of view. The distance between headset and interface can vary fundamentally. Distances of less than half a meter are uncomfortable; distances of over 20 m are perceived as too far away. At long distances, the ability to perceive contrasts and shading diminishes. If the user realigns his or her gaze or headset, the head-up display moves with it without delay. Like a mask, it lies between the user and his virtual environment.Typical computer applications require two-dimensional media that are mainly operated while seated. A three-dimensional medium only develops its full potential when used standing. Head and torso can be moved more flexibly. The information in the entire system is mainly typographical in nature, which is why the size of the integrated font and its legibility play a decisive role. In virtual reality, the font should have a minimum size of 20px to be pleasantly legible [4]. The sans-serif font Roboto was chosen, since especially its figures are well readable in virtual reality (Fig. 3).

Two HUD variants were created and set up with image editing software Adobe Photoshop, Illustrator, InDesign and the development software Unity. They show minor differences regarding the PTZ stage in the center of the field of view and the bordering of the linework in the secondary field of view. The interface is square in the Vive Controller concept and round in the Leap Motion concept. It is based on the appearance of the menu contents and serves a uniform design line. In the center of the primary field of view, the information is arranged according to relevance. They are arranged around the central image section, in which the additional camera image of the PTZ-camera is displayed directly in front of the user. In the center of the image there is a circular element that

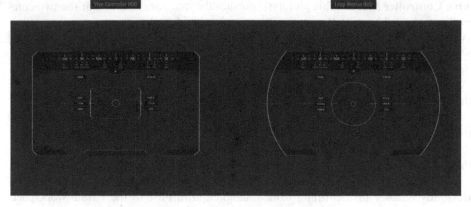

**Fig. 3.** HUD visualization for the interaction concepts

marks the center of the user's field of view. On the left side of the image center, he will find current weather information. On the right side, information regarding visibility and track conditions is arranged. On the left side of the image center, the user finds the identification code of the airport shown and on the right side the coordinated world time (UTC). All information is given without a unit, as the user has the ability to identify it by its value. The intention is to provide him with the minimum amount of necessary information and to keep its mass as concentrated and small as possible. They are retrieved live from the Internet and updated.

A compass is implemented in the upper part of the field of view, which shows the user the full 360° of his environment. On the vertical axis of the field of view, a dynamic value is shown below the compass, indicating the exact degree and cardinal direction of the headset's alignment.

This arrangement is based on the fact that the user is increasingly looking upwards and this information should be quickly visible to him. The cardinal points are also marked on the compass. The entire compass is dynamic and constantly readjusts itself by the orientation of the headset. To ensure a seamless transition, a script in Unity is used to repeat it on the sides as soon as they enter the field of view. The lower part of the compass incorporates colour-coded waypoints of Braunschweig Airport (EDVE), which pilots pass on their approach to the airport. Their identifier, position and distance in kilometres are assigned in the compass and are continuously displayed to the user. These waypoints are intended to make it easier for him to visually locate an aircraft by drawing information on its position from radio traffic and aligning his field of view according to the reference points.

Below the primary field of view, the user is shown process animations, such as the sending of a radio message. All information is framed by a linework that is intended to guide the user's view and assist him in aligning the field of view. All texts are underlaid with a partially transparent surface to ensure their legibility regardless of the background of the video panorama and its lighting conditions. The user experience should be a latency-free, sharp, legible, reduced and clearly structured VR experience.

**Vive Controller Menu.** This prototype pursues the integration of one of the two controllers provided by the VR system. The controller has 4 push buttons, a touch sensitive swipe surface, which can be additionally integrated as a push button and a trigger button, which can be operated as a toggle switch with the index finger. Due to its symmetrical design, the controller can be operated with both hands.

The controller is detected by the Vive System's infrared room tracking. This means that the user must hold the controller in one hand to interact with the system. The controller is equipped with additional user interface elements. These allow the user to make various entries in the system or to call up specific content and functions. The displayed content follows the movement of the controller. The user can therefore hold them closer to his field of view to read them better. This interactive wireless input unit allows the user to perform his work detached from a fixed working position. This flexibility enables a room-filling work situation and full use of the virtual workspace is possible. When observing a concrete situation with the help of the cameras, the user does not have to turn away from what is happening in order to make an input (Fig. 4).

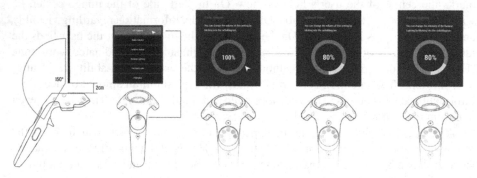

**Fig. 4.** Development of the Controller Menu

New functions have been assigned to the various keys of the controller, which allow the user to navigate through the user interface. The menu button opens the controller menu. The key is located above the touch surface on the front of the controller. Its colour coding communicates to the user that it has been assigned a function. When the menu is opened, a panel appears above the controller, which is fixed upright at an angle of 150°, measured on the vertical axis, and at a distance of 2 cm above its ring element. It has a width of 10 cm and a height of 15 cm. All contents of the menu are displayed on this surface. Via the main menu the user navigates to the different menu contents. These are displayed to him in typographical form as a list. The integration of icons has been refrained from due to poor readability.

It is possible to choose between 6 menu items (from top to bottom: PTZ Camera, Radio Volume, Ambient Sound, Runway Lighting, Contrast Layer and the Flightplan) A cursor is displayed on the panel, which is controlled by the user with the thumb over the touch surface. If the user removes his thumb from the touch surface after an action, the cursor jumps back to its original position. If the user leaves his thumb on the touch surface during operation, he retains constant control of the cursor, which makes it possible to make entries much faster. In addition to the movement of the cursor, the user is also shown the position of the thumb on the touch surface of the virtual controller model. If the cursor moves to a menu item, it is displayed with a coloured background. When a menu item is selected by pressing the touch surface, it is displayed in a lighter colour. Each of these digital buttons therefore has three states.

Within the menu, the user encounters various interactive switching elements. For example, he can activate or deactivate functions via toggle switches. To do this, the user moves the cursor to the switching element and clicks on it with the thumb. With the toggle switch of the menu item Contrast Layer he can contrast the virtual environment. A colored, partially transparent visor is faded in between the head-up display and the virtual panorama environment.

With the other hand, he can adjust the volume of radio traffic and ambient noise as well as the intensity of the railway lighting by means of various colour-coded radial diagrams. To change a radial diagram, it is activated with a click. The input is made by a circular movement with the thumb on the touch surface of the controller. The fill level of the diagram is displayed in real time and the current value is shown as a percentage in the middle of the pie chart. Another click confirms the input and fixes the controller again. To get back to the main menu, the user can click on the left side of the touch surface. The function is located at this point because in familiar computer applications the arrow pointing to the left means "back", confirming the user's mental model. With the index finger he can also operate the trigger button. This is located on the back of the controller. The toggle switch is equipped with the radio function. When deactivated, it is coloured bluish. In action it turns red. To do this, simply hold the index finger button and speak. The radio message is recorded and digitalized via microphones integrated in the headset.

**Leap Motion Menu.** This prototype follows the integration of a motion tracker into the hardware setup, which is mounted in a 3D printed cage on the front of the headset. The hardware is carried with a USB cable connection to the computer via the cable coming from the headset. Using infrared sensor technology, the tracker detects the user's hands when they are within its field of view. His hands are displayed to him in real time in the form of virtual hand models (Fig. 5).

This enables him to operate and change interactive content. He does not need any other physical input devices and has both hands free during use. This also means that all interactions must take place within the field of vision. All buttons within the system are aligned at a horizontal angle of 30° to the user. The average shoulder height of male and female users is between 155 cm and 142,5 cm. Due to the average arm length of 81,5 cm and 75 cm, the menu was placed in the space between the seventh and eighth percentile [5].

The menu is thus located at a height of 110 cm, measured from the floor. The interaction space is also measured by the reach of the arms. The contents are oriented towards the user in such a way that he or she has to make minimal head and eye movements when standing, in order to read the contents well and operate them ergonomically. The motion tracker detects a space that exceeds the average arm length to ensure that everyone can interact fully in virtual reality (Fig. 6).

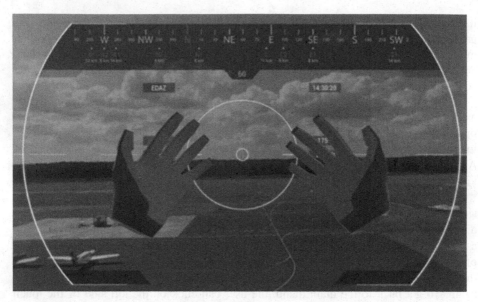

**Fig. 5.** Screenshot from the Leap-Motion Menu (visualisation of the users hands)

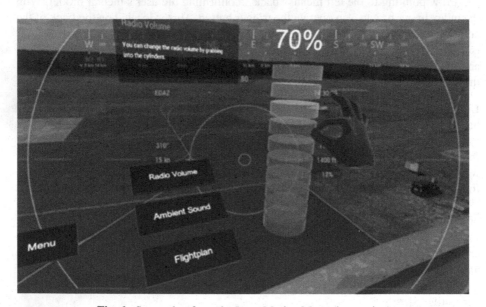

**Fig. 6.** Screenshot from the Leap-Motion Menu (interaction)

All interactive objects are located in close proximity to the main menu button and the physical reach of the user. The user can perform the required user actions in a comfortable upright position while freely looking around in his 360° environment. However, the orientation of the menu is rigid. It provides a clear working position with a view of the runway and taxiway of the displayed airfield. The perceived scene corresponds to the view from the tower of the respective airfield.

The colour design follows the design of the controller menu and its intentions. In contrast to this, a three-dimensional representation of the contents was chosen in this concept. Due to the depth, the objects have a haptic effect and stimulate interaction, which takes place directly between user and object. The user finds a menu button in his environment, which he can use to open the main menu of the system. The interaction takes place by pressing virtual buttons. They are coloured black and have a white label. If the user moves his hand into one of the interaction spaces considered by the buttons, it is coloured in the main colour. This tells the user that he can interact with them. The activation of a button requires the user to press it in to a depth of 2 cm. When pressed, the button turns light green. If the process is interrupted, it reappears in the main colour. If the user leaves the action radius of the button, it appears in its original colour.

When the user opens the main menu, six additional menu items are displayed symmetrically around it. On the left side you will find the settings for the zoom camera, path lighting and the function of the Contrast Layer. The latter is shown as an example in the form of a partially transparent colour area, similar to the Controller menu. On the right side are settings for radio and ambient volume and the function of the flight plan. All buttons have the same reaction behaviour and can be used in the same way. Display panels and three-dimensional controls can be faded in and out via the menu items. If a function is active, it is displayed on another panel so that the user knows which menu item he is in. The system is designed so that only one item can be shown at a time. This is intended to create a space-saving working situation that is also clearly arranged. It is thus possible for several menu items to share the same space for their display. With the buttons for the PTZ camera and the flight plan the user can fade in two-dimensional panels and extract the corresponding information. The flight plan appears on a black transparent background with the colour-coded flight strips described in the Vive concept. The flight plan is facing the user and allows a view without major movements. The size of the flight plan is oriented towards good legibility. The user should also be able to add new flight strips (not implemented yet).

Three-dimensional controls are accessed via the menu contents Runway Lighting, Radio Volume and Ambient Sound. They consist of ten partial cylinders, each representing 10% of the setting. In Unity, these volume bodies were provided with an invisible slider. The user can grasp this slider manually. The current filling level of the cylinder, or the position of the slider, is visually displayed. This adjustment can be made with the palm of the hand closed or open. With the palm open, the user pulls less on the slider, but lifts it much more or pushes it down. All partial cylinders located below the slider are coloured by the system according to the colour coding used in the controller menu for the corresponding menu items. All those above are displayed in transparent white. A dynamic text is placed above the solids, indicating the value of the current setting in percent. This text faces the user vertically by 20°.

The interaction elements must not impair the user's view by obscuring the contents of the video stream. Therefore, the cylinders appear transparent and thus allow the user to see what is behind them.

# 5  Results

The working position of AFISO described here previously represents a very analogue and local human-machine interface. The users interact a lot via the so-called outside view, mostly communicate via radio and use traditional media.

The research project presented here wanted to test a new approach to modernize the working position. The first phase involved an acceptance test. It was not yet the goal to develop a marketable solution or to present a final design concept.

The project team was dealing with very motivated users, but they were still users who had so far had little to do with innovative and experimental interfaces. Accordingly, the present first findings can be seen as the basis for the continuation of the project on a larger scale. Because what is undisputed and has been confirmed by the users, the previous human-machine interaction was accepted, but a potential for improvement was confirmed - both on a technical and an ergonomic level. The usability of the existing set-up is suboptimal for an efficient and secure interaction with the AFISO environment.

The core focus of the test setup was the integration of the binoculars as a well-known interaction element. This interface was assessed as absolutely necessary and was also adopted accordingly - only as an augmented reality artifact. At the beginning of the project, the hypothesis was made that the AFISO would accept virtual binoculars due to the analogy, or would welcome them with AR functions due to the loading.

This hypothesis was confirmed by a qualitative survey. Although it was not yet possible to carry out a representative study on the efficiency and ergonomics of this set-up, initial insights from the tests carried out seem to confirm the hypothesis. AFISO also welcomed the enhancement of the binoculars with zoom displays and additional information – they accepted the 'new' functions' of the device but also were able to use it as their used device - this was confirmed by the very fast and safe interaction with the VR glasses. A very important finding in order to be able to continue the project on a larger scale.

The experimental setup initially served as a first proof-of-concept. The aim of the project was to design and validate new forms of interaction for a previously very analog human-machine interface. The following could be determined in the context of the project:

The system has many advantages through the integration of different technical systems and the use of known interaction types:

- The AFISO always has his main interface at hand
- The system integrates numerous necessary information in the glasses
- The pilot still has a hand free for the radio despite the use of the system
- the use of the system is intuitive and analogous to the well-known HMI (binoculars)

The overall system would take AFISOs' and ATCOs' interaction to a whole new level and the information-gathering workload would be massively reduced.

# 6  Discussion

Currently there is a prototype of an interface that allows full interaction with a VR headset. This system depicts both the outside view of the airport, the additional image of a zoom camera, as well as numerous other interfaces for the control and coordination of air traffic. This overall system is powered by the data of an exemplary air and ground space of an airfield.

With this prototype, the implemented interaction types are checked for validity and optimized. In the next step, the application is planned in the real context (shadow mode). This procedure corresponds to the methodology for the successful implementation of novel HMI in the ATC environment.

# References

1. König, C., Hofmann, T., Bruder, R.: Application of the user-centred design process according to ISO 9241-210 in air traffic control. In: Proceedings of IEA 2012, Recife (2012)
2. Hofmann, T., Bergner, J.: Air Traffic Control HCI - how to handle very special user requirements. In: proceedings of 8th International Conference on Applied Human Factors and Ergonomics, Los Angeles (2017)
3. Hofmann, T., Kamender, T, Stärk, F.: Using virtual reality in the ergonomics-design workflow. In: Proceedings of 8th International Conference on Applied Human Factors and Ergonomics, Los Angeles (2017)
4. Alger, M.: How we design for VR (2018). – web source: https://www.youtube.com/watch?v=49sm52fG0dw&list=PLpDXhkESrNJtNhPKyL0gOfw_Ntl0gL9NI&index=5. Accessed 14 Sept 2019 (2018)
5. Lange, W., Windel, A.: Small ergonomic data collection. 15. updated edition. Cologne. TÜV Media GmbH. [Hrsg.]: Federal Institute for Occupational Safety and Health (2013)

# A Scene Classification Approach
# for Augmented Reality Devices

Aasim Khurshid$^{(\boxtimes)}$, Sergio Cleger, and Ricardo Grunitzki

Sidia Institute of Science and Technology, Manaus, AM, Brazil
{aasim.khurshid,sergio.tamayo,ricardo.grunitzki}@sidia.com

**Abstract.** Augmented Reality (AR) technology can overlay digital content over the physical world to enhance the user's interaction with the
real-world. The increasing number of devices for this purpose, such as
Microsoft HoloLens, MagicLeap, Google Glass, allows to AR an immen-
sity of applications. A critical task to make the AR devices more useful
to users is the scene/environment understanding because this can avoid
the device of mapping elements that were previously mapped and customized by the user. In this direction, we propose a scene classification
approach for AR devices which has two components: i) an AR device
that captures images, and ii) a remote server to perform scene classification. Four methods for scene classification, which utilize convolutional
neural networks, support vector machine and transfer learning are proposed and evaluated. Experiments conducted using real data from an
indoor office environment and Microsoft HoloLens AR device shows that
the proposed AR scene classification approach can reach up to 99% of
accuracy, even with similar texture information across scenes.

**Keywords:** Scene understanding · Convolutional neural network ·
Transfer learning · Augmented reality

## 1 Introduction

Augmented Reality (AR) technology enables users to interact with the physical
world by overlaying digital content on it [6]. Many AR devices are being used for
education and industry applications [4]. Some most commonly used AR devices
include HoloLens [5], MagicLeap [14], GoogleGlass enterprise edition [13] and
Toshiba dynaEdge AR100. Although these AR devices have shown great potential for applications in many fields, existing AR solutions are so limited that it
can not be considered a household gadget.

Most of the existing AR applications require mapping of the scene to overlay
digital content on a scene in the real-world. This mapping is usually done by
locating a flat surface in the scene, which enforces users to map a plane in

This work is partially supported by Sidia institute of science and technology, and
Samsung Eletrônica da Amazônia Ltda, under the auspice of the Brazilian informatics
law no 8.387/91.

the scene to augment [10]. This makes the user experience with the AR device unpleasant, and limit its applicability. Also, once the mapping is done, the digital content is overlayed, this augmented reality is volatile unless the scene is not re-identified next time.

Many methods have been proposed for indoor and outdoor scene classification [15,20,24]. Statistical information about possible inter-dependencies among objects and locations are used in the past [18]. However, these methods performed poorly in discrimination of the complex textural information among the scenes. Most methods utilize texture features for indoor scene classification. Recently, an indoor scene classification method is proposed that utilizes Scale-Invariant Feature Transform (SIFT) features to build a Bilinear Deep Belief Network (BDBN) model using information from a salient region detection technique for indoor scene classification [15]. Similarly, a Structured Constrained Local Model (S-CLM) is used by extracting the Speeded Up Robust Features (SURF) feature of scene image to achieve a high classification rate [24]. In others, local features such as SIFT is used together with supervised classification techniques including Support Vector Machines (SVM) and K-nearest neighbor (K-NN) for scene recognition in an indoor environment [22]. However, such low-level features of images without rich semantic information fails when the number of classes increases. Neural Networks have also been explored for scene classification method and achieved high accuracy for scene category classification [12,20,25].

These methods explained above are tuned for indoor scene *category* classification. For instance, such classification consists of identifying if an environment belongs to a bathroom, bedroom, office or kitchen category. On the other hand, the goal of the proposed approach is to classify among specific scenes (e.g. bedroom A, and bedroom B), and they may belong to the same category (bedroom).

Also, the proposed scene classification approach is specific to AR devices. To the best of the authors' knowledge, scene classification is not explored for AR devices despite its potential to improve the usability of the AR devices. To make AR devices more useful, scene classification and understanding is interesting. After understanding the scene, the smart AR device can facilitate the user to utilize the potential of the AR device. This way, the AR device can also serve as a personal assistant that understands and classify the frequently visited scenes of its owner, and allow personalizing these spaces. As the mapping of the scene is time-consuming, it is useful if the scene can be classified and restored from the previous scene customization of its owner. This can be achieved if the classification is accurate.

In this direction, this paper proposes a scene classification approach for AR devices, which is composed of two components that interact with each other: i) an AR device that contains an application responsible for image acquisition; and ii) a server component that performs scene classification. Four inference models for scene classification are proposed and evaluated. The first, named SURFwithSVM, utilizes SURF to train a classical Support Vector Machine (SVM) model [3,9]. The second (small-CNN) comprises of a CNN with a limited number of layers. To reduce the training effort in the building of the model for scene classification, in the models GoogleNet-TL and AlexNet-TL, we apply transfer learning to retrain two well known existing networks for image classification.

These four models are trained and evaluated with real data, which corresponds to a business building with several indoor scenes with similar texture. In the experiments, we test the effectiveness scene classification on varied number of scenes. Our experimental analysis shows that SURFwithSVM works well for a small number of classes, however, as the number of classes increases, the classification accuracy tends to decrease. Similarly, small-CNN does not work well for scene classification even for a small number of classes. The small number of hidden layers in small-CNN is not able to deal with the inter-class similarity of our indoor environment. The transfer learning-based models (googleNet-TL and AlexNet-TL) perform equally well irrespective of the number of classes. The reason for such behavior is because it learns low-level features using an already-trained network, which was trained on millions of images. This way, it is possible to learn the structure of the image very quickly, and then the newly added layers learn the importance (weights) to discriminate the classes in a more specific context based on the training set. Furthermore, the current experiments are limited to indoor scene classification and utilize HoloLens for dataset collection.

The rest of the paper is organized as follows. Section 2 details the most relevant concepts used in the current experiments. Next, the proposed approach is explained in Sect. 3, followed by the experimental results in Sect. 4. Finally, Sect. 5 provides the conclusions and future work in this direction.

## 2  Fundamental Concepts and Related Work

This Section explains the most relevant concepts about scene classification.

### 2.1  Augmented Reality

Augmented Reality is an interactive experience of the real-world environment, where the real-world objects are enhanced by computer-generated perceptual information. This information could be visual, auditory and/or haptic. The devices used for AR include Eye-glasses, Head-Up Display (HUD), or Contact Lenses. AR devices are being actively used in many applications to help people improve their interaction with the real-world. For example, Google Glass is used as assistive technology for visually impaired people [1]. In [1], Google Glass is used for a specific task of navigation to help blind people. The prospects of the use of AR in education to make the Augmented Remote Laboratory (ARL) is investigated in [2]. For such tasks to be executed automatically, it is interesting for an AR device to classify and understand the scene. It would help, for instance, project the ARL application to the most feasible location in the scene. However, our purpose with scene understanding is to make the AR device a household device, so that it will be used as a customized assistant to its owner. For example, a smart AR device understands the owner's office and home (also sub-scenes such as kitchen, living room, etc.), which can be used to personalize his real-world. When he arrives at the office, the applications that he uses frequently may pop-up and/or important notifications related to the user's day-to-day job. Similarly, it can be used to augment the scenes in the design such as adding a sofa in the living room, or where to place a dining table and so on. More importantly, scene understanding through images can reduce the user effort to map the scenes in AR devices [16].

## 2.2  Speeded Up Robust Features (SURF)

SURF is a scale and rotation-invariant interest point detector and descriptor, which can be computed faster using an integral image technique for image convolution [3]. Firstly, interest points are computed using the determinant of the Hessian matrix ($H_{approx}$) at distinct locations such as corners, blobs, and T-junctions.

$$\det(H_{approx}) = D_{xx}D_{yy} - (\omega D_{xy})^2, \tag{1}$$

where $D_{xx}$ is the approximation for Gaussian second order derivatives with the image $I$ in point $x$ with $\sigma$ variance, and similarly for $D_{xy}$ and $D_{yy}$, whereas $\omega$ balance the relative weights in the expression for the Hessian's determinant. Next, the neighborhood of each interest point is represented as a feature vector, which should be robust and distinctive. The authors of the paper indicated that the distinctive characteristics of the SURF are achieved by mixing of local information and the distribution of gradient related features. For details, we refer to [3].

## 2.3  Support Vector Machines (SVM)

The SVM classifier is a supervised classification technique that utilizes training data to build a linear model to categorize new data observations. SVMs can also perform non-linear classification by implicitly mapping their inputs into a high dimensional space. The goal of the SVM is to design a hyperplane that classifies all the training vectors into their relevant classes. The best choice is the hyperplane that leaves a maximum margin between these classes [9].

## 2.4  Convolutional Neural Network

CNN is a class of Deep Learning algorithms useful for image classification problems [21]. CNN learns filters automatically which can differentiate images by assigning importance (weights) to various aspects of the image. With sufficient training data, CNN has proved very efficient for image data classification problems [8]. Also, CNN requires less pre-processing than other classification algorithms and learn filters automatically. However, the large amount of training data is required for CNN to perform effectively, which also makes the training process time-consuming. For more details on CNN, we refer to [8].

## 2.5  Transfer Learning

To overcome the disadvantage of the CNN, transfer learning approaches are adapted recently to decrease the data dependency and faster training of the

CNNs [17]. Transfer learning is a technique that allows developers to utilize the pre-trained networks and capitalize on the advantage of their training. Many networks such as AlexNet [11], GoogelNet [21] are trained with millions of images such as the ones in ImageNet [19], and can be helpful for general image description. These networks can be used to generate a more general description of the image and use these weights as a cue for a network designed for a more specific task such as scene classification in this work. For more details about transfer learning, we refer to a comprehensive transfer learning survey [23].

## 3   Proposed Approach

The proposed approach is composed of two main components: i) an AR device; and ii) a remote server. The interaction between these two components is illustrated in Fig. 1. The AR device contains an application that continuously acquires images of the scene and sends them to the remote server. Next, the server classifies the scene images according to its inference model and responds with the label (i.e., a scene class) to the AR device.

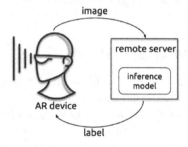

**Fig. 1.** Interaction scheme of the proposed approach.

The communication between the AR device and the remote server is performed via a wireless connection. The motivation for creating a remote server to perform the scene classification is based on the vision to use only minimal computational power and energy from the AR device. The proposed approach creates a minimum load on the AR device.

### 3.1   Inference Models

To find the best model for scene classification, we investigate four inference models explained next.

**SURFwithSVM.** In the proposed scene classification, the SVM is utilized as a classifier to train and test the scene classification. Each scene is considered as a separate class, and an SVM classifier is trained with features obtained using SURF. Consider a training set $X = [x_1, x_2, \ldots, x_N]$, where each $x$ is a scene image of $n \times n$ dimension. The SURF features are computed using the method proposed for each image in the training set as in [3]. Each image SURF feature $z$ is combined into a training feature matrix $Z = [z_1, z_2, \ldots, z_N]$, where $z_i$ is a column vector representing the feature vector of the image $x_i$ for $i = \{1, 2, \ldots, N\}$. The feature matrix $Z$ along with its label vector $Y = \{y_1, y_2, \ldots, y_N\}$ is used to train the SVM classifier similar to [9].

For the classification purpose, the SURF features are computed for the test image and classified using the pre-trained SVM classifier. The SVM classifier output is written as a linear combination of the input vector elements:

$$u^{(i)} = \Omega^T z^{(i)} + \Theta, \tag{2}$$

where $\Omega = [\Omega_1, \Omega_2, \ldots, \Omega_N]$ represents the weight of each element of the input data, and $\Theta$ is a constant acting as a bias to prevent overfitting. The hyperplane is called SVM and is created in a way that each training output $u^{(i)}$ is set to its relevant class label.

**Small-CNN.** This model comprises of a CNN with a limited number of layers, in order to have a model with few hyper-parameters to learn, which during the testing phase, achieves higher speed and has low computational complexity. The architecture of the small is trained with 5 sets of layers to learn hyper-parameters for scene classification.

Apart from the input and output layers, each layer of the small-CNN is composed of multiple two-dimensional planes, and each plane contains an independent neural unit. As shown in Fig. 2, the proposed small-CNN connects the convolutional layer, batch normalization layer, Rectified Linear Units (ReLU) Layer and pooling layer successively from left to right. The batch normalization layer is used to stabilize the learning process and reduce the number of training epochs required to train the small-CNN. Furthermore, the ReLU layer introduces non-linearity and reduces the vanishing gradient problem, which may slow the convergence of the last layers, otherwise. This layer basically changes all the negative activations to 0. Finally, the max-pooling layer is used to reduce the spatial dimensions, which helps decrease the number of hyper-parameters to reduce computational cost and also avoids over-fitting of data. After, these layers, a fully connected layer with a number of classes is used, followed by a softmax Layer and the classification Layer to classify different scenes.

**GoogleNet-TL.** The original GoogleNet model is an Inception based network that allows this network to go deeper without being very slow. It is a particular incarnation of the Inception architecture developed for the ILSVRC 2014 competition [19]. This allows the network to be more accurate because of its depth, and also fast in training and validation.

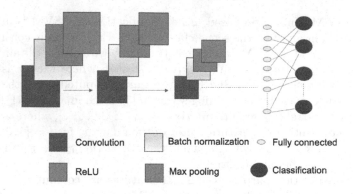

**Fig. 2.** Topology of the small-CNN.

The GoogleNet is a 22 layers network with parameters with five pooling layers, trained and tested on ImageNet [19]. These layers include 2 convolution layers, 5 pooling layers, 5 ReLU layers, and 5 Inception layers. The network is designed with Inception to increase the depth and width of the network. This allows efficient utilization of computing resources inside the network. The number of layers (independent building blocks) has been around 100 if the inception is not used. The exact number of layers depends on how layers are counted by the machine learning infrastructure. In our case, the total number of layers (independent building blocks) is 144, with 170 connections among them.

When GoogleNet is used as a super network, the same architecture is followed. However, the fully connected layer is updated from 1000 connections to 13 connections layer, which is followed accordingly by the Classification Output. Furthermore, to customize the model to our indoor scene classification, the weights of only the first two levels of the convolution layers, and pooling layers are fixed (conv1-7 × 7 s2, conv1-ReLU 7 × 7, pool1-3 × 3 s2, pool1-norm1, conv2-3 × 3 reduce s1, conv2-ReLU-3 × 3 reduce s1, conv2-3 × 3 reduce s1, conv2-ReLU-3 × 3 reduce s1, conv2-norm2, and pool2-3 × 3 s2). Furthermore, the weights of all the Inception layers are learned using the training dataset.

**AlexNet-TL.** This approach is similar to the GoogleNet-TL approach. The original AlexNet [11] contains 25 layers, among which 8 layers are fully connected, 7 dropout layers, 5 convolutional and 05 pooling layers. The network was trained and tested on ImageNet [19] problem.

Due to its promising results in the ImageNet problem, we applied transfer learning to its network to develop a model that can be rapidly trained. The transfer learning model we propose, named AlexNet-TL, reuses the architecture and learned weights of the original AlexNet, and adapt it to our AR scene classification problem.

The transfer learning estimates the irreplaceable layers in AlexNet and freezes their weights. The first 10 layers (2 normalization layers, 2 pooling layers, 2 ReLU layers, and 2 convolutional layers) along with their weights are frozen, and the other layers are allowed to learn their weights. The final fully connected is according to the number of classes in the AR scene classification task.

# 4    Experimental Evaluation

The proposed models are evaluated in a real-world environment, which corresponds to a business building with several indoor scenes. It is noteworthy that the goal of the current experiments is to classify specific scenes, not the scene categories. Therefore, the proposed scene classification system is interesting for an AR device based personal assistant. The AR device used in our experiments is the Microsoft HoloLens. The client application for HoloLens, which is responsible for performing image acquisition, communication to the remote server and display the scene class, is developed in Unity3D[1]. On the server-side, we used a Macbook Pro machine with 2.8 GHz processor corei, 16 GB RAM 2133 MHz LPDDR3. This setup was used for both the training and the evaluation of the models. The classification models were developed using Open Neural Network Exchange[2] format, in Matlab 2019a. Such a format allows utilizing the model across frameworks such as Caffe2. The server application was created in tornado Python[3].

The inference models in the present paper are evaluated according to the following measures: accuracy, training time, and frame rate (number of frames executed per second) during the test. The communication time between the AR device and the remote server is not considered in the current experimentsbecause it may vary according to the AR device and network settings.

## 4.1    Indoor Scene Classification Scenario

For training and testing purposes, the AR scene classification task considered in this paper uses real data collected in an indoor office. This environment is interesting because it may contain similar characteristics such as computers that make scene classification difficult. Some of the examples from the dataset are shown in Fig. 3. In this example, Scene 01 (Fig. 3 (a)) and Scene 03 (Fig. 3 (c)) have very similar texture information, which makes such scene classification very challenging for the classifiers.

The dataset is collected from the Sidia Institute of Science and Technology building[4] using Microsoft HoloLens AR device [5]. The dataset contains a total of 4083 images, of 13 different scenes (classes). Each scene represents one floor of the building. For training and testing, images are resized to 224 × 224 for efficient execution. Table 1 exhibits the details about the dataset distribution among different scenes. The number of images in each class is greater than 300, and a maximum of 370 images. The dataset is distributed into training data

---

[1] https://github.com/Unity-Technologies.

[2] https://onnx.ai/.

[3] https://pypi.org/project/tornado/3.2.1/.

[4] http://www.sidia.com.

(further divided into training and validation set to train the model) and test data. The number of images used for training is 2756, and 1327 images are used for the test set. The training set contains at least 70% of the images from each class from this dataset. Furthermore, the same training set and test set is used for all the methods to make a fair comparison and analysis.

**Table 1.** Data distribution among scenes/classes.

| Label | #Images |
|---------|---------|
| Scene 01 | 312 |
| Scene 02 | 315 |
| Scene 03 | 303 |
| Scene 04 | 305 |
| Scene 05 | 305 |
| Scene 06 | 302 |
| Scene 07 | 315 |
| Scene 08 | 310 |
| Scene 09 | 304 |
| Scene 10 | 318 |
| Scene 11 | 311 |
| Scene 12 | 313 |
| Scene 13 | 370 |

## 4.2  Training Methodology

For SVM training, a linear kernel is used and $\Theta$ is set to 1.1 to prevent over-fitting. Furthermore, to train the small-CNN, softmax is used as an activation function and the batch size is set to 10 samples with a learning rate of $3e-04$ and also allowing a maximum of 6 epochs. Figure 4 shows the training progress of the small-CNN, GoogleNet-TL and AlexNet-TL in terms of accuracy. The training data is divided into a training set and a validation set. Figure 4 shows the training progress of the proposed scene classification inference models, small-CNN, GoogleNet-TL, and AlexNet-TL, on the training and validation sets. Figure 4 indicates that all three architectures fit well on the training set. However, in the validation set, only GoogleNet-TL and AlexNet-TL perform well. This is because the small-CNN overfits on the training data and it is not able to generalize well. Furthermore, the loss functions also show a similar pattern in these networks as shown in Fig. 5. The plots indicate that the transfer learning approaches (GoogleNet-TL and AlexNet-TL) provides much better accuracy and smallest loss, and the plot becomes steady soon. Also, the training time is reduced using the transfer learning approach. The small-CNN takes 253 min to train, however, with transfer learning approaches it is reduced to 80 min. This allows the transfer learning method to retrain cheaply in case of new and updated training data is available for the scenes. This is important because scenes maybe changing characteristics over time, such as objects may be moved among scenes along time.

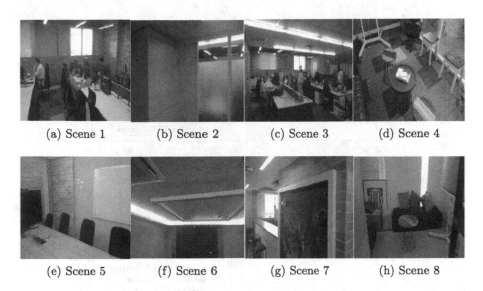

| (a) Scene 1 | (b) Scene 2 | (c) Scene 3 | (d) Scene 4 |

| (e) Scene 5 | (f) Scene 6 | (g) Scene 7 | (h) Scene 8 |

**Fig. 3.** Example of scene classes with interscene similarity. For instance, (a) and (c) has similar texture information, but belongs to different scenes.

### 4.3 Quantitative Evaluation

Table 2 provides the quantitative results of the proposed scene classification on three, five and thirteen scenes, respectively. These results are compared with Resnet152 architecture [7].

In terms of accuracy, the ResNet152 [7] method performs badly for fewer classes which have a similar structure and textural overlap. The small-CNN method performs fairly with a small number of classes. However, when the number of classes increases, the accuracy decreases. Furthermore, the handcrafted features have performed equally well to the transfer learning approaches for a small number of scenes (03 and 05). For the 03 class scene classification problem, handcrafted features outperform the other methods in terms of accuracy and time. However, with an increased number of scenes, the performance of the hand-crafted features deteriorates. On the other hand, the performances of transfer learning approaches remain consistent even when the number of classes increases. Furthermore, the preliminary results show that the transfer learning approach using GoogleNet [21] performs slightly better than AlexNet architecture [11].

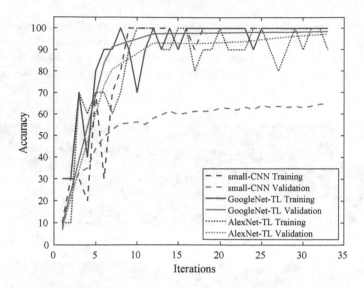

**Fig. 4.** Accuracy vs. time for training and validation process.

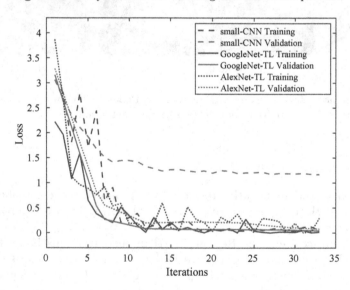

**Fig. 5.** Loss vs. time for training and validation process.

For training time, GoogleNet-TL and AlexNet-TL improved sufficiently the time compared to even a small-CNN. Furthermore, the proposed GoogleNet-TL and AlexNet-TL can classify the scene in 0.08 s or approximately 10 frames per second. This adds to its value in a sense that a user does not have to wait for loading its applications/customization for that particular scene and it performs in an approximate real-time.

**Table 2.** Methods comparison in terms of training time (in minutes), classification accuracy and frame rate in number of frames processed per second (FPS).

| Method | Training time (m) | Accuracy | Frame rate (FPS) |
|---|---|---|---|
| 3 classes | | | |
| ResNet152 [7] | 18.29 | 0.3351 | 18.40 |
| small-CNN | 26.94 | 0.8056 | 12.53 |
| SURFwithSVM | 10.05 | **0.9895** | 5.79 |
| GoogleNet-TL | 24.35 | 0.9826 | 7.84 |
| AlexNet-TL | 18.77 | 0.9861 | 11.57 |
| 5 classes | | | |
| ResNet152 [7] | 30.17 | 0.5870 | 29.39 |
| small-CNN | 83.88 | 0.8492 | 10.57 |
| SURFwithSVM | 4.22 | 0.9876 | 5.37 |
| GoogleNet-TL | 63.79 | **0.9981** | 9.41 |
| AlexNet-TL | 37.09 | 0.9795 | 9.49 |
| 13 classes | | | |
| ResNet152 [7] | 79.15 | 0.6092 | 77.73 |
| small-CNN | 253 | 0.6277 | 10.90 |
| SURFwithSVM | 43.91 | 0.9445 | 7.43 |
| GoogleNet-TL | 79.3 | **0.9895** | 9.53 |
| AlexNet-TL | 80.7 | 0.9880 | 8.23 |

## 5   Conclusions

This work proposes a scene classification approach, which can be utilized to assist Augmented Reality (AR) applications. The proposed scene classification is composed of i) an AR device that continuously captures images about the environment; and ii) a server component that contains an inference model that performs scene classification. For scene classification, four inference models have been evaluated. They were built using methods like hand-crafted features, support vector machine, convolutional neural networks, and transfer learning. The models were trained and evaluated with real data (collected via AR device) from a business building with several indoor scenes with similar textures. The obtained results show that the two transfer learning-based methods achieved up to ≈99% of accuracy in the experiments. Furthermore, even considering that our approach depends on communication between server and AR device, it has proven quite efficient, performing ≈9.5 frames per second in the larger test case, which indicates its effectiveness in the AR environment. In the future, we intend to extend our approach to outdoor scene classification, so that the user of an AR device may customize the outdoor environments as well and also allows inter-AR device classification.

# References

1. Berger, A., Vokalova, A., Maly, F., Poulova, P.: Google glass used as assistive technology its utilization for blind and visually impaired people. In: Younas, M., Awan, I., Holubova, I. (eds.) MobiWIS 2017. LNCS, vol. 10486, pp. 70–82. Springer, Cham (2017). https://doi.org/10.1007/978-3-319-65515-4_6
2. Andujar, J.M., Mejias, A., Marquez, M.A.: Augmented reality for the improvement of remote laboratories: an augmented remote laboratory. IEEE Trans. Educ. **54**(3), 492–500 (2011). https://doi.org/10.1109/TE.2010.2085047
3. Bay, H., Tuytelaars, T., Van Gool, L.: SURF: speeded up robust features. In: Leonardis, A., Bischof, H., Pinz, A. (eds.) ECCV 2006. LNCS, vol. 3951, pp. 404–417. Springer, Heidelberg (2006). https://doi.org/10.1007/11744023_32
4. Bichlmeier, C., Ockert, B., Heining, S.M., Ahmadi, A., Navab, N.: Stepping into the operating theater: ARAV - augmented reality aided vertebroplasty. In: 2008 7th IEEE/ACM International Symposium on Mixed and Augmented Reality, pp. 165–166, September 2008. https://doi.org/10.1109/ISMAR.2008.4637348
5. Evans, G., Miller, J., Pena, M.I., MacAllister, A., Winer, E.: Evaluating the Microsoft HoloLens through an augmented reality assembly application. In: Degraded Environments: Sensing, Processing, and Display 2017, vol. 10197, p. 101970V. International Society for Optics and Photonics (2017)
6. Grubert, J., Langlotz, T., Zollmann, S., Regenbrecht, H.: Towards pervasive augmented reality: context-awareness in augmented reality. IEEE Trans. Visual. Comput. Graph. **23**(6), 1706–1724 (2017). https://doi.org/10.1109/TVCG.2016.2543720
7. He, K., Zhang, X., Ren, S., Sun, J.: Deep residual learning for image recognition. In: 2016 IEEE Conference on Computer Vision and Pattern Recognition (CVPR), pp. 770–778, June 2016. https://doi.org/10.1109/CVPR.2016.90
8. Karpathy, A., Toderici, G., Shetty, S., Leung, T., Sukthankar, R., Fei-Fei, L.: Large-scale video classification with convolutional neural networks. In: 2014 IEEE Conference on Computer Vision and Pattern Recognition, pp. 1725–1732, June 2014. https://doi.org/10.1109/CVPR.2014.223
9. Khurshid, A.: Adaptive face tracking based on online learning (2018)
10. Klein, G., Murray, D.: Parallel tracking and mapping for small AR workspaces. In: 2007 6th IEEE and ACM International Symposium on Mixed and Augmented Reality, pp. 225–234, November 2007. https://doi.org/10.1109/ISMAR.2007.4538852
11. Krizhevsky, A., Sutskever, I., Hinton, G.E.: ImageNet classification with deep convolutional neural networks. In: Advances in Neural Information Processing Systems, pp. 1097–1105 (2012)
12. Liu, S., Tian, G.: An indoor scene classification method for service robot based on CNN feature. J. Robot. **2019**, 1–12 (2019). https://doi.org/10.1155/2019/8591035
13. Mann, S.: Google eye, supplemental material for through the glass, lightly. IEEE Technol. Soc. Mag. **31**(3), 10–14 (2012)
14. Natsume, S.G.: Virtual reality headset. US Patent Application 29/527,040, 07 June 2016
15. Niu, J., Bu, X., Qian, K., Li, Z.: An indoor scene recognition method combining global and saliency region features. Robot **37**(1), 122–128 (2015)
16. Pai, H.: An imitation of 3D projection mapping using augmented reality and shader effects. In: 2016 International Conference on Applied System Innovation (ICASI), pp. 1–4, May 2016. https://doi.org/10.1109/ICASI.2016.7539879

17. Pan, S.J., Yang, Q.: A survey on transfer learning. IEEE Trans. Knowl. Data Eng. **22**(10), 1345–1359 (2010). https://doi.org/10.1109/TKDE.2009.191
18. Pirri, F.: Indoor environment classification and perceptual matching. In: KR, pp. 73–84 (2004)
19. Russakovsky, O., et al.: ImageNet large scale visual recognition challenge. Int. J. Comput. Vision (IJCV) **115**(3), 211–252 (2015). https://doi.org/10.1007/s11263-015-0816-y
20. Santos, D., Lopez-Lopez, E., Pardo, X.M., Iglesias, R., Barro, S., Fdez-Vidal, X.R.: Robust and fast scene recognition in robotics through the automatic identification of meaningful images. Sensors **19**(18), 4024 (2019)
21. Szegedy, C., et al.: Going deeper with convolutions. In: 2015 IEEE Conference on Computer Vision and Pattern Recognition (CVPR), pp. 1–9, June 2015. https://doi.org/10.1109/CVPR.2015.7298594
22. Wang, L., Guo, S., Huang, W., Xiong, Y., Qiao, Y.: Knowledge guided disambiguation for large-scale scene classification with multi-resolution CNNs. IEEE Trans. Image Proc. **26**(4), 2055–2068 (2017)
23. Weiss, K., Khoshgoftaar, T.M., Wang, D.: A survey of transfer learning. J. Big Data **3**(1), 9 (2016). https://doi.org/10.1186/s40537-016-0043-6
24. Wu, P., Li, Y., Yang, F., Kong, L., Hou, Z.: A CLM-based method of indoor affordance areas classification for service robots. Jiqiren/Robot **40**(2), 188–194 (2018)
25. Zhou, B., Lapedriza, A., Xiao, J., Torralba, A., Oliva, A.: Learning deep features for scene recognition using places database. In: Advances in Neural Information Processing Systems, pp. 487–495 (2014)

# Underwater Search and Discovery: From Serious Games to Virtual Reality

Fotis Liarokapis[1]([✉]), Iveta Vidová[2], Selma Rizvić[3], Stella Demesticha[4], and Dimitrios Skarlatos[5]

[1] Research Centre on Interactive Media, Smart Systems and Emerging Technologies (RISE), 1011 Nicosia, Cyprus
f.liarokapis@rise.org.cy
[2] Masaryk University, Brno, Czech Republic
409881@mail.muni.cz
[3] University of Sarajevo, Sarajevo, Bosnia and Herzegovina
srizvic@etf.unsa.ba
[4] University of Cyprus, Nicosia, Cyprus
demesticha@ucy.ac.cy
[5] Cyprus University of Technology, Limassol, Cyprus
dimitrios.skarlatos@cut.ac.cy

**Abstract.** There are different ways of discovering underwater archaeological sites. This paper presents search techniques for discovering artefacts in the form of two different educational games. The first one is a classical serious game that assesses two maritime archaeological methods for search and discovering artefacts including circular and compass search. Evaluation results with 30 participants indicated that the circular search method is the most appropriate one. Based on these results, an immersive virtual reality search and discovery simulation was implemented. To educate the users about underwater site formation process digital storytelling videos were used when an artefact is discovered.

**Keywords:** Serious games · Virtual reality · Underwater archaeology

## 1 Introduction

Underwater archaeological sites are discovered mainly by two different ways. The first one is from an archaeological survey that typically includes remote sensing approaches (i.e. optical, sonar, magnetic devices). The second one is focused on reports by sport-divers or fishermen. In any of the above-mentioned cases, ground-truths dives must be produced in order to verify the reports or the acquired images of the seafloor (remote sensing). In the case of remote sensing, where the site is accurately located by means of a GPS mounted on board the boat, the search may be easier and focused on a specific and limited area. In the other cases, the site needs to be precisely located as the reports may not offer

Supported by iMareCulture EU project.

any accurate location, or there is no precedent knowledge of an existing site. In these cases the search would cover larger areas and be accordingly planned.

Past research has shown that the more types of objects a diver is asked to include in the search, the lower the efficacy is for any single type of object [9]. However, in a typical archaeological search, divers are searching for all available artefacts in the area of interest. There are several techniques that can be used to search for archaeological sites, or to locate a site whose rough location is known. These techniques differ, depending on the type of seafloor and the expected accuracy of the provided location. These are often called search and discovery techniques, and are employed in different fields of expertise, such as archaeology, forensics, law enforcement, search and rescue etc. The concept behind these techniques is to provide the largest coverage of the chosen area with the less time and effort, considering that as with many other underwater activities, time and diver safety are paramount.

The motivation of this research originates from the iMareCulture EU project which focused in raising European identity awareness using maritime and underwater cultural interaction and exchange in Mediterranean Sea [5,19]. This paper presents first a novel serious game that assesses two maritime archaeological methods for search and discovering artefacts including circular and compass search. Evaluation results with 30 participants indicated that the circular search method is the most appropriate one. Based on these results, an immersive virtual reality search and discovery simulation was created. To educate the users about underwater site formation process digital storytelling videos were used when an artefact is discovered. To our knowledge this is the first time that a maritime search and discovery assessment is performed. Also, it is the first time that an immersive VR simulation with digital storytelling for learning and understanding the site formation process is presented.

The rest of the paper is structured as follows. Section 2 presents related work in terms of underwater serious games and VR environments. Section 3, describes the design and implementation of the serious game. Section 4, illustrates the evaluation methodology as well as the experimental results. Section 5, presents the scenarios used for digital storytelling in order to describe the site formation process. Section 6, presents an overview of the VR application and Sect. 7 concludes the paper.

## 2 Related Work

The field of serious games for archaeological purposes has been previously documented [1]. An early approach was the integration of artificial life simulations with serious games focusing on the behavioural representation of species in fragile or long-vanished landscapes and ecosystems [20]. Another approach was the Dolphyn system [4], which consisted of an underwater-computerized display system with various sensors and devices conceived for existing swimming pools and for beach shores, associating computer functions, video gaming and multisensory simulations. More recently, a serious game was developed for seafaring [16] and incorporated probabilistic geospatial analysis of possible ship routes through the re-use and spatial analysis from open Geographical Information System (GIS) maritime, ocean, and weather data. These routes, along with naval engineering and sailing techniques from the period, are used as underlying information for the seafaring game.

In terms of underwater serious games and immersive virtual reality (VR) simulations, there are not a lot of relevant papers. One of the first attempts was the VENUS project [6] that aimed at providing scientific methodologies and technological tools for the virtual exploration of deep underwater archaeology sites. Amphibian presented an immersive SCUBA diving experience through a terrestrial simulator that mimicked buoyancy, drag, and temperature changes through various sensors [10]. Moreover, immersive VR simulator focused on exploration of the Mazotos shipwreck site in Cyprus [12] for raising archaeological knowledge and cultural awareness. Recently, an underwater excavation VR serious game focused on tagging and dredging procedures was presented [11] by implementing and extending an existing voxel-based sand simulation approach and rasterizing it using the marching cubes algorithm on the GPU. Recently, a VR serious game that introduces maritime archaeology students to photogrammetry techniques was presented [7]. Experimental results showed that the system is useful for understanding the basics of underwater photogrammetry and that could be used for learning purposes.

## 3    Design and Implementation of the Serious Game

In this section, firstly the features that were used to create an underwater digital environment, are described. After consultation with maritime archaeologists two techniques were selected for implementation. The circular search is one of the most effective ways for divers to search areas of seabed and to position any objects found during the search. The method does not require any complicated equipment, is easy to learn, is very quick to set up and can be used in poor visibility on any type of seabed. The basic principle of a circular search is for divers to carefully search the seabed in a circular pattern at a known distance around a shot weight that marks the centre of the search area. There are a number of other search methods that have been tried such as a jack-stay and swim-line searches but they are best avoided. These other methods require ropes and weights to be set up on the seabed or co-ordination between many divers underwater, both of which are time-consuming and prone to failure in all but the most ideal conditions (http://www.3hconsulting.com/techniques/TechCircularSearch.html). The second technique that we selected is the compass search, which is also very popular (Fig. 1).

The terrain has width and length set to 1000 units. Boundaries are defined by a transparent wall, to indicate the end of the current game field. All these walls were created from cube game objects with suitably set scales and positions. The color of the wall was set to grey with RGB values equal to (189, 189, 189) and transparency set to 90. A diver model was used as the player's avatar and

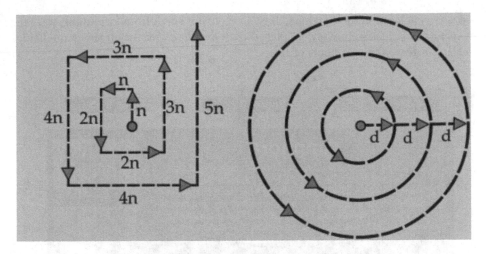

**Fig. 1.** The two search and discovery techniques selected, compass search (left image) and circular search (right image) [22]

animations were used for modeling the swimming of the diver. To achieve the effect of swimming underwater, the diver model was rotated to face the ground and an animation for walking forward was used on it. The scene consists of objects, that the player can interact with, and objects that can't interact with. The objects that the player could pick up and read information about were randomly positioned. The player was able to pick up an object only when was close enough to it. Once, the player is close enough, the name of the object appears above it. This serves as an indicator that the object can be picked up. Figure 2 shows an example of an amphora including some description (https://en.wikipedia.org/wiki/Amphora).

One of the two objectives of the game was to find and collect objects including: five statues, three amphora, three ship wheels, one coin, one axe, and one compass. On the other hand, the objects, that the player couldn't interact with had aesthetic significance and were placed statically in the scene [22]. Many instances of these objects were present in the scene with different scales, positions and rotations. The only exceptions were the objects which were randomly positioned such as: four ships, three different types of rock, five types of shells, fourteen types of plants, four types of wood, a plank, a barrel, two different types of anchors, and one type of fish.

To make the underwater environment more realistic, animated models of fish were added to the scene [22]. The behaviour of the fish is implemented by a flocking algorithm, which is commonly used to model flocking behaviour of animals (e.g. the behaviour of flocks of birds, schools of fishes or herds of mammals). This algorithm is based on three rules: (a) each fish from the group tends to move towards the centre of the group, (b) the whole group is heading towards the same direction, and (c) each fish tries to avoid the other fish, so it will not bump into them. Schools of fish were created according to these rules and were enclosed in an imaginary box, preventing them from swimming too far away. Several of these boxes were placed on different positions on the game field, creating a few smaller schools of fish.

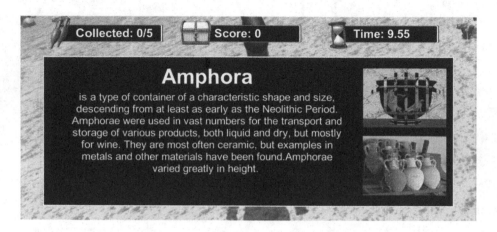

**Fig. 2.** Example of a scene, that provided information about amphora (https://en. wikipedia.org/wiki/Amphora)

Fog was set in each level to model an underwater environment more realistically [22]. It's color was set to dark blue, with the RGB values set to (14, 113, 172). The fog mode was set to exponential squared and the density of the fog was set to different values in each level. In the first level the density was set to 0.015, in the second level it was set to 0.03 and in the third level to 0.02. To make the underwater environment more realistic, the effect of caustics was added to the game. Initially, 16 images of caustic patterns were created and looped through with the frame-per-second (FPS) set to 30.

A mini-map was also created to help the players orient themselves in the scene. In each level, the mini-map shows: (a) the position of the ship, which is represented by a green picture of the ship, (b) the beginning point of the search, which is represented by a yellow dot, and (c) the position of the player represented by a red oval. In the first two levels, the positions of the searched amphora are showed on the mini-map as well. The amphora are represented by green dots. The representation of the maps were created from blurred screenshots of the terrain. An example of the game in action including the mini-map is shown in Fig. 3.

When the game starts an introductory menu is provided containing information about the game and the controllers. After pressing the "Play" button, a screen, that asks the player to enter his name, is shown. This name is used in the following parts to address the player. If the player chooses not to enter any name, the game will address him as a default name (i.e. "Mike"). The game consists of three levels. The first two levels are similar in structure, with the third level being slightly different. A diagrammatic overview of the game is illustrated in Fig. 4.

**Fig. 3.** Example of the game in action as well as the mini-map [22] (Color figure online)

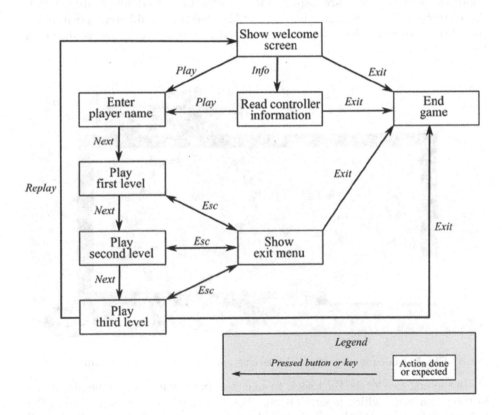

**Fig. 4.** Diagram of the game structure [22]

Each level consists of two parts: (a) dialogue and (b) Collecting or marking of amphora. The dialogue is led between 3 people: the player, another diver (called John) and the captain of the ship. This dialogue is written as a friendly conversation between these three people, where they are discussing the upcoming expedition. In the first two levels, information about the search techniques are provided. The first level provides information about the circular search technique, while the second level provides information about the compass search technique. The dialogues in the first two levels include a part, where the other diver dives and the player, together with the captain, are following him on the monitor of their computer.

During this process, they see a blue circle (which represents the other diver) moving in the direction of the discussed pattern. When the player is close to an amphora, it is also shown on the map as a green dot. The yellow dot represents the point, where the other diver started his dive and the position of the ship is displayed using a green picture of a ship. This map is later used as a mini-map for the player to orient himself at the next part of the level. In the third level, one, of three possible dialogues, is generated. Since the third level is randomized, the dialogue is chosen to match the used technique. These dialogues inform the player about the next steps and are displayed in word bubbles. To differentiate between the characters, each character has his bubble shown on a different position on the screen. An example of the dialogues used in the serious game is shown in Fig. 5.

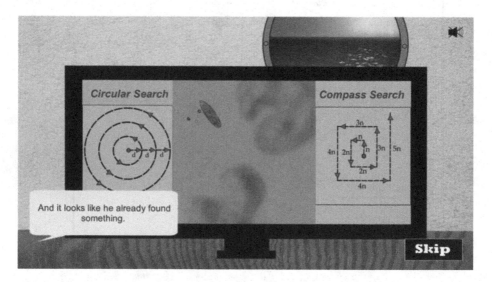

**Fig. 5.** Dialogue example in the serious game [22] (Color figure online)

In the first two levels, the task is to collect as much amphora as the player can in the given time, which is set to 6 min. The player is shown a mini-map in the bottom right corner of the screen to help him orient himself. In the third level, the player does not have the positions of the amphora (which are randomised).

The task is to find them and mark their positions on the map within 10 min. The task is randomly chosen from 3 possible scenarios. These scenarios differ in the technique used to do the exploration. The player should use either the circular or compass search technique, which were explained to him in the previous two levels. In these cases, there is a helping path created at the seafloor as a hint to preserve the pattern. The last possible scenario does not follow any prescribed pattern. The player was allowed to move freely and search. After the third level ends, the player is shown a score.

## 4  Evaluation

### 4.1  Participants and Procedure

33 healthy volunteers took part in the evaluation but three of them were discarded because they did not complete the serious game. The remaining 30 participants (20 male and 10 female) were testing the game in three separate parts. Since the third level in the game is randomized, three separate versions of the game were built, without the randomization of the third level, and each version was tested by an equal number of participants (10 participants). When a participant arrived, was asked to take a seat at any place in the laboratory and fill in a Consent form. After this, a brief description of the experiment was provided, mainly about the structure of the game (number of levels and description of their structure). Participants, who played the versions of the serious game with circular or compass search as the last level, were also asked to adhere the provided path, which served as a hint to preserve the desired pattern. They were also told to freely ask any questions about the game during playing.

### 4.2  Data Collection

Data collection consisted of both questionnaires and log data. The questionnaire consisted of 4 parts. In the first part, demographics were collected. This information was gathered before playing, the rest of the questionnaires were filled in after the game and concerned information about the participants experience with the game. The second part of the questionnaire included 10 custom-made questions about the game. In the third part, participants were asked to sum up their experience using a written open-ended questionnaire. The feedback may serve as an indicator for possible future improvement of the game. The last part included the NASA Task Load Index questionnaire [8] which measures cognitive workload.

During the process of playing the game, log data were collected (see Fig. 6). During one game, 4 text files were created. Three of them contained information about the movement of the players. One text file was created for each level. The last text file contained information about the collected amphora. As soon as the player picked up an amphora, a record about this was made. In the result a list of all collected amphora was created, where they were sorted according to the order, when they were picked up. At the end of the level, the remaining time (in seconds) and a total sum of the collected amphora in that level were recorded as well. The aim was to compare the third level for each of its version. Specifically, the number of amphora collected using each of the 3 methods (circular, compass, free search). In case of circular or compass search, the players were asked to follow the path on the floor, which served as a hint to preserve the pattern. From observing their playing, it was noticeable, that they mostly started their search the correct way, i.e. by following the path and looking around for objects. But many of the participants, who got to the end of the path and saw, that they do not have all the objects, started to search randomly.

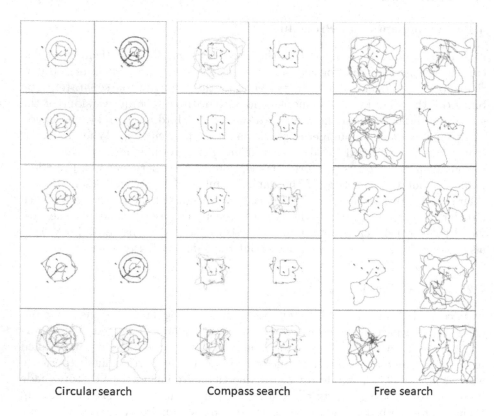

|  Circular search  |  Compass search  |  Free search  |

**Fig. 6.** Comparison of search and discovery techniques

## 4.3   Results

Mann-Whitney analysis [14] was used to analyse the collected data. In terms of interaction, the serious game was considered as easy to get used to. It was reported that it was easy to concentrate on the assigned tasks (i.e. collecting objects and navigating). Overall, participants did not find controlling of the movement of the diver very difficult. Most answers, regarding this question, fell between the values of "not difficult at all" and "somewhat difficult". Only 3 participants found the movement to be rather on the difficult side. This could be explained by the fact, that some of them do not play much games in their free time or not used in playing games in general.

One of the aims of the serious game was to educate players about underwater search and discovery techniques. This was done by the dialogues at the beginnings of each level. Not being sure about the suitability of the length of the dialogues, the players were asked to provide their opinion about this length. With most participants choosing values between 3 and 5, the expected result was met. This result justified the assumption, that the dialogues should be slightly shortened, or made more interactive, for example by giving the player an opportunity to choose his answer and making him this way paying more attention and increase his feeling of being involved in the story (Fig. 7).

| Attribute | What do you think about the length of the dialogues? [22] | How much have you learned about underwater search and discovery techniques? [23] | How interested are you in learning more about underwater exploration techniques? [24] |
|---|---|---|---|
| Mean | 4,133333 | 3,700000 | 2,300000 |
| Std. Deviation | 1,136642 | 1,178836 | 1,512021 |

Fig. 7. Results of the educational part of the serious game [22]

Another question checked whether these participants, who had free search as their final level, really used the mini-map more often than players, who had the other two versions of the level. Although the average values are not very different, it can be seen on Fig. 8, that participants with free search at the end indeed agreed on using the mini-map more often than the rest of the participants. Because the playing part of the game in the first two levels was mainly for fun and for getting the player used to the handling of the game, results of these levels are not going to be analyses in this chapter. The focus is going to be on the third level, which was different in the used techniques.

Furthermore there were observations on whether there were significant differences in success between players of each group. As shown in Fig. 9 the most successful was the group, which had the free search as the final level and the least successful were players with compass search at the end. To find out the significance of these result, the U-Test was performed. This showed that the difference between players, who tried circular search and player who tried the free search is not significant (p = 0,2052). Difference between players, who tried the circular search and player who tried compass search is significant (p = 0,0067). The difference between players with compass search at the end and players with free search at the end was also found to be significant (p = 0,0055).

| Attribute | Circular search | Compass search | Free search |
|---|---|---|---|
| Mean | 1,6000 | 1,2000 | 0,6000 |
| Std. Deviation | 1,3499 | 1,3166 | 0,8433 |

**Fig. 8.** Results of the mini-map usage [22]

| | Circular search | Compass search | Free search |
|---|---|---|---|
| Mean of relevant objects found | 7,9000 | 5,6000 | 8,4000 |
| Std. Deviation of relevant objects found | 1,1001 | 1.95551 | 1.8974 |

**Fig. 9.** Comparison of search and discovery techniques [22]

During the process of playing the game, the log data were collected. During one game, 4 text files were created. Three of them contained information about the movement of the players. One text file was created for each level. The last text file contained information about the collected amphora. As soon as the player picked up an amphora, a record about this was made. At the end of the game, a list of all collected amphora is created, where they were sorted according to the order, when they were picked up. At the end of the level, the remaining time (in seconds) and a total sum of the collected amphora in that level were recorded as well. To assess the different search and discovery techniques a third level of the game was specified.

Specifically, the number of amphora collected using each of the 3 different methods (circular, compass, free search). In case of circular, or compass search, the players were asked to follow the path on the floor, which served as a hint to preserve the pattern. Illustrative results from all the log data from all participants are illustrated in Fig. 6. Moreover, the log data of the game showed that there are significant differences between players who tried the circular search and compass search ($p = 0,0067$). The difference between players with compass search at the end and players with free search at the end is also significant ($p = 0,0055$).

## 5  Site Formation Process Storytelling Scenarios

Digital storytelling is considered to be the narrative entertainment that approaches the audience via media and digital technology [15]. There is a lot of literature that argues in favour of emergent narratives [2,13,21]. Another challenge for digital storytelling is to present the information on immersive VR where the audience can choose their view inside the virtual environment. This includes many challenges [18] and one of them is the Narrative Paradox challenge is specified as a struggle between a user's freedom of choice and the control of the main storyline [3]. Guidelines for introducing the motivation factor as a solution for narrative paradox has been recently presented [17] and followed in this research.

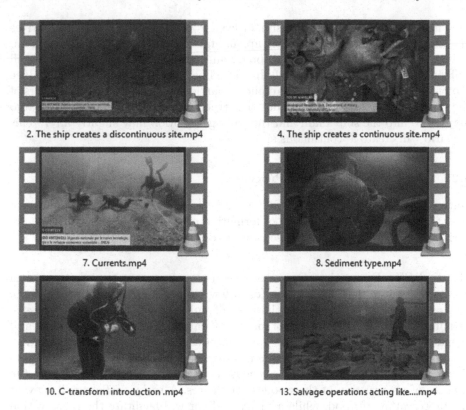

**Fig. 10.** Screenshots from the interactive story about site formation process

After assessing the most appropriate technique for search and discovery (i.e. the circular search), storytelling videos for the VR application were created. They consists of description of the site formation process, which as mentioned before are very important for underwater archaeologists as well as for educating the public. Examples of the storytelling videos are shown at Fig. 10. An interactive story was created using computer animation and voice of the narrator, according to the following scenario illustrated in the next sub-sections.

## 5.1  Sinkage and Impact

Once the ship starts to go down into the water column, the ship itself and the cargo will behave in different ways. These ways will depend on the previous processes (The Wrecking Event), and will end up with different disposition of the ship and cargo on the seafloor. The ship hits a rocky seafloor and brakes apart. Due to the inability to be covered by sand, the ship will remain exposed to eventual currents and wave actions and to the marine wood borers (teredo navalis) which will feast over the wood structure of the ship, slowly eating all the wood. The ship breaks into pieces and the cargo spills out of the hull following different routes to the seafloor ending up spread over a large area on the seafloor.

This creates a 'discontinuous site' or even a 'scattered site'. The ship arrives on a sandy seafloor, bounces slightly up, and falls back hitting the seafloor again. It is possible that this process will break an unbroken hull, and the cargo will spill on one or both sides of the hull. It is also likely that the impacts will produce the breakage of previously intact amphora, particularly those on the top layers of the stacking pile. The ship sinks but the structure remains coherent. The cargo may move into the ship's hull and the stowage position may slightly alter. This scenario will create a 'continuous site'.

## 5.2   Wreck Site Formation Processes

Once the ship has reached the seafloor and has slightly settled, the shipwreck site starts to form. The processes of formation of a shipwreck site over time are various and they are affected by several factors and processes.

**Natural Site Formation Processes** or n-transforms are the environmental agents that cause the deterioration (or in few cases the preservation) of the site and/or the artefacts. Unlike in the case of cultural site formation processes, in natural ones the human factor plays no part. The primary environmental agents which cause n-transforms to underwater archaeological sites are the waves, bottom currents, the sediment, (micro)-organisms and the water itself (i.e. its chemical properties). The Seafloor type is the type of seafloor will influence the chances of preservation of the wooden structures. A rocky seafloor will prevent the preservation of wood, while a sandy seafloor will facilitate the preservation of wood, as the latter will be covered by the sandy sediment and safeguarded in an anoxic environment (low-oxygen environment).

The action of waves and currents can accelerate the process of deterioration of the ship's hull and by oxygenating the water, speed up the decay of wood and the proliferation of wood borers (teredo navalis). Shallow water sites and those close to the shore are the ones most affected by wave actions; not only because of the wave strength, but also because the waves move sediment on the site, constantly exposing and reburying artefacts. Similarly, currents have an effect on the formation of a shipwreck site. Apart from surface currents that may displace floating artefacts at large distances from the wrecking site. Underwater currents move sediment around provoking not only the mechanical deterioration of the wood and fragile organic art3facts, but also by constantly exposing and reburying artefacts. Sand ripples are a testimony of the strength and direction of the currents patterns in the area.

The type of sediment and the bio-chemical composition of the sediment play a role in the preservation or decay of the artefacts (particularly organic materials). The buried, semi-buried or exposed artefacts bear evidence of the site formation processes. On a theoretical base, following the image below, exposed amphora will have traces of marine concretions. Conversely, the semi-buried will have traces only in the exposed portion, while the buried amphora, would not have any concretion. Obviously, on the basis of the effects of wave and currents on the sediments that we have described above, this is not a static and/or unidirectional process, instead it often represents a dynamic process where exposure and burial alternate, leaving evident traces on the artefacts.

Chemical properties of the water Factors influencing survival and condition. The environmental conditions include: Dissolved Oxygen, pH, Temperature, Water Movement, Salinity, Sulphate Reducing Bacteria and Galvanic Coupling. High temperature usually nurture the growth of bacteria and wood borers, while cold waters, low oxygen or high salinity usually limit the growth of these organisms. That's why shipwrecks in the warm Mediterranean waters are more likely to be attached by teredo navalis, while for instance Baltic Sea shipwrecks are usually more well preserved. This is also one of the main reasons why ships sunk in the Black Sea have been preserved to the extent seen here.

**The Cultural Transform** are those processes that involve the intervention of human activity and are developed in the modern Cultural Transformation times, long after the wreckage of the ship. There exists an extensive literature regarding modern salvage activities. With this term archaeologists usually indicate the activities produced in Modern times, before underwater archaeology was established as a discipline. The examples are numerous.

The Nemi shipwrecks in the lake of Nemi (Central Italy) were for example made the subject to several salvage attempts by Leon Battista Alberti in 1446, by the military engineer Francesco De Marchi in 1535 and by Annesio Fusconi in 1827. It is clear how activities intervening long after the shipwreck and long after the wreck has reached an equilibrium with the surrounding environment will create hazards for the preservation of the archaeological remains. Not only these activities have disturbed the equilibrium of the site, they have also disturbed the "coherency" of the archaeological site. Objects may have been removed, while others moved to other positions in order to reach object laying below them and more interesting for the salvage purposes.

Similar to other actions and processes, the salvage operations (ancient or modern) act as both extracting mechanisms, but also scrambling devices, since they move artefacts in order to search underneath them. Trawling (trawl fishing) is a technique that has been criticized from both an environmental and cultural point of view. Trawls are huge nets being dragged by a trawler ship over the seafloor. Clearly the net dragging the seafloor, not only damages the ecosystem but also any eventual cultural heritage site (i.e. shipwreck) encountered on its path. Trawling can not only destroy piles of amphora on the seafloor, but also scatter them over large areas, even far away from their original location and position. This obviously jeopardizes the ability of understanding the site location, the processes of formation and in the last analysis, understand the site.

Another Cultural Formation process to take into account when assessing an underwater site is the possibility of the site being pillaged by divers. For a long time, and still present nowadays although some International measures have been put in place, a common stance among divers has been to believe that what is found underwater is free to anyone to take; what is often called the "Finders

Keepers" attitude. Many sites have been disturbed by divers pillaging artefacts or taking "souvenirs". In terms of its impact, one of the most relevant Cultural Site Formation process is the activity of Treasure Hunters.

The activity is based on private people/companies investing money on enterprises for the discovery, location and recovery of shipwreck cargoes deemed to have high economic value. The investors put their money in this endeavors expecting profit from the selling of artefacts. Their activity is extremely impacting as they usually can employ the most advanced technologies, they do not care much on archaeologically excavating or recording the site, and they tend to use any convenient method to reach the valuable they are aiming at. Clearly this produces several bad outcomes: the disappearance of the archaeological information without proper recording and documentation, the selling of archaeological artefacts into the private market and therefore the complete disappearance of the site and artefacts from the archaeological record.

# 6    Virtual Reality Search and Discovery

The final version of the application contains single VR, where the user can perform search and discovery techniques using the circular search method (Fig. 11). Our goal was to create immersive, believable underwater environment with a light condition and sounds that resembles real diving experience. Therefore, the environment is filled with various static and moving objects, including flocks of fish, seaweed, stones, shipwreck and various small debris - e.g. wooden planks, air bottles, chests, etc. In total, the final environment covers an area of $500 \times 600 \times 50$ m where player can freely move. The environment is "populated" with approximately 200 moving objects and around 1000 static objects.

## 6.1    Scenario

When the player enters the main game underwater environment, the first task is to search the area and find scattered amphora and a shipwreck. When user finds the amphora, they should "tag" this finding. This could be done by pointing at amphora by straight pointer (using the touch-pad on right controller) and pressing the trigger button. The tag with a number appears on amphora. The task of finding and tagging amphora is connected to education videos regarding the site formation process. When user tags the amphora for the first time, a storytelling video is unlocked and starts playing on the panel above the left hand. The video can be replayed later on in the mother base scene. The task is successfully finished when user finds at least 10 amphora out of total 17. After fulfilling the task, it appears green on the task list screen in the base scene. Each amphora also unlock a video. Finding all amphora unlocks all videos in the base scene. Other objects, such as treasure chests or air bottles, are also scattered in the area. User can tag these objects, but this plays no role in finishing the task.

## 6.2   Controls and Interface

Basic control scheme is derived from a common practice of VR applications. Handheld controllers used by users to interact with virtual environment contains only several buttons. The developed application and activities contain a lot of different actions and possible interactions. Therefore, one button can have different meaning based on current environment and activity. In the base scene, the player can get familiarized with the controls. By touching the touch-pad on the left hand, curved pointer appears. When the user presses the touch-pad, can teleport to the point where the end of the curved pointer ends, but only if the target point is in a valid space.

**Fig. 11.** VR application

Figure 12 shows that the left hand is a user interface (UI) menu that can be used to start the VR simulation. The right hand is used for interaction. By pressing the touch-pad, a straight laser pointer appears. Using the laser pointer, the user can interact with the UI on the left hand or objects in the world (i.e. tagging found artefacts). The UI also allows to toggle the locomotion mode between teleportation and swimming. Users who are more likely to get motion sick should use teleportation for more comfortable experience. Swim mode is better for users experience with VR.

## 6.3   Scoring

To engage the player in further exploration of the environment and finish the game, they are motivated by specific task defined in the base scene. The "classic" high score system did not motivate players sufficiently, therefore the "completion" system was developed. The storytelling videos are locked at start and player needs to find all amphora first to unlock them. Each amphora unlocks one video. Although not all videos are needed to fulfil the task.

## 6.4   Desktop and Mobile VR

The VR application is supported by both desktop and mobile configurations. Instead of developing two separate versions of the environment, it was decided

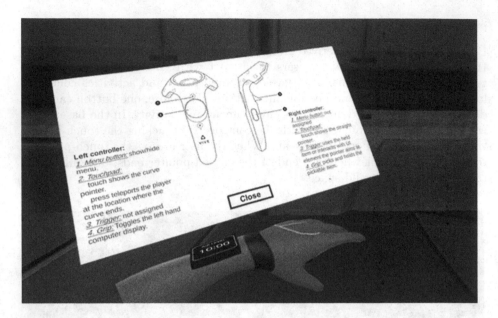

**Fig. 12.** Interface of the VR application

to keep single version usable on both platforms. This approach saves a lot of time otherwise spent by developing separate versions, but also brings a new technological challenges. The mobile VR platforms dispose with much lower computation power compared to desktop computers. Therefore, a scope of the environment, a number of moving objects and a level of detail of individual objects have to be adjusted to avoid overloading of mobile VR platforms. During the development, we continuously tested the application on both desktop and mobile platforms and adjusted technical complexity of environment to ensure smooth user experience on both platforms.

## 7   Conclusions

The focus of this paper was two-fold. First to examine two of the most prominent search techniques for maritime archaeology. This was accomplished by implementing an underwater search and discovery serious game that would assess which of the search techniques is the most appropriate. Log data and questionnaires from 30 participants were used for the analysis. The game consists of 3 levels, with educational dialogues at the beginning of the first two levels. These dialogues serve to educate players about circular and compass search techniques. Results showed that the circular search is the most appropriate maritime method, which also is in line with the feedback from archaeologists.

In the second part of the paper, the most effective technique (the circular search) was ported into an immersive VR environment. Our goal was to create immersive, believable underwater environment with a light condition and sounds that resembles real diving experience. To increase the level of learning, digital storytelling scenarios about the site formation process were used as a motivation for the players. This was used as a reward to the players instead of providing just a numeric score of discovered artefacts.

The main limitation of the paper is that the immersive VR environment was not formally evaluated. Obviously, the evaluation of such environment would be very different than the one performed in the serious game. In the future, we plan to perform a large scale evaluation of the immersive VR simulation and examine typical aspects of VR such as immersion, presence and motion sickness. Another aspect that will be assessed is the player satisfaction as well as the learning aspects of the site formation process.

**Acknowledgements.** This research was part of the i-MareCulture project (Advanced VR, iMmersive Serious Games and Augmented REality as Tools to Raise Awareness and Access to European Underwater CULTURal heritagE, Digital Heritage) that has received funding from the European Union's Horizon 2020 research and innovation programme under grant agreement No. 72715. This research was also partially supported by the project that has received funding from the European Union's Horizon 2020 research and innovation programme under grant agreement No. 739578 (RISE–Call: H2020-WIDESPREAD-01-2016-2017-TeamingPhase2) and the Government of the Republic of Cyprus through the Directorate General for European Programmes, Coordination and Development. Authors would like to thank Milan Dolezal for his VR implementation, Jiri Chmelik for his advice as well as all participants that took place at the user study. A video that illustrates the serious game can be found at: https://youtu.be/P2k5hWBkMcs.

# References

1. Anderson, E.F., McLoughlin, L., Liarokapis, F., Peters, C., Petridis, P., de Freitas, S.: Developing serious games for cultural heritage: a state-of-the-art review. Virtual Real. **14**(4), 255–275 (2010). https://doi.org/10.1007/s10055-010-0177-3
2. Aylett, R.: Narrative in virtual environments-towards emergent narrative. In: Proceedings of the AAAI Fall Symposium on Narrative Intelligence, pp. 83–86 (1999)
3. Aylett, R.: Emergent narrative, social immersion and storification. In: Proceedings of the 1st International Workshop on Narrative and Interactive Learning Environments, pp. 35–44 (2000)
4. Bellarbi, A., Domingues, C., Otmane, S., Benbelkacem, S., Dinis, A.: Underwater augmented reality game using the DOLPHYN. In: Proceedings of the 18th ACM Symposium on Virtual Reality Software and Technology, VRST 2012, pp. 187–188. Association for Computing Machinery, New York (2012). https://doi.org/10.1145/2407336.2407372
5. Bruno, F., et al.: Development and integration of digital technologies addressed to raise awareness and access to European underwater cultural heritage. an overview of the H2020 i-MARECULTURE project. In: OCEANS 2017 - Aberdeen, pp. 1–10 (2017)
6. Chapman, P., et al.: Venus, virtual exploration of underwater sites. In: The 7th International Symposium on Virtual Reality, Archaeology and Cultural Heritage (VAST), pp. 1–8. The Eurographics Association (2006)

7. Doležal, M., Vlachos, M., Secci, M., Demesticha, S., Skarlatos, D., Liarokapis, F.: Understanding underwater photogrammetry for maritime archaeology through immersive virtual reality. Int. Arch. Photogramm. Remote Sens. Spatial Inf. Sci. **XLII-2/W10**, 85–91 (2019). https://doi.org/10.5194/isprs-archives-XLII-2-W10-85-2019. https://www.int-arch-photogramm-remote-sens-spatial-inf-sci.net/XLII-2-W10/85/2019/

8. Hart, S.G., Staveland, L.E.: Development of NASA-TLX (Task Load Index): results of empirical and theoretical research. In: Hancock, P.A., Meshkati, N. (eds.) Advances in Psychology, Human Mental Workload, vol. 52, pp. 139–183. North-Holland, Amsterdam (1988). https://doi.org/10.1016/S0166-4115(08)62386-9. http://www.sciencedirect.com/science/article/pii/S0166411508623869

9. Henke, S.E.: The effect of multiple search items and item abundance on the efficiency of human searchers. J. Herpetol. **32**(1), 112–115 (1998). http://www.jstor.org/stable/1565489

10. Jain, D., et al.: Immersive terrestrial Scuba diving using virtual reality. In: Proceedings of the 2016 CHI Conference Extended Abstracts on Human Factors in Computing Systems, CHIEA 2016, pp. 1563–1569. Association for Computing Machinery, New York (2016). https://doi.org/10.1145/2851581.2892503

11. Kouřil, P., Liarokapis, F.: Simulation of underwater excavation using dredging procedures. IEEE Comput. Graph. Appl. **38**(2), 103–111 (2018)

12. Liarokapis, F., Kouřil, P., Agrafiotis, P., Demesticha, S., Chmelík, J., Skarlatos, D.: 3D modelling and mapping for virtual exploration of underwater archaeology assets. Int. Soc. Photogramm. Remote Sens. **42**, 425–431 (2017). https://doi.org/10.5194/isprs-archives-XLII-2-W3-425-2017

13. Louchart, S., Aylett, R.: Solving the narrative paradox in VEs – lessons from RPGs. In: Rist, T., Aylett, R.S., Ballin, D., Rickel, J. (eds.) IVA 2003. LNCS (LNAI), vol. 2792, pp. 244–248. Springer, Heidelberg (2003). https://doi.org/10.1007/978-3-540-39396-2_41

14. Mann, H.B., Whitney, D.R.: On a test of whether one of two random variables is stochastically larger than the other. Ann. Math. Stat. **18**(1), 50–60 (1947). http://www.jstor.org/stable/2236101

15. Miller, C.H.: Digital Storytelling: a Creator's Guide to Interactive Entertainment. Taylor and Francis, Hoboken (2004)

16. Philbin-Briscoe, O., et al.: A serious game for understanding ancient seafaring in the Mediterranean sea. In: 2017 9th International Conference on Virtual Worlds and Games for Serious Applications (VS-Games), pp. 1–5 (2017)

17. Rizvic, S., et al.: Guidelines for interactive digital storytelling presentations of cultural heritage. In: 2017 9th International Conference on Virtual Worlds and Games for Serious Applications (VS-Games), pp. 253–259. IEEE (2017)

18. Schoenau-Fog, H.: Adaptive storyworlds. In: Schoenau-Fog, H., Bruni, L.E., Louchart, S., Baceviciute, S. (eds.) ICIDS 2015. LNCS, vol. 9445, pp. 58–65. Springer, Cham (2015). https://doi.org/10.1007/978-3-319-27036-4_6

19. Skarlatos, D., et al.: Project iMARECULTURE: advanced VR, iMmersive serious games and Augmented REality as tools to raise awareness and access to European underwater CULTURal heritagE. In: Ioannides, M., et al. (eds.) EuroMed 2016. LNCS, vol. 10058, pp. 805–813. Springer, Cham (2016). https://doi.org/10.1007/978-3-319-48496-9_64

20. Stone, R., White, D., Guest, R., Francis, B.: The virtual scylla: An exploration of "serious games", artificial life and simulation complexity. Virtual Real. **13**(1), 13–25 (2009)

21. Temte, B.F., Schoenau-Fog, H.: Coffee Tables and Cryo Chambers: a comparison of user experience and diegetic time between traditional and virtual environment-based roleplaying game scenarios. In: Oyarzun, D., Peinado, F., Young, R.M., Elizalde, A., Méndez, G. (eds.) ICIDS 2012. LNCS, vol. 7648, pp. 102–113. Springer, Heidelberg (2012). https://doi.org/10.1007/978-3-642-34851-8_10
22. Vidová, I.: Underwater search and discovery serious game. Master's thesis, Faculty of Informatics, Masaryk University, Brno, Czech Republic (2018)

# Emergent Behaviour of Therapists in Virtual Reality Rehabilitation of Acquired Brain Injury

Henrik Sæderup[1,2], Flaviu Vreme[4], Hans Pauli Arnoldson[4], Alexandru Diaconu[1,2,3], and Michael Boelstoft Holte[1,2,3(✉)]

[1] 3D Lab, University Hospital of Southern Denmark, Esbjerg, Denmark
{Henrik.Saederup,Alexandru.Diaconu,
Michael.Boelstoft.Holte}@rsyd.dk
[2] Department of Oral and Maxillofacial Surgery, University Hospital of Southern Denmark, Esbjerg, Denmark
[3] Department of Regional Health Science, University of Southern Denmark, Esbjerg, Denmark
[4] Department of Architecture, Design and Media Technology, Aalborg University, Esbjerg, Denmark
flaviu.vreme@gmail.com, mail@hanspauli.com

**Abstract.** This study investigates how therapists are able to adopt a virtual reality toolset for rehabilitation of patients with acquired brain injury. This was investigated by conducting a case study where the therapists and their interactions with the system as well as with the patients were in focus. A tracked tablet gives the therapist a virtual camera and control over the virtual environment. Video recordings, participant observers and field notes were the main sources for data used in an interaction analysis. Results reveal emergent behaviour and resourcefulness by the therapists in utilizing the virtual tools in combination with their conventional approaches to rehabilitation.

**Keywords:** Acquired brain injury · Rehabilitation · Virtual reality · Emergent behaviour

## 1 Introduction

The use of Virtual Reality (VR) for rehabilitation of various disabilities was deemed viable as early as in the mid 90's [1], and in recent years, it has been applied and researched extensively with promising results. Using VR in rehabilitation of upper limb mobility as well as cognitive functions seems especially useful [2], and combining traditional methods with VR might be more effective than either method by itself [3]. Studies on VR in rehabilitation tend to focus on how a given VR system compares with traditional methods in relation to the improvements of the user's condition. This often leaves out the role of the therapist who should be the expert when it comes to facilitating rehabilitation sessions, be it traditional methods, VR methods or a combination of both.

This study builds on top of research the authors have done in the field of VR rehabilitation. The earliest study [4] investigated what aspects to consider when designing

VR experiences for disabled people, while in the most recent study [5] the focus was on how to make a flexible system that can be used by people with various degrees of disabilities caused by Acquired Brain Injury (ABI).

The aim of this study is to explore how therapists at Lunden, a rehabilitation centre located in Varde, Denmark, interact with a VR system designed to give the facilitator control over the virtual environment (VE) and the virtual artefacts within it, during a rehabilitation session. The design is based on how the therapists at Lunden use their resourcefulness when performing traditional therapy and is meant as an addition to the tools they already use rather than a replacement. Furthermore, a flexible system is needed as the patients at Lunden have individual needs depending on the nature and severity of their condition. The system should let the therapists use their general expertise, as well as their experience with the individual patients, to fine-tune the system and use it in combination with the existing tools and methods they have.

To frame this study, the following research questions were formulated:

- What tools and methods do therapists utilise in rehabilitation exercises with patients?
- Which aspect of rehabilitation could be enhanced by VR?
- Which aspect of rehabilitation should not be substituted by VR?
- What can a computerized VR system offer to rehabilitation that traditional methods cannot?
- How does the conventional approach of therapists transfer to a virtual reality toolset?

## 2 Background

### 2.1 Acquired Brain Injury

ABI is one of the most common causes of disability in adults, with ischaemic stroke and Traumatic Brain Injury (TBI) being the most common forms of ABI [6, 7]. The World Health Organization (WHO) defines stroke as the "rapid onset of [...] cerebral deficit, lasting more than 24 h [...], with no apparent cause other than a vascular one" [7]. The WHO definition for TBI is that of injury to the brain caused by an external mechanical force [7]. This definition for TBI is also independently formulated later by Menon et al. [8].

In terms of rehabilitation, the authors follow the definition given by Wade and Jong [9]. Notably, rehabilitation services contain a multidisciplinary unit of people, are reiterative, focused on assessment and intervention, and aim to "maximise the participation of the patient in his or her social setting" while minimising "the pain and distress experienced by the patient" [9] (p. 1386). Turner-Stokes et al. [10] offer an insight into the multidisciplinary aspect of ABI rehabilitation services, by conducting a review on the topic, regarding ABI in adults of working age. The reviewers note that the rehabilitation goals of non-retired adults differ from those of older populations, as they must cope with the effects of disability for most of their life. As such, any opportunity they may have to recover independence is worthwhile. The conclusions of the review indicate that multidisciplinary rehabilitation services conducted by experts do improve the effects after ABI and that faster and improved recovery is possible if these services are offered at higher intensity.

## 2.2  VR in Healthcare

Several studies outline the potential of VR and how it can improve healthcare by adding to and replacing parts of medical practice [11], distraction from pain during surgery [12], alternative to surgery [13], assessment of trait paranoia [14], supplementing cognitive therapy [15], treatment of clinical syndromes [16] and art therapy [16].

More specifically Henderson et al. [17] conduct a review on the effectiveness of VR for motor upper limb rehabilitations of stroke patients. At the time of the review, the relevant research was more limited than today, and the review concludes that the existing evidence was promising enough to advocate for further trials on the matter. However, it is their definition for virtual environments that will be used in this paper. Henderson et al. [17] classify VR applications either as fully immersive, or as non-immersive. The former category includes head-mounted displays (HMDs), large screen projections and other systems in which the environment is projected on a concave surface. The latter category includes any other application displayed on a computer screen.

## 3  Related Works

In more recent ABI rehabilitation research, VR is commonly used in post-stroke and post-TBI therapy, for both motor and cognitive rehabilitation.

In a review of 11 studies on immersive virtual reality rehabilitation for patients with traumatic brain injury, Aida et al. [18] conclude that TBI therapy was improved by the use of immersive VR. The review indicates that immersive VR within the context of TBI rehabilitation is mostly used for gait and cognitive deficits. Additionally, TBI rehabilitation has the potential of becoming more engaging because of the increased availability of immersive VR, but the currently available studies are limited, which leaves a degree of uncertainty regarding the utility of immersive VR in the field. Notably, the review exclusively selects studies that use immersive VR in their approach. According to the reviewers, many studies label any computer-generated environment as VR, and by the definition used in this paper it is understood that Aida et al. use 'VR' to mean 'immersive VR'. In terms of side effects, the usage of VR was reported to be well tolerated, but it did include motion sickness, discomfort, fatigue and frustration, which draws focus on users with a precedent of motion sickness [18].

A different perspective on the topic is brought by Laver et al. [3], who conduct a review on 72 studies on VR based rehabilitation for stroke patients. The review is conducted from the perspective of comparing the effectiveness of VR rehabilitation and video gaming methods with that of traditional means, and with a combination of both. The review primarily targets upper limb function, but secondary conclusions are also drawn on adverse effects, gait and balance. According to the reviewers, stroke patients did not show significant improvement in upper limb function when undergoing VR based rehabilitation, compared to the traditional methods. However, the effects were significantly more positive when the two approaches to rehabilitation were combined. Additionally, gait and balance VR rehabilitation effects were not statistically different from traditional method effects, and adverse effects were reported to be sparse and mild.

Pietrzak et al. [2] cover 18 studies in their review on various methods available for VR rehabilitation for patients with ABI. The reviewers note that VR based systems offered positive results in before/after studies, but no significant difference in when compared to conventional methods. In this case, VR systems ranged from commercially available platforms to large, custom installations. The immersion, cost, and availability of these systems varied. Regarding motor rehabilitation, the results indicated improvements in the areas of balance and upper extremity functions. Overall, the review indicates that VR is particularly well suited toward rehabilitation of upper limb mobility and cognitive function. Furthermore, the prevailing attitude of the patients towards the technology is positive, and there is potential in the lower cost and accessibility of VR systems.

**Commercial VR Hardware in Rehabilitation.** As hardware evolve and become more widely available, using commercially available gaming technology for rehabilitation purposes often has great benefits including cost, accessibility, flexibility and data logging. The following examples all use commercially available hardware, not limited to immersive VR, for rehabilitation purposes.

Anderson et al. [19] develop a system called Virtual Wiihab that uses Wii remotes and the Wii balance board (WBB) for rehabilitation. Their system does data logging, and has a wide variety of customizability, feedback cues and motivation factors. The therapist can specify the type, duration and intensity of these feedback cues. The system allows for patient-therapist or patient-patient interaction to complete the tasks. According to Anderson et al. [19], this is done to improve patient motivation and conformity.

Gil-Gómez et al. [20] also use a WBB in their system, Easy Balance Virtual Rehabilitation (eBaViR), developed for balance rehabilitation for patients with ABI. A therapist can configure the difficulty by adjusting the speed of the exercises and number of goals on screen. According to their evaluation, the WBB-based system indicated significant improvement in static balance of patients compared to traditional therapy. Furthermore, they report significant improvement in dynamic balance over time, but not significantly better than the traditional therapy group. The results bolster the use of such systems for VR rehabilitation.

Cuthbert et al. [21] study the feasibility of commercially available VR gaming products in balance rehabilitation. In a randomized controlled trial, the study has patients perform rehabilitation exercises for balance training using conventional methods and by playing commercially available Wii games. This is the notable aspect of the study, where no custom software or hardware is used for the rehabilitation exercises. Although the VR training group did not have significantly increased results from the traditional therapy group, the study supports commercially available semi-immersive VR hardware and software as having potential in the treatment of TBI victims in need of balance training.

Holmes et al. [22] combine several commercially available devices to form a VR rehabilitation system called the Target Acquiring Exercise (TAGER). TAGER is designed to augment physiotherapy with engaging exercises tailored to the user. Their system uses a Microsoft Kinect and a Leap Motion controller to track input from the upper body and arms, a Myo armband for data logging and an Oculus Rift DK1 HMD for viewing. The tasks were perceived to be easier with the HMD on, and the Leap Motion controller was found enjoyable. The main result was that object acquisition was improved with visual cues.

Few studies focus on the therapist, but Christensen and Holte [23] examined how the interaction between patients and therapist played a role in VR rehabilitation. Using a Leap Motion, HMD and custom built task simulator software, they found that therapists without prior VR rehabilitation experience unknowingly change their behaviour in facilitating rehabilitation exercises when there is a VR system in between the patient and themselves. Therapists with experience in VR rehabilitation show a clear understanding of limitations of the system, and do not feel hindered in regular physical and verbal guidance together with motivating words in patient-therapist interaction. Christensen and Holte [23] concludes that with proper training in VR rehabilitation, many of the barriers of such a system would be eliminated and could be used as any other rehabilitation tool.

### 3.1 Previous Work

**Multisensory Virtual Environment.** In Diaconu et al. [4], the authors designed a multisensory virtual environment (MSVE) based on the sensing room concept. The participants were all residents of a facility for the developmentally disabled and some had physical disabilities as well. Data gathering was done in the form of a case study that focused on how the participants interacted with the MSVE and how the therapists/pedagogues facilitated the experience. A set of preliminary guidelines for designing facilitated VR experiences aimed at people with disabilities was established:

- "The facilitator should be both familiar with the individual user and well-versed in the full extent of the possibilities the system offers. This would ensure the best mediated experience.
- Extra attention is necessary in making the controls as simple and intuitive as possible to ensure that the controls are understandable, usable, and memorable for the user regardless of their developmental level.
- The system must be reliable and flexible in order to rapidly adapt during use to the wide range of needs created by the larger context. Disruptions should not require a session to stop or restart." [4] (p. 11).

**Requirements for Designing VR Experiences for the Disabled.** The most recent work preceding this study is the exploratory case study by Arnoldson et al. [5]. The study included a total of five visits to a rehabilitation centre where one point of interest was to determine to what extent a consumer available HTC Vive VR system could be used by people with ABI. The system was designed as a toolbox for the facilitator to use for whatever exercise might be suitable and relevant for the specific user.

The facilitator had several options to change the controls and environment so that the experience could be customized to address the current user's unique condition. Adjustments include virtual table height, controller button mapping and resetting the virtual environment. With at least one therapist as a facilitator during the sessions, the participants interacted with different kinds of virtual objects and with different types of controls. The therapists were free to come up with tasks for the users as the session went along and to customize the experience. Based on an interaction analysis of the video recordings and the rest of the data, a tentative model with three categories of requirements for the users were created (see Table 1).

**Table 1.** Model of user requirements for VR ABI rehabilitation. Dexterity refers to hand and finger strength. Mobility indicates range of upper body motion including the head. Playstyle describes preferred interaction method with the virtual environment and the facilitator [5].

| Requirement categories | | | |
|---|---|---|---|
| User | Dexterity | Mobility | Playstyle |
| W | Low | High | Sandbox |
| Gi | High | High | Social/guided |
| B | High | High | Social/guided |
| Ge | High | Low | Social/guided |
| St | High | High | Goal-oriented |
| F | High | High | Sandbox |
| K | Low | Low | Social/guided |
| Si | Low | High | Sandbox |
| M | High | High | Goal-oriented |
| R | High | Low | Goal-oriented |
| T | High | High | Social/guided |
| Mo | High | High | Social/guided |

Based on the model, appropriate solutions for each dexterity, mobility and playstyle were implemented to fit the different users.

Overall the study addressed what can be done in terms of software to make an out-of-the-box VR system useful for people with various limitations, and how the facilitator can play a role by guiding the 0075sers in performing activities that might be useful in improving their condition.

# 4 Methods

This study used a variety of data gathering, data analysis and development/design methods to achieve its goal. This section describes why specific methods were chosen and how they were used.

## 4.1 Stakeholder

Lunden is a rehabilitation centre located in Varde, Denmark. Their mission is to provide specialized rehabilitation for patients suffering from ABI. Intensive therapy and training, both individual cognitive and physical training, but also involvement in ordinary activities, is part of everyday life for a patient at Lunden. Lunden has a staff of around 120 people; among them are social and healthcare assistants, physiotherapists, occupational therapists, nurses and support staff.

The collaboration happened over the course of four months, where the researchers visited Lunden four times. In total one section leader, three therapists and eight patients participated in the study. The visits were a mix of interviews with the section leader and therapists, gathering requirements for designing the VR toolset/environment and testing sessions with patients.

### 4.2  System Development

Through iterative design and with the requirements set forth by the therapists in this study, the research group developed the virtual environment and software solution that was used in the sessions where both therapists and patients were present. The final system consists of:

- *Hardware*: HTC Vive VR system with HMD, two tracked controllers and base stations, one additional Vive tracker that was mounted on an iPad, one wireless router and two 24″ monitors connected to a desktop computer PC.
- *Software*: A custom built VE that uses the HMD and two monitors. Monitor 1 displays the facilitator view while Monitor 2 displays a duplicate of the HMD. The iPad runs an open source game streaming application and that duplicates Monitor 1 (Fig. 1).

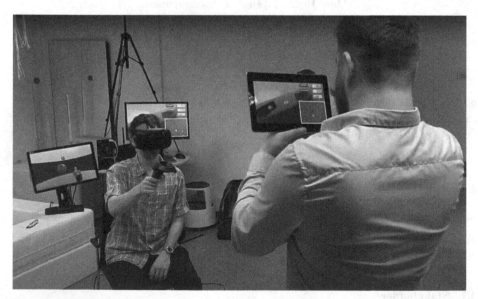

**Fig. 1.** System setup at Lunden. Foreground: researcher with tracked tablet and participant sitting on a chair with the HMD on. Background: two monitors with HMD and tablet view.

### 4.3  Case Study

This study follows what Hancock and Algozzine [24] define as an explanatory case study. This type of case study aims to establish a causality - to determine which factors affect a certain outcome. In the case of this study, the aim was to determine how the VR system affects the practice of therapists (doing ABI rehabilitation), and vice versa. Specific tools for data gathering were: participant observers, questionnaires, field notes, documents/archival research, video and audio recordings [24, 25]. A case study approach was chosen in lieu of other methods due to the complex nature of the subject being researched [24–26]. When the context (be it social, geographical, temporal, etc.) cannot be divorced from the phenomenon being studied, quantitative methods are ill suited due to their reliance on controlled setups and minimal variables, where a case study is not [24, 25]. On the contrary, a case study treats all such internal and external factors as a core part of the research. For this study, the context of Lunden cannot be ignored: Both the therapists and the patients are strongly dependent on the location of, the social environment at, and their history with Lunden as an institution.

Furthermore, the way in which rehabilitation therapy sessions occur are inherently unique to Lunden (as would be the case with any other similar rehabilitation centre), thus warranting the use of a case study approach.

### 4.4  Interaction Analysis

This paper makes use of a method known as interaction analysis (IA) to break down the events captured on video during the testing sessions at Lunden.

Video recordings are the primary source of data for IA due to being repeatedly reviewable to check for inconsistencies and to ensure the conclusions drawn accurately reflect events [27]. IA also makes use of other data collection methods such as field notes, observers (passive and participant), interviews, questionnaires, and other ethnographic practices [27]. These are used to supplement the primary data source with context by relaying events that occur before or after video recordings, or that take place outside the field of view of the camera(s).

Due in part to its firm origin in ethnography, IA works best when paired with participant observers [27]. This type of observation complements IA well because of its inherent aim to get close to the target group in a social context and learn by participating in their regular activities. Furthermore, IA is best done by researchers who have first-hand knowledge of the social setting and the participants due to their role as participant observers in the field [27]. All this makes IA the preferred choice for this study: the four researchers acted as participant observers in Varde for the duration of the study, they have in-depth knowledge of the location, the social context, and are familiar with the participants (both therapists and patients). The four researchers individually coded the available data focusing their analysis on these key aspects: Interactions of therapists with the system (positive and negative); Novel/emergent behaviours between therapists and the system; and interactions between therapists and end users. Common themes were extracted from the IA that will be explained in the results section.

## 4.5 Ethical and Safety Concerns

Ethical considerations were taken into account and applied throughout the whole process. Written consent forms for video and audio recording to be used only in relation to the study were filled out by all participants. At no point were a patient in VR without a therapist being close by. Any participant in the test sessions, patient and therapist alike, could at any point stop without further ado. Patients were continually asked by the therapists how they felt in the HMD, and if there were any sign of nausea or discomfort, the test session would immediately stop. Best practices of VR experiences were carefully followed (no unwarranted movement of the player, correct human scale, continuous monitoring of system functionally, etc.) to avoid discomfort for the patients.

## 5 Results

This study gathered a lot of data though the various methods mentioned in Sect. 4. Methods. What follows is a selection of the analysed results.

### 5.1 Interaction Analysis

As described in Sect. 4. Methods, interaction analysis was used to interpret video footage recorded during visits at Lunden. This section presents the common themes observed by the authors across the key aspects (positive/negative interactions of the therapists with the system, emergent interactions of therapists with the system and interactions between the therapists and the patients).

**Fig. 2.** Two iterations of the reach exercise both featuring the therapist duck avatar.

**Positive Interactions of Therapists with the System**

- Therapists observing the interaction from various angles, both in the real world and through the tablet.
- Therapists observing the interaction directly, on the tablet, and on the desktop monitors (1st person from HMD and roaming camera from tablet tracker). In most cases, there is at least one way for the therapists to see into VR, depending on the placement of desktop monitors and tablet.
- Therapists using UI controls on tablet to customize parameters of the experience before play session. After several times being exposed to that, therapists use UI controls on tablet to customize parameters of the experience during the play session.
- System easily accommodates additional props and techniques used in the rehabilitation process: tables, low friction cloth, foam pads.

**Negative Interactions of Therapists with the System**

- Therapists encountering difficulties with the limited input mode of the tablet (touchpad behaviour instead of direct touchscreen).
- Physically navigating in the VE is difficult due to mirrored perspective when looking at the HMD view on the monitor, often leading to walking the wrong way around the patient with the tablet
- Pointing, smiling and other visual cues are lost, due to the patient wearing the HMD. New cues emerged from this limitation, see next key aspect.
- When a therapist sets down the tablet to physically help the patient with the exercise, the avatar is now static for the patient, increasing the likelihood of patient accidentally bumping into the therapist
- Placement of monitors might cause the therapist to lose their view into the VE. A second therapist is often needed in current iteration of the setup.

**Emergent Interactions between Therapists and System**

- The avatar duck (Fig. 2) became a central tool in the activity of the therapists, who used it to replace the missing visual cues they give to the patient when their vision into the room is not obscured by the HMD. The avatar is used to grab and direct attention of the patient, although initially developed only to act as an indicator for the position of the therapist using the tablet.
- The avatar duck is used to gauge the limits of the area of neglect of patients.
- In some cases, the second controller is used by the therapist instead of the avatar to grab and direct attention of the patient.
- The view on the tablet is shared with other observers/therapists.

- Randomized colour of targets, implemented as a customization for visual preference, is used as a tool to check whether patient with neglect actually sees the target instead of guessing that the target is in their area of neglect.
- Software bugs can also lead to useful interactions. One of the patients was encouraged to keep playing to train the impaired arm when a software bug in the 'reach' exercise caused multiple targets to appear.
- Flowers spawning as a visual indicator of a reached goal were used as interactables for physical training of impaired arm.
- Second version of wall reach exercise was configured by therapists to work as a different exercise (flex).

**Interactions between Therapists and Patients**

- Plenty of verbal communication between therapist and patient, especially praise and instruction. Casual conversation too.
- Common instructions are to use the impaired arm, to direct attention towards side with neglect, to direct attention towards target, to switch to other arm.
- Other communication is to enquire about state of patient (e.g. tired, happy) and state of interaction (e.g. difficulty, fun).
- Often the therapists would touch the patient as a mode of ensuring the quality of movement, correcting posture, drawing attention to neglected side, instruction and praise.
- Common touches are used to set the limb in the correct position in preparation for the exercise, to correct the posture when the patient is compensating with the rest of the body, to correct the movement when the limb is straying. Also common are touches to draw and direct attention, especially in patients with neglect. Sometimes pats and rubs are given for praise.

## 6  Discussion

For this study, the implemented system is designed as a tool for the therapist, the main stakeholder of the work. This is distinguished from the end user, which is the patient. The focus lies on the therapist using the tools available in the VR system to address the requirements encountered in each unique session with their patients.

The software used in this context should be flexible enough to accommodate the requirements of therapists in their work. The therapists at Lunden are applying their knowledge and skills in a holistic manner as part of the rehabilitation process, building on theoretical expertise from their education with additional material throughout their development in their profession. This process includes looking at the overall progress of the patient, measuring specific markers of mobility, and using their ingenuity to perform a tailored form of therapy suited to each individual's needs. As such, there is no recipe to follow that is generalizable to all therapy sessions, because these sessions are not entirely what the theory makes them to be.

Using novel approaches to rehabilitation is nothing new for the therapists. However, using new methods and tools usually requires learning how to use a whole new system for it to be effective for the desired purpose. Observations throughout this study show that the therapists use their resourcefulness and turn everyday objects into tools for rehabilitation exercises. The flexibility and robustness of the VR system allows for this by handling a few objects occluding the trackers.

## 6.1 Addressing the Research Questions

**What Tools and Methods do Therapists Utilise in Rehabilitation Exercises with Patients?** As the main focus of this study was to provide the therapists with a set of virtual tools to use in rehabilitation sessions, a good starting point was to find out what physical tools they already use, and for what type of exercise. Some of the exercises initially described by the therapists did not rely on any tools other than a table to rest an arm on. Such was the repetitive wrist rotation exercise where the patient sit at a table with an arm resting on it, rotating the wrist back and forth. The same with the arm extension exercise where the user pushes and pulls their arm back and forth, sometimes with a physical object to connect with in front of them. These kind of exercises are well suited for adaptation to a VR system.

Another exercise that the therapists explained was meant to get patients with neglect to practice searching for objects on the neglect side. The exercise consisted of several coloured targets taped to a wall, and the patient would be asked to locate and touch a target with a specific colour. There being no other feedback than that from the therapist means that it can become a boring task to do after a short time. Adapting this exercise for VR not only had the potential to make it more engaging and rewarding for the patient, but it also gave the therapist a much more convenient way of adjusting the exercise in various ways, some of which would not be possible with a physical, paper and wall setup.

**Which Aspect of Rehabilitation Could be Enhanced by VR?** According to the answers from the questionnaire that was given to the therapists, VR could enhance several aspects of rehabilitation. Motivation, feedback cues and statistics were all mentioned. The idea of masking the exercises as a fun game has roots in gamification research and makes good sense in this context. The therapists at Lunden observed that some of their patients would spend more time on a task in VR than they would otherwise, indicating a potential in increasing focus. However it is unclear if it was due to repetitive exercises in VR actually increasing focus, or if the observer effects and novelty factors played a role. For further development of the system, implementing statistics should be considered. Therapists reported that data logging and scoring systems to monitor a patient's performance over time, could be useful and it ties in with other research in this field [19].

**Which Aspect of Rehabilitation Should not be Substituted by VR?** The therapists all agreed that the aspect with the greatest impact on the effectiveness and validity of the rehabilitation training is the therapists themselves. Knowing the individual patient and their condition, what to look for during training to make sure the patient does the right movements and how to structure the whole rehabilitation process, is something only a therapist knows and should not be replaced by machines.

**What Can a Computerized VR System Offer to Rehabilitation that Traditional Methods can not?** One of the major benefits of using VR in this context is the different perspectives that can be available for the facilitator. In this study the therapist had a portable camera into the VR world that could be viewed through the tablet or one of the desktop monitors. On the same screen was a top down view of one of the exercises. Additionally the possibility to have a first person view of the patient is something that would never really occur in traditional rehabilitation. Many parameters of the virtual environment can be adjusted on the fly, for deeper customization of exercises and more gradual difficulty adjustments. Furthermore, not only does the VR system not interfere with the props of traditional rehabilitation, but it can complement them. There is also big potential for computerized systems to automatically log much more information that therapists can manually, in order to review improvements at a later stage.

**How does the Conventional Approach of Therapists Transfer to a Virtual Reality Tool-Set?** Using VR technology in rehabilitation brings some of the limitations of the medium into the therapy session. By not being able to have eye contact with the patient, natural visual cues between therapist and patient are lost. It seems, however, that the therapists quickly employ the new tools that come with the medium, especially the avatar on the tracker to mitigate this limitation and introduce new communication channels to the patients. They make use of the avatar and controllers to reach into the VR play area of the patient and redirect their attention towards the goal.

Guidelines are required for such situations, where the alliance between therapist and patient depends on implicit and explicit communication. Both the therapist and the patient would benefit from constant cues to remind them of the way VR works. The chaperone system integrated with SteamVR and the avatar for the therapist are just two solutions. At least in the case of Lunden, therapy sessions involve a single therapist for a patient during a session. Throughout this study, however, multiple therapists took part in the play sessions, perhaps in part due to the novelty aspect. Nonetheless, with the new devices, perspectives and controls introduced by the VR system, its handling without sacrificing attention to the patient was in some cases only possible with the intervention of at least two therapists. If this type of system changes the nature of the sessions at this fundamental level, it is desirable to do so in the direction of minimizing the workload on the therapists instead.

## 6.2 Weaknesses of Study

Due to privacy concerns, the researchers were not given more than short descriptions of each patients symptoms, range of movement and general limitations and could therefore not develop specific tools guided towards specific conditions. Another weakness of this study is the difficulty in generalizing any conclusions. This stems from not only the use of a case study approach, but also from the inherent nature of therapy sessions at Lunden, which are unique. That is to say that each combination of rehabilitation centre (like Lunden), therapist, and patient create a unique context which must be treated as a separate case. Common patterns can be discerned, and a limited set of conclusions can be drawn from this, but generalizing further would not be possible without additional research using other methods.

# 7   Conclusion

The therapist, and how they interact with a VR rehabilitation system, is not often the main focus of similar studies in this field. Looking into how therapists use a collaborative VR rehabilitation system revealed how they use their resourcefulness and years of experience in traditional rehabilitation to continuously shape the given system and its limitations, into the needed tools for performing specific exercises. Results indicate emergent behaviours can overcome design flaws and discovered the usefulness of giving the therapist a better view into, and more control over, the virtual environment. Having established this, future work can move past the weaknesses of case studies by employing long term, more controlled research methods in collaborative and flexible VR systems for ABI rehabilitation.

**Acknowledgements.** The authors would like to acknowledge Lunden Rehabilitation Centre for their collaboration throughout this study.

# References

1. Kuhlen, T., Dohle, C.: Virtual reality for physically disabled people. Comput. Biol. Med. **25**, 205–211 (1995). https://doi.org/10.1016/0010-4825(94)00039-S
2. Pietrzak, E., Pullman, S., McGuire, A.: Using virtual reality and videogames for traumatic brain injury rehabilitation: a structured literature review. Games Health J. **3**, 202–214 (2014). https://doi.org/10.1089/g4h.2014.0013
3. Laver, K.E., Lange, B., George, S., Deutsch, J.E., Saposnik, G., Crotty, M.: Virtual reality for stroke rehabilitation. Cochrane Database Syst. Rev. (2017). https://doi.org/10.1002/146 51858.CD008349.pub4
4. Diaconu, A., et al.: An interactive multisensory virtual environment for developmentally disabled. In: Brooks, A.L., Brooks, E., Sylla, C. (eds.) ArtsIT/DLI - 2018. LNICST, vol. 265, pp. 406–417. Springer, Cham (2019). https://doi.org/10.1007/978-3-030-06134-0_44
5. Arnoldson, H.P., Vreme, F., Sæderup, H.: Requirements for designing virtual reality experiences for the disabled (2019)
6. Feigin, V.L., Barker-Collo, S., Krishnamurthi, R., Theadom, A., Starkey, N.: Epidemiology of ischaemic stroke and traumatic brain injury. Best Pract. Res. Clin. Anaesthesiol. **24**, 485–494 (2010). https://doi.org/10.1016/j.bpa.2010.10.006
7. World Health Organization: Neurological disorders: public health challenges. World Health Organization, Geneva (2006)
8. Menon, D.K., Schwab, K., Wright, D.W., Maas, A.I.: Position statement: definition of traumatic brain injury. Arch. Phys. Med. Rehabil. **91**, 1637–1640 (2010). https://doi.org/10.1016/j.apmr.2010.05.017
9. Wade, D.T., de Jong, B.A.: Recent advances in rehabilitation. BMJ **320**, 1385–1388 (2000). https://doi.org/10.1136/bmj.320.7246.1385
10. Turner-Stokes, L., Pick, A., Nair, A., Disler, P.B., Wade, D.T.: Multi-disciplinary rehabilitation for acquired brain injury in adults of working age. Cochrane Database Syst. Rev. (2015). https://doi.org/10.1002/14651858.CD004170.pub3
11. Wiederhold, B.K., Riva, G.: Virtual reality therapy: emerging topics and future challenges. Cyberpsychol. Behav. Soc. Netw. **22**, 3–6 (2019). https://doi.org/10.1089/cyber.2018.291 36.bkw

12. Mosso Vázquez, J.L., Mosso Lara, D., Mosso Lara, J.L., Miller, I., Wiederhold, M.D., Wieder-hold, B.K.: Pain distraction during ambulatory surgery: virtual reality and mobile devices. Cyberpsychol. Behav. Soc. Netw. **22**, 15–21 (2019). https://doi.org/10.1089/cyber.2017.0714

13. Li, X., et al.: Intermittent exotropia treatment with dichoptic visual training using a unique virtual reality platform. Cyberpsychol. Behav. Soc. Netw. **22**, 22–30 (2018). https://doi.org/10.1089/cyber.2018.0259

14. Riches, S., et al.: Using virtual reality to assess associations between paranoid ideation and components of social performance: a pilot validation study. Cyberpsychol. Behav. Soc. Netw. **22**, 51–59 (2018). https://doi.org/10.1089/cyber.2017.0656

15. Ferrer-Garcia, M., et al.: A randomized trial of virtual reality-based cue exposure second-level therapy and cognitive behavior second-level therapy for bulimia nervosa and binge-eating disorder: outcome at six-month followup. Cyberpsychol. Behav. Soc. Netw. **22**, 60–68 (2019). https://doi.org/10.1089/cyber.2017.0675

16. Hacmun, I., Regev, D., Salomon, R.: The principles of art therapy in virtual reality. Front. Psychol. **9**, 2082 (2018). https://doi.org/10.3389/fpsyg.2018.02082

17. Henderson, A., Korner-Bitensky, N., Levin, M.: Virtual reality in stroke rehabilitation: a systematic review of its effectiveness for upper limb motor recovery. Top. Stroke Rehabil. **14**, 52–61 (2007). https://doi.org/10.1310/tsr1402-52

18. Aida, J., Chau, B., Dunn, J.: Immersive virtual reality in traumatic brain injury rehabilitation: a literature review. NeuroRehabilitation **42**, 441–448 (2018). https://doi.org/10.3233/NRE-172361

19. Anderson, F., Anett, M., Bischof, W.F.: Lean on Wii: physical rehabilitation with virtual reality Wii peripherals. Stud. Health Technol. Inform. **154**, 229–234 (2010)

20. Gil-Gómez, J.-A., Lloréns, R., Alcañiz, M., Colomer, C.: Effectiveness of a Wii balance board-based system (eBaViR) for balance rehabilitation: a pilot randomized clinical trial in patients with acquired brain injury. J. NeuroEng. Rehabil. **8**, 30 (2011). https://doi.org/10.1186/1743-0003-8-30

21. Cuthbert, J.P., Staniszewski, K., Hays, K., Gerber, D., Natale, A., O'Dell, D.: Virtual reality-based therapy for the treatment of balance deficits in patients receiving inpatient rehabilitation for traumatic brain injury. Brain Inj. **28**, 181–188 (2014). https://doi.org/10.3109/02699052.2013.860475

22. Holmes, D.E., Charles, D.K., Morrow, P.J., McClean, S., McDonough, S.M.: Usability and performance of leap motion and oculus rift for upper arm virtual reality stroke rehabilitation. In: Virtual Reality, vol. 11 (2016)

23. Christensen, D.J.R., Holte, M.B.: The impact of virtual reality training on patient-therapist interaction. In: Brooks, A.L., Brooks, E., Vidakis, N. (eds.) ArtsIT/DLI -2017. LNICSSITE, vol. 229, pp. 127–138. Springer, Cham (2018). https://doi.org/10.1007/978-3-319-76908-0_13

24. Hancock, D.R., Algozzine, R.: Doing Case Study Research: A Practical Guide for Beginning Researchers. Teachers College Press, New York (2017)

25. Yin, R.K.: Case Study Research: Design and Methods. SAGE, Los Angeles (2014)

26. Baxter, P., Jack, S.: Qualitative case study methodology: study design and implementation for novice researchers. Qual. Rep. **13**, 544–559 (2008)

27. Jordan, B., Henderson, A.: Interaction analysis: foundations and practice. J. Learn. Sci. **4**, 39–103 (1995). https://doi.org/10.1207/s15327809jls0401_2

# Improving Emergency Response Training and Decision Making Using a Collaborative Virtual Reality Environment for Building Evacuation

Sharad Sharma(✉) (iD)

Department of Computer Science, Bowie State University, Bowie, MD, USA
ssharma@bowiestate.edu

**Abstract.** Emergency response training is needed to remember and implement emergency operation plans (EOP) and procedures over long periods until an emergency occurs. There is also a need to develop an effective mechanism of teamwork under emergency conditions such as bomb blasts and active shooter events inside a building. One way to address these needs is to create a collaborative training module to study these emergencies and perform virtual evacuation drills. This paper presents a collaborative virtual reality (VR) environment for performing emergency response training for fire and smoke as well as for active shooter training scenarios. The collaborative environment is implemented in Unity 3D and is based on run, hide, and fight mode of emergency response. Our proposed collaborative virtual environment (CVE) is set up on the cloud and the participants can enter the VR environment as a policeman or as a civilian. We have used game creation as a metaphor for developing a CVE platform for conducting training exercises for different what-if scenarios in a safe and cost-effective manner. The novelty of our work lies in modeling behaviors of two kinds of agents in the environment: user-controlled agents and computer-controlled agents. The computer controlled agents are defined with preexisting rules of behaviors whereas the user controlled agents are autonomous agents that provide controls to the user to navigate in the CVE at their own pace. Our contribution lies in our approach to combine these two approaches of behavior to perform emergency response training for building evacuation.

**Keywords:** Virtual reality · Immersive VR · Building evacuation · Collaborative virtual environment

## 1 Introduction

During emergencies and disasters, there is a need to minimize the potential impact on life and property. As a result, emergency response teams are organized to serve as pre-emergency preparedness function such as planning, training, and exercising. Training is needed for all emergency response tasks because of the uncertainties involved in it and a need for urgency in response [1]. Thus emergency response training is needed to prepare

© Springer Nature Switzerland AG 2020
C. Stephanidis et al. (Eds.): HCII 2020, LNCS 12428, pp. 213–224, 2020.
https://doi.org/10.1007/978-3-030-59990-4_17

an emergency response team and first aid responders to promptly detect the onset of an emergency, assess the situation, and respond effectively to the situation. Emergency response training is a learning process that can lead to the development of individual and team expertise. Expertise can be outlined as the achievement of consistently superior performance through experience and training. Expertise can be built through guidance programs which include formal training programs, on-job activities, and other learning experiences. Expertise can be classified as the breadth of expertise and depth of expertise. The breadth of expertise focuses on the diversity of training and learning gain as part of a career within an organization. On the other hand, the depth of expertise focuses on skill-building that is needed to build expertise. For emergency response training and decision making there is a need to understand the depth of learning as well as the content or breadth of learning activities. Depth of expertise consists of three components: (a) knowledge that is highly procedural zed and principled, (b) mental models that are well organized and structured, (c) self-regulatory systems that are well developed [2].

Training exercises are needed to explore possible responses to emergency events and explore different what-if scenarios. Thus, training exercises are planned to address the cognitive skills of the user to respond to unexpected events and scenarios. The purpose and scope of these training scenarios for unexpected events include training of emergency response teams, evaluation of new evacuation guidelines, building trust on trainees, and testing the emergency operation plans (EOP). One of the possible ways to accomplish this type of training is to create a collaborative virtual reality environment (CVE) where the training concepts are attractive to the users and allow them to be fully immersed in the environment without being exposed to any dangers. The CVE can help in improving the planning of security tasks and procedures as well as for performing virtual evacuation drills. The development of such CVE is very critical for the training of different types of scenarios in emergency events to enhance the security agent's skills (Fig. 1).

**Fig. 1.** View of security personnel training module for active shooter events

With the recent advances in technology, CVE based training incorporates real-world scenarios and creates a "sense of presence" in the environment. The collaborative virtual environment offers a considerable advantage for performing real-time evacuation drills for different what-if scenarios. The use of CVE allows us to run virtual evacuation drills, eliminates the risks of injury to participants, and allows for the testing of scenarios that could not be tested in real life due to possible legal issues and health risks to participants. This paper presents a CVE platform for improving emergency response training and decision making for building evacuations. The CVE will aid in developing an experimental setup to train emergency response personnel as well as first aid responders for decision-making strategies and what-if scenarios. The emergencies can be a result of fire and smoke as well as active shooter events. This work also verifies the feasibility of designing a CVE for training security personnel and occupants of the building in active shooter events. The proposed CVE is implemented in Unity 3D and is based on run, hide, and fight mode for emergency response. The user can enter the CVE as a policeman or as an occupant of the building. Our proposed CVE offers flexibility to run multiple scenarios for evacuation drills for emergency preparedness and response. The modeling of such an environment is very critical in today's life because of the need to train for emergency events. This paper presents a CVE platform where experiments for disaster response can be performed in CVE by including 1) Artificial Intelligent (AI) agents: defined by pre-existing rules, 2) User-Controlled agents: to navigate as autonomous agents.

The rest of the paper is structured as follows. Section 2 briefly describes the related work for CVE, active shooter response, and disaster evacuation drills in CVE. Section 3 describes the collaborative virtual reality environment for building evacuation. Section 4, describes the implementation of the CVE in three phases. Section 5 presents the simulation and results. Finally, Sect. 6 discusses the conclusions.

## 2   Related Work

The need for evacuation training goes beyond training participants to evacuate a building. It is also useful in training airline pilots to conduct airplane evacuations [3]. An example of a successful evacuation training program can be seen in [4] where a pre-test and a post-test assessing the participant's knowledge gain in regards to proper evacuation methods revealed a positive knowledge gain among participants. Wagner et al. [5] have argued that evacuees are more confident and less hostile while evacuating with an evacuation assistant than evacuating without an evacuation assistant. Alemeida et al. [6] have built a VR environment that studied user's compliance with safety warnings regarding potential safety hazards in the environment. Sharma et al. [7, 8] have created a CVE for emergency response, training, and decision making in a megacity environment. They have also developed a CVE for a real-time emergency evacuation of a nightclub disaster [9] and an active shooter training environment [10]. Their CVE includes user-controlled agents as well as computer-controlled agents. computer-controlled agents include behaviors such as hostile agents, non-hostile agents, leader-following agents, goal-following agents, selfish agents, and fuzzy agents. On the other hand, user-controlled agents are autonomous agents and have specific roles such as police officer, medic, firefighter, and swat officials. Sharma et al. [11] have also used VR as a theme-based game tool for training and education tools. They have conducted virtual evacuation drills for an aircraft evacuation [12], a building evacuation [13], a subway evacuation [14], a university campus evacuation [15], and a virtual city [16].

Lindell, Prater and Wu [17] have argued that preparation and warning times during evacuation time analysis are very important to a successful evacuation. Lindell and Perry [18] have emphasized the importance of evacuation planning in emergency management. Similarly, Bowman et al. [19, 20] have also stressed on the use of VR gaming approach and VR evacuation drills. Musse [21, 22] has provided a real-time communication between users and virtual agents by defining three levels of autonomy: 1) Autonomous crowd 2) Guided crowd 3) Programmed crowd. The above three levels of autonomy are represented in ViCrowd using two kinds of interfaces: scripted and guided interface. The scripted interface incorporates script language where action, motion, and behavioral rules are defined for crowd behaviors.

# 3  Collaborative Virtual Reality Environment for Building Evacuation

Our proposed collaborative VR environment for building evacuation includes: 1) immersive (use of Oculus Rift), and 2) non-immersive environments (Desktop Environment). The participant can enter the CVE setup on the cloud and participate in the emergency evacuation drill, which leads to considerable cost advantages over large-scale, real-life exercises. We hypothesize that the "sense of presence" provided by the CVE will allow running training simulations and conducting evacuation drills. Virtual evacuation drills are:

- More Cost Effective
- Take Less Setup Time
- Able to Simulate Real Dangers
- Improved Response Time.

## 3.1  Non-immersive Environment

The non-immersive environment or a desktop environment was developed to interact with the VR environment using mouse and keyboard. Photon Unity Networking (PUN) is a built-in networking asset in Unity 3D to create a multi-user environment on the cloud. The photon cloud has PCs running photon server on them. The cloud of servers offers the flexibility to host CVE or multiplayer games. Figure 2 shows that the movement of the avatar (client), can be described in 5 stages:

1. The user initiates the action to create a CVE on photon cloud network (PUN)
2. The action is submitted to the photon cloud of servers.
3. A server validates the action and assigns a room ID for the CVE.
4. Any Change done by one client in CVE is propagated to all the other client avatars.
5. All client avatars can see the motion and updated actions of the other ava avatars.

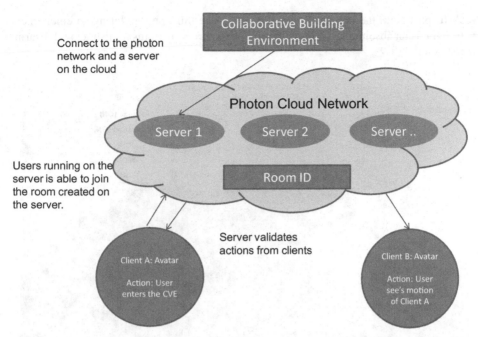

**Fig. 2.** System architecture diagram for CVE using photon networking asset tool in Unity 3D.

Figure 2 shows the proposed system architecture diagram for CVE on the photon cloud network. The photon cloud network has multiple servers on the cloud. The host player connects to the photon network to a server inside the cloud with a valid room ID. The created room on the server is available to all the client's avatars who would like to join the room. The clients also have the option to communicate via chat with the other client avatars who enter that room. When the connection is established the server validates actions from clients and calculates the new CVE environment according to the updated action.

### 3.2   Immersive Environment (Oculus Rift)

An oculus Rift head-mounted display was used to allow the user to be immersed in the CVE. The environment was designed in such a way that the user can enter the environment as policemen or as a building occupant. If the user entered as a building occupant the training module is based on run, hide, and fight mode of emergency response. It offers the flexibility to train building occupants for different what-if scenarios. During an emergency, the first option for the user is to run and escape from the building. This can be done by running away from the active shooter by leaving behind their belongings. During this option, it becomes important to help others to escape and warn other building occupants from entering the building. The second option during an emergency is hide. This can be done by getting out of the active shooter's view and staying quite. It becomes important to silence cell phones and lock doors and windows. Thus staying quite at a locked space until the law enforcement arrives. The third option during an emergency is to fight as an absolute last resort. This involves throwing items to distract and disarm the shooter.

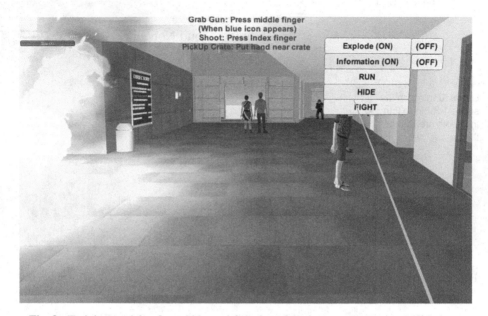

**Fig. 3.** Training module of run, hide, and fight for a fully immersed user-controlled player

Active shooter response for run, hide and fight is shown in Fig. 3. The user can use the laser pointer triggered through the oculus controllers to choose the correct option from the menu in the CVE. The menu is used to trigger the explosion as well as to navigate in the CVE. The user can grab the objects in the environment through an oculus touch controller and throw them at the active shooter.

## 4    Implementation of the CVE

The implementation of a collaborative virtual reality environment for building evacuation for improving emergency response training and decision making was done in four phases.

### 4.1    Phase 1: The Modeling Process

Phase 1 consists of modeling the building on campus using 3D Studio Max and Google Sketch Up. The building was modeled to scale and incorporated real-time textures. The environment also includes adding 3D models for furniture, table, chairs, computers, etc. in the room, lecture halls, and meeting places. Figure 4 shows the view of the first floor of the building in Google Sketch-Up.

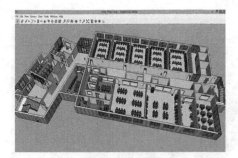

**Fig. 4.** Google Sketch Up 3D scene view of the model.

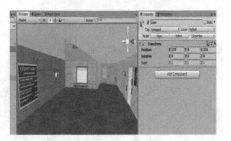

**Fig. 5.** Unity 3D scene view of the game.

## 4.2 Phase 2: Unity 3D and Photon Cloud Setup

Phase 2 consisted of exporting the environment from Google Sketch Up to the Unity 3D gaming engine as shown in Fig. 5. Unity 3D tools for animating avatars were incorporated to give each agent in the CVE the necessary behavior for navigation in the environment. There were two kinds of agents implemented in the CVE. User-controlled agents and computer-controlled AI agents. Both the agents were given functionalities for walking, running, and jumping. Besides, C# scripts were added to the user-controlled agents to give users the ability to communicate with the AI agents and grab the objects in the environment. Multiple C# scripts were added for triggering events in the environment. For example, user-controlled agents were given the ability to communicate with the menu using a laser pointer for selection as shown in Fig. 3. Similarly, C# scripts to implement the Photon server/client networking system were developed to allow collaboration and communication in the CVE.

## 4.3 Phase 3: Oculus Integration and Controller Hand Simulation

Phase 2 consisted of incorporating Oculus Rift and the Oculus Touch controllers in Unity 3D. The left and right controllers were used for integrating the menus and laser pointer for selecting the objects from the menu. The menu design includes options for triggering the explosion which resulted in fire and smoke. The menu also included options for triggering run, hide, and fight modes of interaction for active shooter events. Figure 4 shows the menu option to trigger smoke and fire in the modeled environment using the two oculus touch controllers. With the use of Oculus Rift and the Oculus Touch controllers, users can navigate the environment and interact with objects. The Oculus Touch headset allows users to experience full immersion in the CVE. Oculus Touch controllers also give haptic feedback to the user when using grab option. The trigger button situated on the right controller is used to grab and place the objects in their preferred locations by selecting, holding, dragging, and releasing the trigger as shown in Fig. 6.

**Fig. 6.** Unity 3D scene showing the grab option with Oculus Rift S touch controllers.

## 5  Simulation and Results

One of the goals of this project is to demonstrate the feasibility of conducting emergency response training and virtual evacuation drills in a building using a CVE. We have tested our approach using 10 clients running at the same time on the photon cloud. The environment incorporates smoke and fire. The users can use Oculus Rift to immerse themselves in the CVE and are can turn around 360°.

We have modeled two kinds of threat scenarios such as bomb threat and an active shooter gunman threat in the CVE for a building. Figure 7 shows a scenario where a policeman (as a user-controlled agent) is responding to the active shooter threat. We have modeled behaviors based on pre-defined rules for computer-controlled agents. Through this experimental approach in CVE, emergency personnel and building occupants can be trained on how to respond to emergencies safely and securely by following proper procedures. The platform can also be used in building teamwork for decision-making strategies to follow in case of emergencies. User-controlled agents can enter the CVE to interact with the computer-controlled characters as well as the objects (grab) present in the environment. Multiple users can interact with one another and train on how to respond to active shooter events using the run, hide, and fight option for different what-if scenarios. The collaborative environment is implemented in Unity 3D and is based on run, hide, and fight mode of emergency response as shown in Fig. 8.

**Fig. 7.** Active shooter response for policeman training

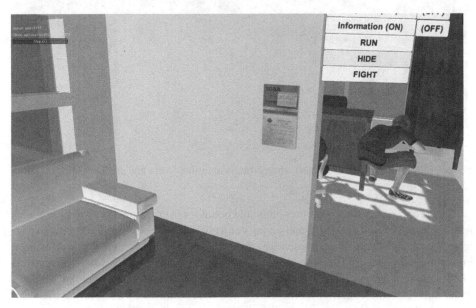

**Fig. 8.** Active shooter response for hide for building occupant's training module

## 6   Conclusions

In conclusion, this CVE system will act as a platform to allow emergency response training to police offices/security personnel and building occupants to follow protocol in an emergency situation. Unity 3D was used to develop the collaborative server/client-based virtual environment that runs on the photon cloud. Our proposed CVE incorporates artificial agents with simple and complex rules that emulate human behavior by using AI.

We have presented a hybrid (human-artificial) platform where experiments for disaster response and training can be conducted using computer-controlled (AI) agents and user-controlled agents. We hope our proposed CVE will aid in visualizing emergency evacuation time and training for different what-if scenarios that are difficult to model in real life. CVE can also act as a training and educational tool for strategizing emergency response and decision-making strategies. Future work will involve the implementation of more behaviors such as altruistic behavior and family behavior for AI agents in the environment. Figure 9 shows a bomb blast scenario where a policeman (as a user-controlled agent) is responding to the threat. An explosion is triggered at the building leading to fire and smoke.

**Fig. 9.** User triggers a bomb blasts resulting in fire and smoke.

The proposed CVE incorporates emulated behavior in emergencies by defining rules for computer-controlled agents, and giving controls to the user-controlled agents to navigate the environment in an immersive environment (head-mounted display) or non-immersive environment (desktop monitor, mouse, keyboard) in real-time. The analysis of these combination controls make it possible to better understand human behavior under such extreme conditions. Knowledge and data obtained by performing the evacuation drill and training exercises will facilitate a more efficient evacuation procedure. Emergency responders will benefit from education on time management during an emergency evacuation drill by better managing and reducing response time to save more lives and casualties. Our contribution in knowledge will complement and improve the skills of emergency responders to handle the traditional emergency response and evacuation drill in real life more efficiently.

**Acknowledgments.** This work is funded in part by the NSF award #1923986 and NSF award #2032344. The author would like to acknowledge Mr. Manik R Arrolla and Mr. Sri Teja Bodempudi who were involved in the development of the project using Unity 3D and Oculus Rift S.

# References

1. Lindell, M.K., Perry, R.W.: Behavioral Foundations of Community Emergency Planning. Hemisphere Publishing, Washington, D.C. (1992)
2. Ford, J.K., Kraiger, K.: The application of cognitive constructs to the instructional systems model of training: implications for needs assessment, design, and transfer. Cooper, C.L., Robertson, I.T. (eds.) International Review of Industrial and Organizational Psychology, p. 1. Wiley, Chichester (1995)
3. O'Connell, K.M., De Jong, M.J., Dufour, D.M., Millwater, T.L., Dukes, S.F., Winik, C.L.: An integrated review of simulation use in aeromedical evacuation training. Clin. Simul. Nurs. **10**(1), e11–e18 (2014)
4. Alim, S., Kawabata, M., Nakazawa, M.: Evacuation of disaster preparedness training and disaster drill for nursing students. Nurse Educ. Today **35**(1), 25–31 (2015)
5. Wagner, V., et al.: Implications for behavioral inhibition and activation in evacuation scenarios: applied human factors analysis. Procedia Manuf. **3**, 1796–1803 (2015)
6. Almeida, A., Rebelo, F., Noriega, P., Vilar, E., Borges, T.: Virtual environment evaluation for a safety warning effectiveness study. Procedia Manuf. **3**, 5971–5978 (2015)
7. Sharma, S, Devreaux, D., Scribner, D., Grynovicki, J., Grazaitis, P.: Artificial intelligence agents for crowd simulation in an immersive environment for emergency response. IS&T International Symposium on Electronic Imaging (EI 2019), in the Engineering Reality of Virtual Reality, Hyatt Regency San Francisco Airport, Burlingame, California, pp. 176-1–176-8, 13–17 January 2019
8. Sharma, S., Devreaux, P., Scribner, P., Grynovicki, J., Grazaitis, P.: Megacity: a collaborative virtual reality environment for emergency response, training, and decision making. In: IS&T International Symposium on Electronic Imaging (EI 2017), in the Visualization and Data Analysis, Proceedings Papers, Burlingame, California, pp. 70–77, 29 January–2 February 2017. https://doi.org/10.2352/ISSN.2470-1173.2017.1.VDA-390
9. Sharma, S, Frempong, I.A., Scribner, D., Grynovicki, J., Grazaitis, P.: Collaborative virtual reality environment for a real-time emergency evacuation of a nightclub disaster. In: IS&T International Symposium on Electronic Imaging (EI 2019), in the Engineering Reality of Virtual Reality, Hyatt Regency San Francisco Airport, Burlingame, California, pp. 181-1–181-10 (2019)
10. Sharma, S, Bodempudi, S.T., Scribner, D., Grazaitis, P.: Active shooter response training environment for a building evacuation in a collaborative virtual environment. In: IS&T International Symposium on Electronic Imaging (EI 2020), in the Engineering Reality of Virtual Reality, Hyatt Regency San Francisco Airport, Burlingame, California, 26–30 January 2020
11. Sharma, S., Otunba, S.: Virtual reality as a theme-based game tool for homeland security applications. In: Proceedings of ACM Military Modeling & Simulation Symposium (MMS11), Boston, MA, USA, pp. 61–65, 4–7 April 2011
12. Sharma, S., Otunba, S.: Collaborative virtual environment to study aircraft evacuation for training and education. In: Proceedings of IEEE, International Workshop on Collaboration in Virtual Environments (CoVE - 2012), as part of the International Conference on Collaboration Technologies and Systems (CTS 2012), Denver, Colorado, USA, pp. 569–574, 21–25 May 2012
13. Sharma, S., Vadali, H.: Simulation and modeling of a virtual library for navigation and evacuation. In: MSV 2008 - The International Conference on Modeling, Simulation and Visualization Methods, Monte Carlo Resort, Las Vegas, Nevada, USA, 14–17 July 2008
14. Sharma, S., Jerripothula, S., Mackey, S., Soumare, O.: Immersive virtual reality environment of a subway evacuation on a cloud for disaster preparedness and response training. In: Proceedings of IEEE Symposium Series on Computational Intelligence (IEEE SSCI), Orlando, Florida, USA, pp. 1–6, 9–12 December 2014. https://doi.org/10.1109/cihli.2014.7013380

15. Sharma, S., Jerripothula, P., Devreaux, P.: An immersive collaborative virtual environment of a university campus for performing virtual campus evacuation drills and tours for campus safety. In: Proceedings of IEEE International Conference on Collaboration Technologies and Systems (CTS), Atlanta, Georgia, USA, pp. 84–89, 01–05 June 2015. https://doi.org/10.1109/cts.2015.7210404

16. Sharma, S.: A collaborative virtual environment for safe driving in a virtual city by obeying traffic laws. J. Traffic Logist. Eng. (JTLE) 5(2), 84–91 (2017). ISSN:2301-3680. https://doi.org/10.18178/jtle.5.2.84-91

17. Lindell, M.K., Prater, C.S., Wu, J.Y.: Hurricane evacuation time estimates for the Texas gulf coast. Hazard Reduction & Recovery Center, Texas A&M University, College Station, TX (March 2002)

18. Perry, R.W., Lindell, M.K., Greene, M.R.: Evacuation Planning in Emergency Management. Hemisphere Pub., Washington, D.C., pp. 181–196 (1992)

19. McMahan, R.P., Bowman, D.A., Zielinski, D.J., Brady, R.B.: Evaluating display fidelity and interaction fidelity in a virtual reality game. IEEE Trans. Vis. Comput. Graph. 18(4), 626–633 (2012)

20. Ragan, E.D., Sowndararajan, A., Kopper, R., Bowman, D.A.: The effects of higher levels of immersion on procedure memorization performance and implication foe educational virtual environments. Presence: Teleoperators Virtual Environ. 19(6), 527–543 (2010)

21. Musse, S.R., Thalmann, D.: Hierachical model for real time simulation of virtual human crowds. IEEE Trans. Vis. Comput. Graph. 7(2), 152–164 (2001)

22. Musse, S.R., Garat, F., Thalmann, D.: Guiding and interacting with virtual crowds in real-time. In: Proceedings of the Eurographics Workshop on Computer Animation and Simulation 1999, September 7–8, pp. 23–34 (1999)

# Text Entry in Virtual Reality: Implementation of FLIK Method and Text Entry Testbed

Eduardo Soto and Robert J. Teather[✉]

Carleton University, Ottawa, ON, Canada
easm93@gmail.com, rob.teather@carleton.ca

**Abstract.** We present a testbed for testing text entry techniques in virtual reality, and two experiments employing the testbed. The purpose of the testbed is to provide a flexible and reusable experiment tool for text entry studies, in such a way to include studies from a variety of sources, more specifically to this work, from virtual reality text entry experiments. Our experiments evaluate common text entry techniques and one novel one that we have dubbed the Fluid Interaction Keyboard (FLIK). These experiments not only serve as a way of validating the text entry test-bed, but also contribute the results of these studies to the pool of research related to text entry in virtual reality.

**Keywords:** Virtual reality · Text entry · FLIK · Words per minute · Head-mounted displays

## 1 Introduction

According to Ko and Wobbrock, text is a ubiquitous form of verbal communication [8]. Text entry refers to the process of creating messages composed of characters, numbers, and symbols using an interface between a user and a machine. Text entry has been extensively studied in the field of human-computer interaction (HCI) over the past few decades, especially for use in desktop and mobile contexts. Efficient text entry is important because it directly impacts the user's ability to write documents, send messages, and communicate effectively when verbal communication is not available.

Virtual reality (VR) is defined as the use of computer technology to create simulated environments. It often employs head-mounted displays (HMD) along with spatially tracked controllers or hand trackers as input devices. VR has become increasingly popular in recent years; increased consumer demand means the interaction effectiveness of VR systems affects more people than ever. Our work focuses on text entry in VR, such as entering text in search bars or sending text messages. In such use cases, it is important to include an effective option to enter text in VR systems as well as an effective method to evaluate novel text entry techniques and compare with existing techniques.

To type long-form text in a VR environment is not yet feasible or likely even desirable. However, there are many examples where text input of quick messages or short notes is important, even in VR. For example, consider playing a multiplayer online VR game, where communicating in written form may be preferable to speech communication. For

© Springer Nature Switzerland AG 2020
C. Stephanidis et al. (Eds.): HCII 2020, LNCS 12428, pp. 225–244, 2020.
https://doi.org/10.1007/978-3-030-59990-4_18

gaming scenarios, one might find the need to type a quick SMS to a friend inviting them to come online, without interrupting the gameplay. In scientific fields or even in architecture, it might be useful to be immersed in a 3D environment and be able to annotate different components of the environment. In a VR conference call, a user might like to take notes and potentially send back-channel messages to other attendees.

While speech-to-text technologies could work well for some purposes, there are cases where some form of typing in VR might be preferable to voice recognition. Consider, for example, being immersed in VR whilst in a loud environment, or when you need to maintain quiet. For example, VR use has increased in areas such as office work, collaboration, and training, and education. For these applications, inputting text is an essential part of these experiences and more often than not, precise text entry is required rather than using a speech-to-text method, which may disturb other people. Modification of existing text is unreliable and inaccurate as compared to more refined and direct text entry techniques.

The first contribution of this paper is a text entry testbed. A common issue for text entry research is consistency and better standardization of methodological practice in evaluating text entry systems [1]. Methodological consistency is important in text entry research since usually, analysis is comparative, i.e., involves empirically comparing text entry techniques to determine which is most efficient. Employment of similar methods and metrics to ensure comparability between studies is motivating factor of the text entry testbed. The text input testbed is a tool for conducting text entry studies which supports desktop, mobile, VR, and other text input methods, that adheres to standard practice in HCI text input research. The testbed provides a consistent platform to perform text entry comparisons regardless of the techniques used.

The second contribution is a novel VR text entry technique called Fluid Interaction Keyboard (FLIK). FLIK uses two 6-degree of freedom (6DOF) controllers (e.g., Oculus Touch controllers) together with a simple interface for fast and intuitive text entry. It operates by selecting characters from a virtual soft keyboard by directly touching the characters. When eliminating physical button presses on the controllers, users are enabled to type faster using this technique and achieve fluid hand motions to increase the speed of character selection. FLIK was inspired by another text entry technique, Cutie Keys [2, 19] and was designed to improve on existing practice in VR text entry. The third and final contribution is a formal evaluation of FLIK, comparing it to existing VR text entry techniques through two user studies.

## 2 Related Work

Text entry has been studied for decades. In typical text entry experiments, the time to enter a phrase is recorded while transcribing text to provide a measure of entry speed, while the transcribed text is compared with the original text to measure errors. Entry speed is a key metric for text entry since the goal of text entry techniques is most commonly to offer fast ways of entering text into a system. However, entry speed cannot be looked at by itself; error rate (e.g., number of mistyped characters) is another metric that is used in text entry studies along with entry speed to find a balance between fast text entry speed and acceptable error rates.

Methodology for conducting text entry studies is detailed by Mackenzie [10]; he focuses on the evaluation of text entry techniques and laid out strategies in order to do so. A typical text entry study starts with a working prototype. Users must be able to enter text, and have it displayed as a result. Once this prototype is implemented, substantial pre-experimental testing can begin. The first step is to get a rough idea of the entry speed possible with the technique in question. Next is to decide what the evaluation task will consist of. For text entry studies, the task is commonly to transcribe a set of phrases as quickly and accurately as possibly using some text entry technique as the typing method.

For the methodology described above, a set of phrases is needed to conduct the study. Mackenzie provides such a phrase set for text entry experiments [11] which provides a consistent metric for text entry research independent of technique used, technology involved, or even the main researcher conducting a study. Mackenzie's phrase set consists of 500 phrases to be used in such studies; this provides easy access to much needed source material to be used in text entry.

Wobbrock et al. [20] describe measures of text entry performance in which they include words per minute as a text entry speed measure as well as the minimum string distance as an error rate measure. We now summarize key metrics of text entry performance, several of which we employ in our experiments.

Entry speed, in words per minute, is calculated as seen in Eq. 1, as described by Boletsis et al. [2].

$$wpm = \frac{|T| - 1}{S} \times 60 \times \frac{1}{5} \tag{1}$$

where $S$ is the time in seconds from the first to the last key press and $|T|$ is the number of characters in the transcribed text (i.e., the phrase length). The constant '60' corresponds to the number of seconds in a minute, and the factor of one fifth corresponds to how many characters compose an average word, which is defined to be 5 characters in text entry experiments. This definition makes results more generalizable and consistent across varying phrase sets used for input. We subtract 1 from the length since timing starts as soon as participants enter the first key, hence the first key is not timed, necessitating reducing the phrase length by 1.

An important metric for measuring text entry accuracy is the minimum string distance (MSD) [12, 14–16]. MSD instead is based on an algorithm used in DNA analysis, phylogenetics, spelling correction, and linguistics for measuring the distance between two strings. As an example, consider two strings 'abcd' and 'acbd'. To calculate the error between these two strings you could take the transcribed string and delete the c, then, insert a c after the b. This requires two actions to make the strings identical, so MSD = 2. Minimum String Distance is denoted MSD(A, B) where A is the presented text and B is the transcribed text. With this in mind, MSD(A, B) = 0 means A and B are identical. Further refinement for error analysis [12] notes the difficulty of analyzing errors in text entry systems, in that the displacement of letters in certain text entry errors might make correctly typed letters incorrect.

Text entry techniques are the input methods for entering text into a system. These can include physical keyboards, touchpads, voice recognition, and even pencils. There has been extensive research done in the area of text entry using physical keyboards [5–7]. This work typically focuses on conducting text entry experiments using standard keyboards to enter text into a computer system, comparing entry speed and accuracy measures. The main focus lies on physical keyboard text entry studies in the context of VR use. One prominent topic in the use of physical keyboards in VR includes hand visualizations when typing on physical keyboards [5–7], and comparing different hand representations in VR text entry. For example, Grubert et al. [6] made use of the Logitech Bridge SDK [18], which was used to track the physical keyboard and hands, and display a digital hand representation in VR. The hand representations included no hands, realistic hands, finger tips only, and VideoHand, where a video pass-through allows the user to see their hands in VR. They found no significant results on entry speed; however, participants preferred VideoHand.

VR text entry techniques usually do not use a physical keyboard, as standard keyboards tend to restrict the user's mobility, while VR systems often require users to physically move around. Thus, the two most common techniques are controller pointing and bimanual text entry. Controller pointing requires spatially tracked controllers to point at and select keys on a soft (i.e., virtual) keyboard, typically via ray-casting. Bimanual text entry presents a soft keyboard split into two sides and requires two touchpads or joysticks to move the cursor on each side of the keyboard to select keys. Each thumb controls one of the two cursors on the corresponding side of the virtual keyboard [13]. Bimanual entry offers entry speeds as high as 15.6 WPM, but on average 8.1 WPM. Similar techniques have been included in several studies, which consistently report low entry speeds between 5.3 and 10 WPM [2, 17].

We finish our discussion of past VR text entry techniques by mentioning BigKey by Faraj et al. [4], as we employ it in our second experiment, and it is a feature supported by our text entry testbed. BigKey resizes keys on the virtual keyboard in real-time. It uses the probability of a letter being typed *next* to increase the size of the predicted next key, and decrease the size of less probable keys. On average, this can increase entry speed, since as described by Fitts' law, larger targets (at otherwise equal distance) offer faster selection times [9].

## 3   Text Entry in Virtual Reality

This section describes the implementation of our text entry testbed, BigKey, word disambiguation, and the input techniques evaluated in the studies.

### 3.1   Text Entry Testbed

Noting the need for an experimental framework to conduct text entry studies in VR, we first developed a text entry experiment testbed. The design and implementation of the text entry testbed was targeted as a flexible text entry experiment framework developed with Unity 3D 2018.3.7f1. It was developed for conducting user studies on text entry in VR for the purpose of this work. However, it doubles as a general purpose text entry experiment tool since it can be adapted for use with any (i.e., non-VR) text entry technique. Researchers can develop their text entry techniques and easily integrate it with the text entry testbed for initial performance test, up to full scale user studies.

Starting the testbed, the experimenter is presented with the welcome screen. The experimenter enters a participant ID, the number of phrases that will be presented to the participant in each block, and the number of blocks (i.e., repetitions of all experiment conditions). For example, if 10 phrases are chosen, and 3 blocks are chosen, 30 phrases will be typed overall split into 3 blocks.

Upon proceeding, the experimenter is then prompted to select a text entry method on the next screen. The input technique selection screen presents a list of options which is customized by the experimenter depending on what text entry techniques they include in the evaluation. If the study requires multiple levels of independent variables to choose from, additional options appear allowing the experimenter to choose each condition. Once everything is set and an initial text entry method to test is chosen, the system will start with the first block of the chosen method.

When starting the first block and any subsequent block, a new unique log file is created. In the scene, there are three main components: The phrase to be transcribed, the text entered so far, and the text entry technique are the main components that the participant is able to visualize in the scene. The testbed currently uses Mackenzie's phrase set of 500 phrases as the default phrases, which are commonly used in text entry experiments [11], but these can be changed by replacing a file called 'phrases.txt' with any other phrase set as long as it follows the same format as Mackenzie's phrase set text file. At this point, the main experiment begins, and is seen in Fig. 1.

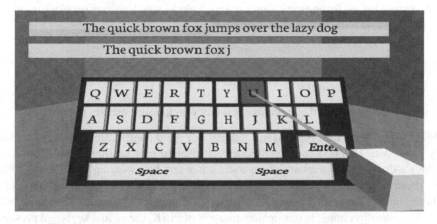

**Fig. 1.** Elements presented to participants during text entry experiment

As each trial commences, the participant sees the target phrase, the text they have entered so far in a separate text field, and any components that need to be visualized required for a given text entry technique. The testbed randomly selects a phrase from the phrase set; this is the "target" phrase, and it is presented to the participant in the phrase to transcribe section and will wait for the participant to enter the first letter, this is repeated for the number of phrases selected by the experimenter for each block. When the participant enters the first character, a timer starts. The timer stops once the participant hits ENTER or an equivalent key on the virtual keyboard being used. The timer measures how long it took the participant to enter the phrase. While the participant is entering text, the system records the raw input stream, which corresponds to all characters selected in order; this includes space, backspace, and modifiers.

When the participant completes a phrase and hits ENTER, the testbed records all relevant details to a log file (see Fig. 2). In the log file, each trial appears as a single row, and includes experiment setup details such as the participant ID, text entry technique, current condition, block number, the phrase to be transcribed, the text that is actually transcribed by the participant, the raw input stream, time, the calculated entry speed in words per minute, and error rate in minimum string distance [1].

| Participant ID | Text Entry Technique | Condition | Block | Phrase | Transcribed | Keystrokes | Time | MSD | Length Phrase | Length Trans | wpm | Error Rate % |
|---|---|---|---|---|---|---|---|---|---|---|---|---|
| 1 | Ray | None | 1 | bad for the environment | bad for the environment | bad-for-the-environment | 16.1202 | 0 | 23 | 23 | 16.37696803 | 0.00% |
| 1 | Ray | None | 1 | we park in driveways | we park in driveways | we-park-in-driveways | 19.9295 | 0 | 20 | 20 | 12.04469215 | 0.00% |
| 1 | Ray | None | 1 | what a monkey sees a monkey will do | what a monkey sees a monkey will do | what-a-monkey-sees-a-monkey-will-do | 29.9984 | 0 | 35 | 35 | 13.60072537 | 0.00% |
| 1 | Ray | None | 1 | world population is growing | world population is growing | world-population-is-growing | 21.8406 | 0 | 27 | 27 | 14.41201741 | 0.00% |
| 1 | Ray | None | 1 | what a lovely red jacket | what a lovely red jacket | what-a-lovely-red-jacket | 17.1598 | 1 | 24 | 25 | 16.78341239 | 4.00% |
| 1 | Ray | None | 1 | an excellent way to communicate | an excellent way to communicate | an-excellent-way-to-communicate | 24.5003 | 0 | 31 | 31 | 14.81483192 | 0.00% |
| 1 | Ray | None | 1 | he is still on our team | he is still on our team | he-is-still-on-our-team | 17.9646 | 0 | 23 | 23 | 15.03023126 | 0.00% |
| 1 | Ray | None | 1 | not quite so smart as you think | not quite so smart as you think | not-quite-so-smart-as-you-think | 27.0699 | 0 | 31 | 31 | 13.26937994 | 0.00% |
| 1 | Ray | None | 2 | my bike has a flat tire | my bike has a flat tire | my-bike-has-a-flat-tire | 16.3452 | 0 | 23 | 23 | 17.20407684 | 0.00% |
| 1 | Ray | None | 2 | effort is what it will take | effort is what it will take | effort-is-what-it-will-take | 22.0915 | 0 | 27 | 27 | 14.12307901 | 0.00% |
| 1 | Ray | None | 2 | she wears too much makeup | she wears too much makeup | she-wears-too-much-makeup | 18.5272 | 0 | 25 | 25 | 15.54471264 | 0.00% |
| 1 | Ray | None | 2 | this equation is too complicated | this equation is too complicated | this-equation-is-too-complicated | 20.7602 | 0 | 32 | 32 | 17.91800252 | 0.00% |
| 1 | Ray | None | 2 | I took the rover from the shop | I took the rover from the shop | I-took-the-rover-from-the-shop | 20.9074 | 0 | 30 | 30 | 16.64462432 | 0.00% |
| 1 | Ray | None | 2 | join us on the patio | join us on the patio | join-us-on-the-patio | 19.6099 | 0 | 20 | 20 | 11.62078035 | 0.00% |
| 1 | Ray | None | 2 | an inefficient way to heat a house | an inefficient way to heat a house | an-inefficient-way-to-heat-a-house | 28.6977 | 2 | 34 | 34 | 13.70901525 | 5.88% |
| 1 | Ray | None | 2 | meet tomorrow in the lavatory | meet tomorrow in the lavatory | meet-tomorrow-in-the-lavatory | 27.8705 | 0 | 29 | 29 | 12.05575788 | 0.00% |
| 1 | Ray | None | 3 | companies announce a merger | companies announce a merger | companies-announce-a-merger | 22.192 | 0 | 27 | 27 | 14.11636956 | 0.00% |
| 1 | Ray | None | 3 | the stock exchange dipped | the stock exchange dipped | the-stock-exchange-dipped | 21.308 | 0 | 25 | 25 | 13.51605031 | 0.00% |
| 1 | Ray | None | 3 | this is a very good idea | this is a very good idea | this-is-a-very-good-idea | 17.0346 | 0 | 24 | 24 | 16.20231764 | 0.00% |
| 1 | Ray | None | 3 | so you think you deserve a raise | so you think you deserve a raise | so-you-think-you-deserve-a-raise | 20.5906 | 0 | 32 | 32 | 18.06649636 | 0.00% |
| 1 | Ray | None | 3 | look in the syllabus for the course | look in the syllabus for the course | look-in-the-syllabus-for-the-course | 26.7766 | 0 | 35 | 35 | 15.23718471 | 0.00% |
| 1 | Ray | None | 3 | protect your environment | protect your environment | protect-your-environment | 14.0739 | 0 | 24 | 24 | 19.80324732 | 0.00% |
| 1 | Ray | None | 3 | be persistent to win a strive | be persistent to win a strive | be-persistent-to-win-a-strive | 19.4204 | 0 | 29 | 29 | 17.29901283 | 0.00% |
| 1 | Ray | None | 3 | a big scratch on the tabletop | a big scratch on the tabletop | a-big-scratch-on-the-tabletop | 20.7282 | 0 | 29 | 29 | 16.21214763 | 0.00% |

**Fig. 2.** Sample log file for one participant covering 3 blocks with 8 phrases per block using one text entry technique using a particular modifier condition.

## 3.2   Text Entry Techniques

This subsection describes the main four text entry techniques used in our studies.

**FLIK (Fluid Interaction Keyboard).** One of our main contributions is the FLIK text entry technique. This text entry technique was inspired by CutieKeys [19]. CutieKeys employs a drumstick metaphor, where users enter text by swinging the VR controller like a drumstick to hit pads corresponding to keys organized in the standard QWERTY keyboard layout. However, unlike CutieKeys, FLIK supports contacting keys from both top and bottom directions. In our implementation of FLIK, the virtual keyboard presents keys as spheres, with each positioned far enough from its neighbor to prevent overlap or mistaken keystrokes due to accidental collisions between the cursor and the key sphere. FLIK uses bimanual 3D tracked controllers. A selection cursor sphere is attached to the tip of each controller; these selection spheres are what the user manipulates, and intersects with the key spheres to issue a specific keystroke. Intersection between the selection cursor and the key sphere simultaneously selects *and* confirms the key input; no further button presses are required with FLIK. Upon intersecting a key sphere, the user can move in arcs in both direction, and loop back through the keyboard to the desired key. In contrast, CutieKeys requires that the user move the controller back (upwards) from below upon striking a key, then move back down *again* from above to strike the next key; FLIK effectively cuts the total movement to half that required by CutieKeys.

Although this difference between FLIK and CutieKeys is subtle, we anticipate that this minor difference will improve entry rates due to the continuous movement of the hands. In contrast, with CutieKeys, the user has to tap on a key, and then reverse the direction of movement to stop selecting the key. See Fig. 3.

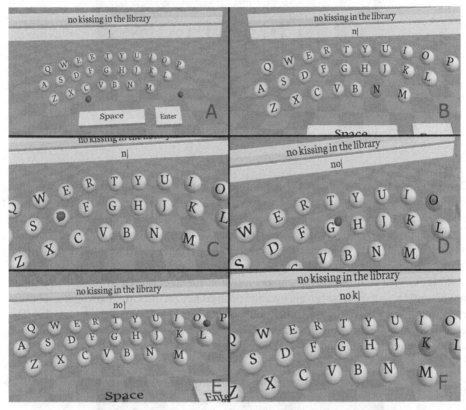

**Fig. 3.** Fluid intersection keyboard (FLIK). (A) Shows the phrase to be transcribed before the user has entered any text. (B) The user selects the character 'n' by moving his right hand through it so that the red selection cursor touches and goes through the character as seen in (C) where the sphere cursor is now behind (and partially occluded by) the 'n' character key sphere. (D) Again, shows the same right hand having looped around the bottom of the keyboard and selecting the 'o' character key sphere from beneath. (E) As the right hand exits the character 'o', the left hand now moves through the spacebar to select a space character. (F) User continues in this manner selecting the next character.

**Controller Pointing.** Controller pointing requires two 6DOF controllers, each of which emits a selection ray. Participants can bimanually (using both controllers, one in each hand) remotely point at a virtual (i.e., soft) keyboard presented in front of them. Individual keystrokes are issued upon pressing the primary thumb button on the controller. Visual feedback is provided by changing the colour of the intersected key. Vibrotactile feedback and auditory feedback (a "click" sound) are also provided when the button is pressed. Both rays pointing at the same key would not have any side-effects. See Fig. 4.

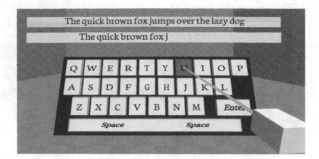

**Fig. 4.** Images showing the Controller Pointing keyboard

**Continuous Cursor Selection.** With continuous cursor selection, the virtual keyboard is divided in left and right halves to support bimanual entry. Keys on the right half of the keyboard are selected via a cursor controlled by the right hand and vice versa. A 2D selection cursor is initially placed at the midpoint on each keyboard half. These selection cursors are controlled by moving the corresponding (right or left) joystick (e.g., with an Oculus Touch controller) or sliding a finger/thumb on the controller's touchpad (e.g., on an HTC Vive controller). The cursors move in the direction of joystick or touchpad movement. Once the cursor hovers over the desired key on the virtual keyboard, a button press on the corresponding controller confirms selection and issues the keystroke. The same feedback mechanisms described above for the controller pointing technique are employed with this technique as well. See Fig. 5.

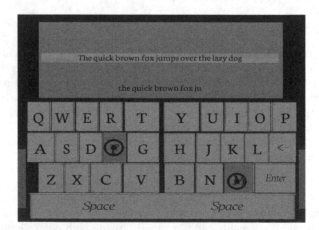

**Fig. 5.** Continuous Cursor Selection. By dragging each thumb on their respective touchpads, the cursors move on their respective half in order to select characters from the virtual keyboard.

**Physical Keyboard.** Lastly, as a baseline condition, we included a tracked physical keyboard with virtual hand representations via the Logitech Bridge API [18] and an HTC Vive individual tracker mounted on a keyboard. The hand representations are provided by the Logitech Bridge API via the HTC Vive's integrated camera and Logitech's proprietary computer-vision hand recognition techniques. This technique behaves the same as using a physical keyboard except that the user sees a virtual representation of the keyboard (and a video image of their hands) rather than the physical keyboard itself (Fig. 6).

**Fig. 6.** Logitech Bridge virtual keyboard with hand representation

### 3.3 Text Entry Aids

The testbed supports the following add-on features that can be applied with any of the above text entry techniques (or any other that could be added). This demonstrates the flexibility of the testbed in providing a way to add different techniques with varying levels of complexity. The testbed currently supports two such aids, BigKey and word disambiguation, described below.

**BigKey:** BigKey [4] employs a custom algorithm which analyzes the input stream of the current word. The higher the probability a letter will be typed next, the bigger the size of its key on the virtual keyboard. The algorithm starts when an initial character for a new word is entered (i.e., following press of the space key). Prior to entry of the initial character for a new word, all keys are the same average size.

Following the first keystroke for the new word, the algorithm searches the phrase set for all possible words that start with entered character. Since the testbed included only a fixed phrase set size, there are a limited number of possible words that can be entered, so the overall computing time for finding matches is short. Prior to entering the next character, the system calculates the frequency of all subsequent next characters. This is done by counting the occurrences of each character in sequential order, starting with 'a', 'b', 'c', and so on, that directly follow the entered character for all words in the array. After this calculation is complete, all letters with 0 frequency (i.e., those that *never* follow the initial character) are scaled to the smallest size, while the letter or letters with highest frequency, are scaled up to the maximum size. Letters with frequencies between 0 and the highest frequency are scaled by a factor multiplier which starts with the scaling factor of the letter with the highest count. The highest ranked letter would be scaled up by a factor of 1.75, the second by a factor of 1.7, and so on in decrements of 0.05. This ensures that all characters that *might* follow the initial character receive some scaling multiplier, with the more probable ones being scaled proportionally larger. See Fig. 7.

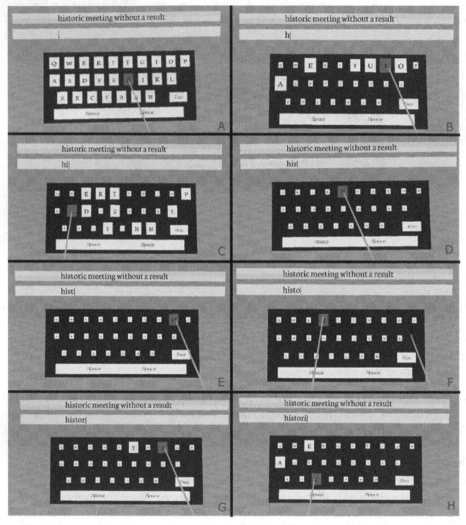

**Fig. 7.** Image showing keys rescaling on virtual keyboard based on BigKey algorithm. In every slide, a new character is selected, and the rest of the characters are resized based on the words found with those characters as an option to be typed next.

**Word Disambiguation:** In every language, there are frequencies of how often each word appears. From these frequencies, one can implement a general Word Disambiguation system that can be used with most text entry scenarios. The testbed implements this in the form of suggestions, every time the user enters a key, the system computes the most likely word to be typed in order of ranking, then it displays the top three ranked words as selection options to complete a word in one selection rather than finishing the whole word.

This system works in a similar way to how key prediction is implemented for BigKey. The main difference is that with disambiguation, the ranking for every word in the phrase set is used. Along with Mackenzie's phrase set, a list of word rankings is provided. This list contains all words used in the phrase set along with their ranking. The rankings are in the form of a simple integer number, representing the frequency of use of this word.

## 4   User Study 1

This section presents the first user study using the text entry testbed comparing four VR text entry techniques. This first experiment consists of a performance comparison of the four text entry techniques for VR: Controller Pointing, Continuous Cursor Selection, FLIK, and Physical Keyboard.

### 4.1   Participants

This study included 24 participants, 14 male, 10 female, and aged 18–30 ($SD = 2.96$). All participants stated that they are comfortable with basic or advanced computer use.

### 4.2   Hardware

The experiment was conducted using a VR-ready laptop with an Intel core i7-6700HQ quad core processor, an Nvidia Geforce GTX 1070 GPU, and 16 GB of RAM, running the latest build of Microsoft Windows 10. The Logitech G810 was used as the physical keyboard due to integration requirement with the Logitech Bridge SDK. We used the HTC Vive HMD with its two touchpad-enabled 6DOF controllers. The HTC Vive provides a $1080 \times 1200$-pixel resolution per eye, 90 Hz refresh rate, and 110 degrees of field of view.

### 4.3   Software

The study employed the text entry testbed described above. It was developed in Unity 3D and it integrates easily with any text entry technique. The study included four text entry techniques: Controller Pointing, Continuous Cursor Selection, FLIK, and Physical Keyboard. Each text entry technique used in this study was developed in Unity independent of the text entry testbed and was then integrated with the testbed in order to use the full functionality of the system.

## 4.4  Procedure

Upon arrival, participants were informed of their right to withdraw from the experiment at any point without any obligation to finish the experiment if at any time they felt uncomfortable or nauseous. Participants were then asked to try the HMD, and were presented with their first text entry technique (based on counterbalancing order). They were given around two minutes to practice with the technique, to become familiar with the general system and the first text entry technique they would use.

The task involved transcribing presented phrases using the current text entry technique. Participants performed 3 blocks with each of the four text entry techniques, where each block consisted of 8 phrases to be transcribed. When all three blocks were finished, the next text entry technique was chosen and the process repeated. Participants were instructed to transcribe the phrases presented as quickly and accurately as possible.

## 4.5  Design

The experiment employed a within-subjects design, with two independent variables, text entry technique with four levels (FLIK, controller pointing, physical keyboard, and continuous cursor selection) and block with three levels (block 1, 2, and 3). The order of text entry technique was counterbalanced using a Latin square.

We recorded three dependent variables (entry speed, error rate, and NASA-TLX scores). Entry speed was measured in words per minute (WPM), and calculated as seen in Eq. 1. Error rate was based on the MSD, see Eq. 2.

$$Error\ Rate\ \% = \frac{100 \times MSD(P, T)}{\max(|P|, |T|)} \tag{2}$$

TLX scores are the overall results from the NASA-TLX questionnaire. Across all 24 participants, this yielded *3 blocks × 8 phrases × 4 text entry techniques × 24 participants = 2304 phrases transcribed.*

## 4.6  Results

**Task Performance.** We first present quantitative results in terms of task performance, computed per participant, for each text entry technique, and averaged per block. Entry speed ranged from 12.32 WPM for Continuous Cursor Selection to 49.56 WPM for Physical Keyboard. See Fig. 8 for full results.

Repeated-measures analysis of variance revealed that the effect of text entry technique on entry speed was statistically significant ($F_{3, 69} = 54.886, p < .0001$), as was the effect of block on entry speed ($F_{2, 46} = 88.432, p < .0001$). A Post Hoc Bonferroni-Dunn test at the $p < .05$ level revealed pairwise significant differences between some text entry techniques, represented as horizontal bars on Fig. 8.

Error rates ranged from 1.83% error for physical keyboard to 4.03% for continuous cursor selection. Analysis of variance revealed that the effect of text entry technique on error rate was not statistically significant ($F_{3, 69} = 0.431$, ns), nor was the effect of block ($F_{2, 46} = 1.681, p > .05$). See Fig. 9.

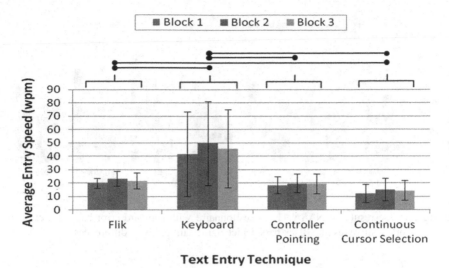

**Fig. 8.** Text entry speed average comparison chart. Error bars show standard deviation. Black horizontal bars show pairwise significant differences between text entry techniques.

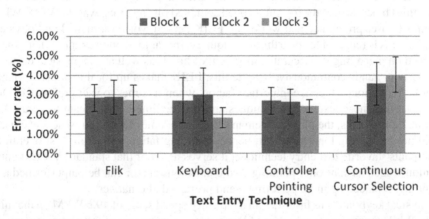

**Fig. 9.** Average error rate (%) comparison chart, error bars show standard deviation.

**User Preference.** User workload was assessed using Hart and Staveland's NASA Task Load Index (TLX). NASA TLX. A lower overall score signifies less workload and thus greater overall preference to a technique. Results were analyzed using a Friedman test, with Conover's test as a posthoc. Significant pairwise differences ($p < .05$) are represented on Fig. 10 by horizontal bars between each text entry technique.

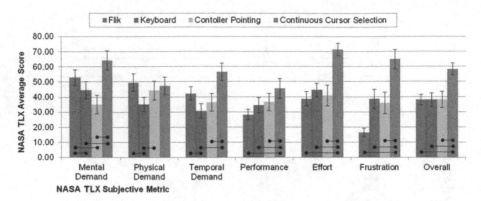

**Fig. 10.** Average results of NASA TLX questionnaire with standard error bars. Lower scores represent more positive results. Black bars depict significant pairwise differences.

## 4.7 Discussion

The physical keyboard condition did best overall, however, this was expected. After that, FLIK did about as well as controller pointing, with an average entry speed of 23 WPM on the third block, where the top entry speed achieved on any block was 28 WPM. When it comes to user preference and workload, FLIK scored worse on mental and physical demand. This is reasonable since this technique is similar to playing drums, where focus is placed on managing the accurate movement of hands as well as the physical effort of moving both hands continuously. Despite this, FLIK makes up for this in performance and low frustration, which are important factors when it comes to user preference. Due to the fact that these text entry techniques are, for the time being, meant to be used in short text entry tasks, the physical demand aspect becomes less of an issue since users would not be writing long form text, hence, reducing fatigue. Ranking first overall as participants' favorite text entry technique, it serves as proof that standard VR text entry techniques such as controller pointing, and continuous cursor can be outperformed and potentially replaced by more performant and preferred alternatives.

Physical keyboard was fastest, with a top entry speed score of 49.6 WPM on the third block. While there were some cases of low scores among some of the participants, with one participant scoring the lowest of 15.7 WPM, this is far offset with the proficiency of participants using regular keyboards, with most being able to type without looking at the keyboard. Cases where scores were low could be attributed to poor tracking, appearing as the virtual hands not aligning perfectly with real world hands. Such occurrences were rare and the mismatch between the virtual and real hands was not extreme. While the physical keyboard offered performance double that of the other techniques, we again note that the physical keyboard is, in many ways, a sub-optimal choice as a VR text entry technique due to several issues unrelated to performance. For example, the extra hardware (specific Logitech keyboard, attachment piece for the keyboard, and external HTC Vive sensors) and software required is not readily available with out of the box VR headsets. This decreases the adoption rate of such a technique, at least until these components become an integrated part of a HMD package. The most negative factor is that physical keyboards are not very portable. VR users tend to be standing and walking, rotating in place, and generally changing their position more often than not. Carrying a full-sized keyboard does not lend itself very well for this type of activity since it would need to be stationary.

Controller pointing is currently the most commonly used technique in commercial VR products, and was shown to offer good performance as well. However, with text entry speed results coming in below FLIK (although not significantly worse), and being overall ranked second as participants' favorite choice of text entry technique, there is room for improvement when it comes to the standard method of entering text in VR. The continuous cursor technique, which is also commonly available in commercial products, was by far the least preferred and worst performing technique, further demonstrating that the industry VR software developers have gravitated towards several sub-optimal techniques.

## 5 User Study 2

We present a second user study comparing performance between Controller Pointing and FLIK with the addition of text entry aids described earlier, BigKey and word disambiguation. To further demonstrate the flexibility of the text entry testbed, we used an Oculus Rift CV1 in this study instead of the HTC Vive, which have different setup requirements and APIs.

### 5.1 Participants

This second study consisted of 24 participants, 13 male, 11 female, and aged 18–55 (*SD* = 9.01). None of these participants completed the first user study.

### 5.2 Hardware

The experiment employed same hardware as the first study, with the exception of using an Oculus Rift CV1 HMD with two Oculus Touch 6DOF controllers instead of the HTC Vive. The Oculus Rift CV1 provides a $1080 \times 1200$-pixel resolution per eye, a 90 Hz refresh rate, and 110 degrees of field of view.

### 5.3 Software

The text entry testbed was again used for this study. We used a subset of the text entry techniques from the previous study. The only new additions to the software were the BigKey and word disambiguation aids.

### 5.4 Procedure

Upon arrival, participants were told that if at any time they felt uncomfortable or nauseous that they were free to stop the experiment without finishing. Participants were then asked to try the HMD, followed by presenting them with their first text entry technique with the first condition (no aids, BigKey, or word disambiguation), where they were free to try it out and practice for around two minutes. Once they felt comfortable, the study began with that text entry technique.

This study employs the same procedure as the first study, but with the techniques and aids described below.

## 5.5  Design

The experiment employed a within-subjects design, with three independent variables:

*Text entry technique*: FLIK, controller pointing.
*Aid*: no aid, BigKey, Word Disambiguation.
*Block*: 1, 2, and 3.

The text entry techniques and aids were counterbalanced using a Latin square. With all 24 participants, this resulted in *3 blocks * 8 phrases * 2 text entry techniques * 3 aids * 24 participants = 3456 phrases transcribed*.

## 5.6  Results

**Task Performance.** The slowest condition was Controller Pointing with no aid, at 18.12 WPM. The fastest condition, at 27.8 WPM was FLIK+BigKey. See Fig. 11.

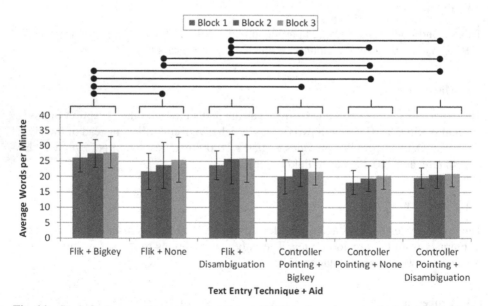

**Fig. 11.** Study 2 average entry speed comparison table. Error bars show standard deviation. Black horizontal bars show pairwise significant differences between some text entry techniques

Repeated-measures ANOVA revealed that the effect of text entry technique on entry speed was statistically significant ($F_{1,23} = 67.705$, $p < .0001$), as was aid ($F_{2,46} = 41.098$, $p < .0001$) and block ($F_{2,46} = 107.446$, $p < .0001$). The interaction effects for text entry technique-block ($F_{2,46} = 5.338$, $p < .01$) and aid-block were also statistically significant ($F_{4,92} = 6.179$, $p < .0005$). Post Hoc Bonferroni-Dunn test at the $p < .05$ level results are represented as horizontal bars showing individual significant results for each text entry technique + aid combination.

Error rates ranged from 0.73% error for Controller Pointing + BigKey to 4.11% for FLIK + word disambiguation. Repeated-measures analysis of variance revealed that the effect of text entry technique on error rate was statistically significant ($F_{1,23} = 6.456$, $p < .05$). The effect of aid on error rate was statistically significant ($F_{2,46} = 23.412$, $p < .0001$). The text entry technique-block interaction effect was statistically significant ($F_{2,46} = 13.855$, $p < .0001$). The aid-block interaction effect was significant ($F_{4,92} = 12.067$, $p < .0001$). A post hoc Bonferroni-Dunn test at the $p < .05$ level showed individual significant results on error rate for each of the text entry technique + aid combination and are represented by horizontal bars on Fig. 12.

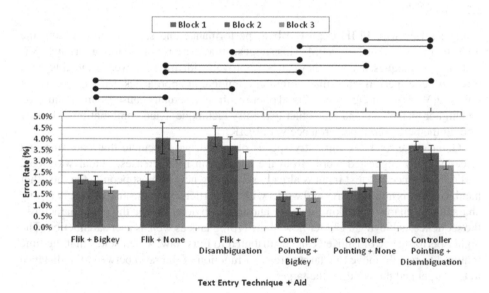

**Fig. 12.** Study 2 average error rate (%) comparison table. Black horizontal bars show pairwise significant differences between some text entry techniques.

**User Preference.** As seen in Fig. 13, FLIK + None and Controller Pointing + None scored similarly to corresponding conditions in study 1. Friedman post hoc pairwise comparison using Conover's F test results are represented by horizontal bars between each text entry technique.

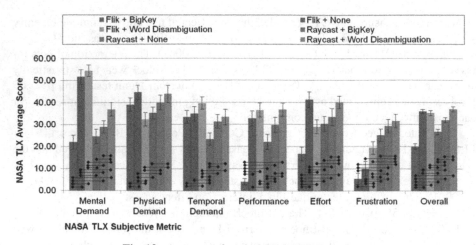

**Fig. 13.** Averages of study 2 NASA TLX results

## 5.7 Discussion

Using BigKey with FLIK improved both performance and satisfaction, yielding the highest entry speed at 27.9 WPM. This suggests that there is potential in alternative VR text entry techniques, and how they can be improved with different tweaks and aids, even offering better performance than common commercial approaches, such as controller pointing. Word disambiguation offered poor results and actually caused some frustration and higher mental demand than other text entry technique + aid combinations. BigKey offered the best overall performance with FLIK.

Given the error rates and entry speeds observed, we speculate that choosing and selecting fully suggested words from the suggestion list increases mental workload, particularly since VR text input is already something new to most people. On the other hand, BigKey is non-intrusive and works with the participant in the task of purely typing character by character, making it easier and more obvious to type the next character in the sequence, achieving a greater flow in the typing task as well as user satisfaction. The higher performance of BigKey can be attributed to Fitts' law, which states that the time required to rapidly move to a target area is a function of the ratio between the distance to the target and the width of the target.

## 6    Conclusion and Future Work

We developed a general-purpose text entry testbed focused on VR text entry experiments. Using the testbed, we conducted two comparative studies that evaluate different text entry techniques.

Our results for controller pointing and for continuous cursor selection are comparable to results provided by Speicher et al. [17]. While our scores for controller pointing were slightly higher than either Speicher et al. [17] or Boletsis and Kongsvik [3], our results reflect the same relative difference between controller pointing and continuous cursor selection. Boletsis and Kongsvik [2] evaluation of CutieKeys, which is the basis of our technique FLIK, reveals similar performance between the two techniques, and the same relative difference between these techniques and others. Table 1 summarizes these results for easy comparison.

**Table 1.** Text entry speed (WPM) comparison to similar studies.

|  | Our studies | Speicher [17] | Boletsis [2] | Grubert [6] | Knierim [7] |
|---|---|---|---|---|---|
| Controller pointing | 19.13 | 15.44 | 16.65 | – | – |
| Continuous cursor selection | 13.9 | 8.35 | 10.17 | – | – |
| FLIK | 21.49 | – | *21.01 | – | – |
| Physical keyboard | 45.64 | – | – | 38.7 | 40 |

*Used CutieKeys technique, which is the most similar to Flik.

Grubert et al. [6] reported an entry speed of 38.7 WPM text entry speed on their VideoHand technique, which is what our physical keyboard technique is based on. Table 1 again shows that these related papers report similar results on keyboard entry speeds in VR. These papers also report high standard deviation with keyboard-based input, likely due to including participants with varying typing proficiency, like our studies.

For future iterations of the testbed, a main focus is to improve extensibility, to require less programming ability to add new text-input techniques and/or keyboard layouts. This could use, for example, XML-based approaches to define new techniques/layouts.

Other future work could explore different kinds of text entry experiments using this software. From plain desktop text entry techniques to virtual or augmented reality techniques, as well as other types of human-computer interactions other than these such as wearable devices (smart watches, smart glasses…), brain-computer technologies such as EEG's like EMOTIV devices [3]. Conducting studies comparing different types of HMDs and controllers is also an important study to perform since this could potentially be a significant factor when it comes to text entry in VR.

**Acknowledgments.** Thanks to all participants for taking part in the experiments. Thanks to Aidan Kehoe of Logitech for providing the Logitech Bridge hardware. This work was supported by the Natural Sciences and Engineering Research Council of Canada.

# References

1. Arif, A.S., Stuerzlinger, W.: Analysis of text entry performance metrics. In: IEEE Toronto International Conference on Science and Technology for Humanity (TIC-STH), pp. 100–105 (2009). https://doi.org/10.1109/tic-sth.2009.5444533
2. Boletsis, C., Kongsvik, S.: Controller-based text-input techniques for virtual reality: an empirical comparison. Int. J. Virtual Real. **19**(3), 2–15 (2019). https://doi.org/10.20870/ijvr.2019.19.3.2917
3. EMOTIV: EMOTIV. https://www.emotiv.com/. Accessed 10 Feb 2020

4. Al Faraj, K., Mojahid, M., Vigouroux, N.: BigKey: a virtual keyboard for mobile devices. In: Jacko, J.A. (ed.) HCI 2009. LNCS, vol. 5612, pp. 3–10. Springer, Heidelberg (2009). https://doi.org/10.1007/978-3-642-02580-8_1

5. Grubert, J., Witzani, L., Ofek, E., Pahud, M., Kranz, M., Kristensson, P.O.: Text entry in immersive head-mounted display-based virtual reality using standard keyboards. In: Proceedings of IEEE Conference on Virtual Reality and 3D User Interfaces, pp. 159–166 (2018)

6. Grubert, J., Witzani, L., Ofek, E., Pahud, M., Kranz, M., Kristensson, P.O.: Effects of hand representations for typing in virtual reality. In: Proceedings of IEEE Conference on Virtual Reality and 3D User Interfaces, pp. 151–158 (2018)

7. Knierim, P., Schwind, V., Feit, A.M., Nieuwenhuizen, F., Henze, N.: Physical keyboards in virtual reality: analysis of typing performance and effects of avatar hands. In: Proceedings of ACM Conference on Human Factors in Computing Systems, pp. 1–9 (2018). https://doi.org/10.1145/3173574.3173919

8. Ko, A.J., Wobbrock, J.O.: Text entry. User Interface Software and Technology. https://faculty.washington.edu/ajko/books/uist/text-entry.html. Accessed 18 June 2020

9. MacKenzie, I.S.: Fitts' law. In: Norman, K.L., Kirakowski, J. (eds.) Handbook of Human-Computer Interaction, pp. 349–370. Wiley, Hoboken (2018). https://doi.org/10.1002/9781118976005

10. Mackenzie, I.S.: Evaluation of text entry techniques. In: MacKenzie, I.S., Tanaka-Ishii, K. (eds.) Text Entry Systems: Mobility, Accessibility, Universality, pp. 75–101 (2007). https://doi.org/10.1016/b978-012373591-1/50004-8

11. MacKenzie, I.S., Soukoreff, R.W.: Phrase sets for evaluating text entry techniques. In: Extended Abstracts on Human Factors in Computing Systems - CHI 2003, pp. 754–755 (2003). https://doi.org/10.1145/765891.765971

12. MacKenzie, I.S., Soukoreff, R.W.: A character-level error analysis technique for evaluating text entry methods. In: Proceedings of the Nordic Conference on Human-Computer Interaction - NordiCHI. 2002, p. 243 (2002). https://doi.org/10.1145/572020.572056

13. Sandnes, F.E., Aubert, A.: Bimanual text entry using game controllers: relying on users' spatial familiarity with QWERTY. Interact. Comput. 19(2), 140–150 (2007). https://doi.org/10.1016/j.intcom.2006.08.003

14. Soukoreff, R.W., MacKenzie, I.S.: Measuring errors in text entry tasks. In: Extended Abstracts on Human Factors in Computing Systems - CHI 2001, pp. 319–320 (2001). https://doi.org/10.1145/634067.634256

15. Soukoreff, R.W., MacKenzie, I.S.: Recent developments in text-entry error rate measurement. In: Extended Abstracts on Human Factors in Computing Systems - CHI 2004, pp. 1425–1428 (2004). https://doi.org/10.1145/985921.986081

16. Soukoreff, R.W., MacKenzie, I.S.: Metrics for text entry research - an evaluation of MSD and KSPC, and a new unified error metric. In: Proceedings of ACM Conference on Human Factors in Computing Systems - CHI 2003, pp. 113–120 (2003). https://doi.org/10.1145/642611.642632

17. Speicher, M., Feit, A.M., Ziegler, P., Krüger, A.: Selection-based text entry in virtual reality. In: Proceedings of ACM Conference on Human Factors in Computing Systems - CHI 2018, pp. 1–13 (2018). https://doi.org/10.1145/3173574.3174221

18. Tucker, V.: Introducing the logitech BRIDGE SDK (2017). https://blog.vive.com/us/2017/11/02/introducing-the-logitech-bridge-sdk. Accessed 18 June 2020

19. Weisel, M.: Cutie keys. https://github.com/NormalVR/CutieKeys/. Accessed 18 June 2020

20. Wobbrock, J.O.: Measures of text entry performance. In: MacKenzie, I.S., Tanaka-Ishii, K. (eds.) Text Entry Systems: Mobility, Accessibility, Universality, pp. 47–74 (2007)

# Appropriately Representing Military Tasks for Human-Machine Teaming Research

Chad C. Tossell[(✉)] ⬥, Boyoung Kim, Bianca Donadio, Ewart J. de Visser⬥, Ryan Holec, and Elizabeth Phillips⬥

Warfighter Effectiveness Research Center (DFBL), USAF Academy,
Colorado Springs, CO 80840, USA
Chad.tossell@usafa.edu

**Abstract.** The use of simulation has become a popular way to develop knowledge and skills in aviation, medicine, and several other domains. Given the promise of human-robot teaming in many of these same contexts, the amount of research in human-autonomy teaming has increased over the last decade. The United States Air Force Academy (USAFA), for example, has developed several testbeds to explore human-autonomy teaming in and out of the laboratory. Fidelity requirements have been carefully established in order to assess important factors in line with the goals of the research. This paper describes how appropriate fidelity is established across a range of human-autonomy research objectives. We provide descriptions of testbeds ranging from robots in the laboratory to higher-fidelity flight simulations and real-world driving. We conclude with a description and guideline for selecting appropriate levels of fidelity given a research objective in human-machine teaming research.

**Keywords:** Human-machine teaming · Autonomy · Robots · Simulation · Fidelity

## 1 Introduction

Machines already play an integral role in defense. Across military services, operators team with machines to perform important tasks such as diffusing explosive ordinance or safely flying an aircraft. As automated technologies have advanced, so has the military's reliance on them. For example, remotely piloted aircraft (RPAs) extend the reach of humans and have kept numerous pilots out of harm's way. Newer automated systems are being fielded at higher levels of automation [1]. Instead of relying on pilots to shift to autopilot, the Auto Ground Collision Avoidance System (Auto-GCAS) on many fighter aircraft take control of the aircraft based on its own calculations. If the system detects the distance to ground and trajectory of the aircraft are unsafe, it will take control of the aircraft from the pilot and fly to safer airspace. Auto-GCAS has already been credited with saving seven lives and is a good example of how automation can team effectively with humans [2].

This is a U.S. government work and not under copyright protection
in the U.S.; foreign copyright protection may apply 2020
C. Stephanidis et al. (Eds.): HCII 2020, LNCS 12428, pp. 245–265, 2020.
https://doi.org/10.1007/978-3-030-59990-4_19

Autonomous systems have also been utilized in defense since World War II [3]. Autonomous systems differ from automated systems because autonomous systems can independently determine courses of action based on their knowledge of itself and the environment [4, 5] whereas automated systems are more restricted to execution of a set of scripted pre-determined sets of actions. Autonomous systems are expected to use this knowledge to achieve goals in situations that are not pre-programmed. Simply because they can operate independently in these unanticipated environments does not mean they will operate independent of humans. Indeed, most concepts of operations for military autonomous systems have these systems teamed with humans across warfighting and peacetime contexts. AI-based systems are already used in drone swarms to learn patterns based on their observation. Other forecasted applications of AI that are currently in development include search and rescue techniques and exoskeleton suits [6]. AI technologies will continue to penetrate battlefields to help human operators perceive complex battlespaces, fight effectively, and stay safer across military domains. One challenge to this end is to enable human-autonomy interactions that are effective and natural across a wide range of military tasks. Studies to improve trust, shared situational awareness, social norms, and collaboration in human-autonomy systems are actively being conducted to facilitate these interactions [7]. The United States Air Force (USAF), in particular, has provided a vision for humans teaming with autonomy [8]:

*In this vision of the future, autonomous systems will be designed to serve as a part of a collaborative team with airmen. Flexible autonomy will allow the control of tasks, functions, sub-systems, and even entire vehicles to pass back and forth over time between the airman and the autonomous system, as needed to succeed under changing circumstances. Many functions will be supported at varying levels of autonomy, from fully manual, to recommendations for decision aiding, to human-on-the-loop supervisory control of an autonomous system, to one that operates fully autonomously with no human intervention at all. The airman will be able to make informed choices about where and when to invoke autonomy based on considerations of trust, the ability to verify its operations, the level of risk and risk mitigation available for a particular operation, the operational need for the autonomy, and the degree to which the system supports the needed partnership with the airman. In certain limited cases the system may allow the autonomy to take over automatically from the airman, when timelines are very short for example, or when loss of lives are imminent. However, human decision making for the exercise of force with weapon systems is a fundamental requirement, in keeping with Department of Defense directives.*

This paper describes the approach being used to help develop these capabilities within the Warfighter Effectiveness Research Center (WERC) at the United States Air Force Academy (USAFA). Realistic simulations have been developed to explore trust, workload, human-robot interaction (HRI), social norms, and performance across a wide range of military tasks. Importantly, the fidelity requirements of the intelligent agents, tasks, and environments have been carefully designed according to research goals and applications to future battlefields.

## 2   Fidelity in Human-Machine Teaming Research

Cadets studying at USAFA are involved in this research as part of the research team and as participants for experiments. As part of the research team, select firstie (i.e., senior) cadets majoring in human factors (HF) engineering and behavioral sciences learn important concepts in HF (e.g., trust, workload, and situation awareness [9, 10]) by helping to design experiments in HMT. Another learning goal is to afford opportunities for these cadets to critically think about how intelligent systems might be involved in future warfighting. Cadets learn quickly that the effective integration of these systems is important for military tasks with life-or-death consequences. Thus, in determining fidelity requirements for HMT studies, we oftentimes aim for higher degrees of realism.

To achieve this goal, different levels of fidelity have been used in research environments to represent a variety of Air Force tasks. Fidelity has been defined as the degree to which a simulation replicates reality [11]. When a simulation more closely mimics the real world, it is higher fidelity. Simulations lower in fidelity are more artificial without as many matching elements in the real world. In medical and aviation training, high-fidelity simulations have included full-body manikins programmed to provide realistic physiological responses to care and 360°, full-motion flight simulators to prepare pilots for live flight. Low-fidelity trainers in the same domains include patient vignettes read from a sheet of paper to test medical students and chair flying [12, 13]. Fidelity in these contexts have generally referred to elements of the simulation environment (e.g., graphics, haptics, etc.) and labeled *physical fidelity* [14].

Fidelity has also been characterized beyond simply the physical features of the environment. To create immersive experiences in gaming environments and realistic behaviors in psychological experiments, fidelity has been considered based on human elements. *Conceptual fidelity* measures the degree to which the narrative/scenario elements in a simulation are connected and make sense to humans in the loop. Similarly, *psychological/emotional fidelity* is the extent to which the task mimics the real-world task to provide a sense of realism [12]. *Cognitive fidelity* has also been used to measure the level of human engagement with simulations [15]. Fidelity has thus been more broadly defined to capture the extent to which the environments and other elements of the simulation come together to elicit the intended emotional, cognitive, and behavioral responses from humans. Even simulations low in physical fidelity can create visceral human reactions and realistic responses to stimuli (e.g., crying in response to reading sad vignettes on a piece of paper [16]). Games can provide humans a very immersive experience when the narrative, gameplay, and graphics are combined in consistent ways and not based on synchronization or the level of physical fidelity [17–24].

Experiments in HMT, like most psychology experiments, are (among other things) designed to examine specific questions of interest by measuring effects through partitioning the sources of variability. Studies using artefacts that are low in physical fidelity have been able to examine antecedents and consequences of mind perception toward robots. Tanibe and colleagues read vignettes of robots sustaining damage in a kitchen to participants before obtaining perceptions of mind via questionnaires [27]. High-fidelity simulations have also been used in HMT research. In an elegantly designed study of mistrust, Alan Wagner had participants work in a building where a simulated fire with smoke started. Participants were then instructed to follow a robot out of the building.

Even though the robot navigated incorrectly to a room with no exit (by design), participants still frequently followed the robot. Participants seemed to experience real risk and relied on the robot despite it making an obvious error. Studies across the fidelity spectrum have attempted to maximize internal, external, and/or ecological validity to test research hypotheses and improve the precision of the results.

In HMT studies at USAFA, we have set up controlled, yet realistic, environments to maximize internal, external, and ecological validity. Cadets, many of whom will be users of highly automated and autonomous systems, are the participants in these studies. The simulations used in our experimental settings were developed to maximize the human-centered types of fidelity at low cost. We have used well established methods such as Wizard of Oz (WoZ) for both high-fidelity and low-fidelity simulations to try and mimic future functions of machine agents [28, 29]. Additionally, we have developed novel tasks in higher-fidelity settings to study trust in autonomous tools (e.g., a Tesla). Across these experimental settings, cadet researchers work with military operators and faculty members to study HMT in future warfighting scenarios [25]. We describe a subset of these settings along with examples of studies below (Table 1).

**Table 1.** Research testbeds described in this paper

| Testbed | Simulation technology | Ecological target | Example studies |
|---|---|---|---|
| A. Autonomous Flight Teaming (AFT) | F-35 Flight Sim (Prepar3d, COTS) | Flying an F-35 with 4 autonomous F-16s | Workload on Trust, SA in Multitasking in Air Combat [30] |
| B. Human Automation Research in a Tesla (HART) | Tesla Model X | Trust in auto systems in risky tasks | Trust in Risky Autopark [31–33] |
| C. Gaming Research Laboratory | Games (e.g., Overcooked) | AI agent teaming in interdependent tasks | Teaming with AI vs Human on Task Performance [26] |
| D. Social Robotics Laboratory | Robot APIs (Pepper, NAO, Aibo) | Robotic teammates using natural language | Social norms with robots [34, 35] |

# 3   Human-Machine Teaming Research Settings at USAFA

## 3.1   Testbed A: Autonomous Flight Teaming (AFT)

As mentioned above, one vision for future flight operations is for autonomous aircraft to seamlessly integrate with human F-35 pilots [8]. To explore factors and evaluate designs to facilitate this integration, we adapted a flight simulator to allow human pilots to fly with virtual autonomous F-16s (Fig. 1). The AFT consists of three features that must work together to provide an environment for human participants to operate their simulated aircraft in a team of autonomous systems: Hardware/software, flight scenarios, and measurement systems.

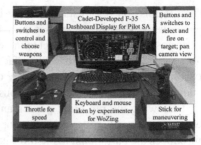

**Fig. 1.** AFT displays and controls

### 3.1.1 Hardware and Software for Visualization and Control

As shown in Fig. 1, the AFT was designed as a high-fidelity simulator to mimic future autonomous flight teaming. Participants fly in the simulator using the F-35 Hand on Throttle and Stick (HOTAS), which is similar to the stick and throttle used in the actual F-35 Lightning II. The displays found in F-35 simulators used in USAF training are overlaid with new interfaces and models designed by cadets with input from subject matter experts (SMEs) and faculty at USAFA.

The flight simulator integrates three, 72-in. monitors into the testing environment with an additional visual display for prototyping an F-35 pilot's dashboard display. The integration of the four monitors into the testing environment, each with their own information stream, allows for more robust cognitive fidelity in the experiment that reflects the heavy workload and strained SA of actual pilots. A VR system can also be integrated into the AFT to further immerse the participant in the scenario or to study Wizard of Oz (WoZ) teaming scenarios.

Other artifacts are also integrated on a case-by-case basis. For example, pilots in live flight are continually checking their kneeboard and monitoring the gauges, radar screens, and the Heads-Up Display (HUD). The SMEs helped develop a "cheat sheet" that is used to mimic a pilot's kneeboard. This kneeboard includes the names of the targets at each SAM site, how to make a radio call, the call-signs of the autonomous F-16s, and the scenario-dependent flight parameters.

### 3.1.2 Scenario Development

Like gaming systems, our scenario development focused on blending the narrative, graphical elements, and physics of the simulation to create an immersive experience for participants. Our goal is in these scenarios is to replicate future operations in autonomous flight to a level where participants are highly engaged and motivated to succeed with their virtual autonomous F-16s. The general sequence is outlined here:

1. After pre-brief, take off from airfield and identify the SAM sites
2. Conduct an orientation flight to learn basic flight skills
3. Determine how to destroy SAM sites: self and/or autonomous F-16s
4. Neutralize targets and be on lookout for other threats
5. Have F-16s form up on F-35 to complete mission

The pre-mission brief establishes the importance of the mission and how participants can do well. Most of our studies do not offer real incentives for performing well in our scenarios. However, we attempt to increase motivation through artificial incentives (e.g., "to succeed in this mission, all surface-to-air missiles must be destroyed without any losses to your flight team"). We rely on cadets' competitiveness in these activities to study ways they trust their autonomous wingmen.

Following the pre-brief, participants fly in three different scenarios; first, a familiarization scenario, followed by two operational scenarios. The familiarization scenario introduces the participant to the information streams of the four screens and guides them through using each of the flight controls. The participants also practice making radio calls and learn how to engage their autonomous F-16s based on the goals of the study. For example, one study assessed three different ways to communicate with the autonomous wingmen using supervisory control methods and a "play calling" technique [10].

With input from SMEs (i.e., experienced Air Force pilots), we have developed a range of scenarios at different levels of difficulty and workload in order to assess different ways participants trust, communicate, and team with autonomous wingmen. One scenario requires participants to attack enemy sites that each contain many surface-to-air missiles (SAMs). The participant leads all three of the autonomous F-16s in this mission. The participant can engage each of the targets individually (and likely fail) or rely on the autonomous wingmen to assist. New displays (e.g., Fig. 2) and methods to communicate (e.g., voice versus supervisory control on a gaming controller). Workload and difficulty levels are increased systematically by introducing air threats and/or increasing radio traffic. When air threats are introduced, participants must multitask with their autonomous wingmen to neutralize the air threat in addition to targeting ground threats with a limited number of missiles.

**Fig. 2.** Visual display that shows statuses of F-16s.

### 3.1.3 Measurement

We have examined questions involving workload, trust, and performance in teaming with autonomous systems via passive means using physiological sensors, telemetric performance measures, and experimenter/SME observations (Table 2). To avoid disrupting the scenario and influencing the immersive experience, the scenarios are not interrupted for measurements. The Tobii Pro Glasses system is worn like any other pair of glasses, and collects data on eye fixation, saccade patterns, and pupil dilation. This eye-tracking system provides the research team with clear behavioral metrics of the participant's attention, which research suggests can be linked to trust [9].

Additionally, overall team performance is assessed based on mission success (e.g., number of enemy targets destroyed, number of hits received, etc.). The built-in telemetry system allows for real-time recording of flight data (e.g., missiles fired, successful hits, location of wingmen, etc.), which augments behavioral observations and allows for quantifiable performance metrics (including their adherence to the flight parameters and time on target) to be analyzed post-experiment. Depending on the study, other measures can be collected (Table 2).

**Table 2.** Examples of data collected in the AFT.

| Type | Construct | Metric |
| --- | --- | --- |
| Behavioral | Performance, Attention | Telemetry, Eye-Tracking |
| Physiological | Cognitive Workload, Stress | EEG, ECG & GSR |
| Subjective | Workload, SA, Trust/Self-Confidence | TLX, SART, Post-Q |
| Observed | Performance | Mission Success Rate (# of Targets Destroyed) |

### 3.1.4 Testbed Validation

Validation of the AFT has been conducted in proof-of-concept studies [30]. In a recent unpublished study, eight cadets engaged in risky behaviors because of the lack of real-world consequences. They relied less on their teammates and more on themselves and explicitly acknowledged flying too low or too high, aggressively going after SAMs, and performing advanced maneuvers during the mission that were not necessary. Thus, even though our simulation was higher in physical fidelity than other laboratory-based multi-tasking tasks (e.g., MAT-B), participants still recognized it was an artificial environment and their behaviors were reported as unrealistic.

The risky behaviors were minimized in follow-on (also unpublished) studies using higher-fidelity scenarios. Adding tasks such as keeping an altitude, not exceeding airspeed levels, and neutralizing air threats reduced the frequency of observed risky behaviors. Our current study (on hold due to COVID-19) is evaluating trust and performance as a function of workload and experience level. Across scenarios and experience levels, we are seeing variance in reliance on the autonomous wingmen and overall performance. Our measures are sensitive to the changing dynamics of the scenario and trust levels. For example, at the individual level, EEG has been captured and correlated to task load and trust levels (Fig. 3). All measures are time-synchronized and correlated for a more complete understanding of stress, workload, and SA. Even though there are no real consequences for mission failure, our goal of creating an environment to study future teaming concepts, trust, SA, and other phenomena in USAF tasks has provided an important look into HMT beyond more basic laboratory tasks.

### 3.2 Testbed B: Human Automation Research in a Tesla (HART)

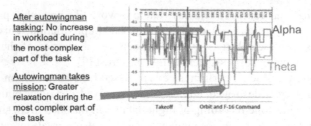

After autowingman tasking: No increase in workload during the most complex part of the task

Autowingman takes mission: Greater relaxation during the most complex part of the task

**Fig. 3.** Alpha and theta waves recorded from a participant during takeoff and initial task allocations with autonomous wingmen.

Given the lack of real consequences in testbeds such as AFT, the HART testbed has allowed us to evaluate trust in real-world and potentially risky environments. While driving is obviously a different task than flying, there are similarities in trusting autonomous systems in both environments. The goal for HMTs is to engender calibrated trust where the expected performance of automation matches the actual performance

**Fig. 4.** The Tesla Model X - Air Force Version

of automation [36, 37]. Research has shown that humans may have a propensity to over-trust robots in realistic emergency scenarios [38]. This over-trust could lead an individual to underestimate the risk associated with using an intelligent agent or even foster misplaced reliance on technological teammates [39]. While there has been recent work attempting to develop an "adaptive trust calibration" system, which would help with issues of over-trust and under-trust [40], such a system has not been used and tested on a variety of technological agents in high risk environments with a focus on high physical and cognitive fidelity.

To address the need for assessing trust calibration in high risk environments, the WERC has established a mobile research laboratory known as HART (Human-Automation Research in a Tesla) mobile lab. This mobile lab environment is set up in a 2017 Tesla Model X (Fig. 4), equipped with various automated features which include lane-following, adaptive cruise control (ACC), and automated parking. Within the HART mobile lab are five distinct pieces of technology for data recording that do not impede the participant's experience, therefore maintaining psychological fidelity during the task. Unifying all this technology is its mobility, which allows researchers to collect a multitude of data in the most ecologically valid way in a dynamic, naturalistic environment. Like the AFT, we can collect data throughout the study on eye fixations via eye-tracking, stress fluctuations via GSR, and heart rate readings via ECG with the use of the Tobii Pro Glasses system and NeuroTechnology's BioRadio. The third piece of technology used in the HART is the Advanced Brain Monitoring B-Alert X24 Mobile electroencephalography (EEG). This EEG will allow us to measure workload and attention [41]. Additionally, cameras mounted inside the car will capture the interior and exterior environment but also the participant's face to analyze their emotional and cognitive states based on Ekman and Friesen [42] action unit measurements of facial muscle movement. The final piece of technology being used is a RaceCapture telemetry system, which will allow for the real-time recording of vehicle data (i.e. acceleration, braking, and steering).

### 3.2.1 Testbed Validation

We have conducted a series of studies to examine how trust in real autonomous systems develops [31–33]. An initial study evaluated driver intervention behaviors during an autonomous parking task (Fig. 5). While recent research has explored the use of autonomous features in self-driving cars, none has focused on autonomous parking. Recent incidents and research have demonstrated that drivers sometimes use autonomous features in unexpected and harmful ways.

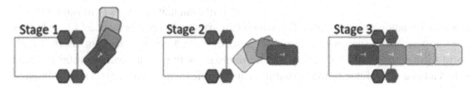

**Fig. 5.** The three distinct stages of the Tesla's autoparking feature. The Tesla is represented by the solid blue rectangles with the "T" on them. Time is represented by those rectangles transitioning from a lighter to a darker shade of blue. The trash cans are represented by the hexagons. (Color figure online)

Participants completed a series of autonomous parking trials with a Tesla Model X and their behavioral interventions were recorded. Participants also completed a risk-taking behavior test and a post-experiment questionnaire which contained, amongst other measures, questions about trust in the system, likelihood of using the autopark feature, and preference for either the autonomous parking feature or self-parking. Initial intervention rates were over 50%, but declined steeply in later trials (Fig. 6). Responses to open-ended questions revealed that once participants understood what the system was doing, they were much more likely to trust it. Trust in the autonomous parking feature was predicted by a model including risk-taking behaviors, self-confidence, self-reported number of errors committed by the Tesla and the proportion of trials in which the driver intervened. Using autonomy with little knowledge of its workings can lead to high degree of initial distrust and disuse. Repeated exposure of autonomous features to drivers can greatly increase their use. Simple tutorials and brief explanations of the workings of autonomous features may greatly improve trust in the system when drivers are first introduced to autonomous systems.

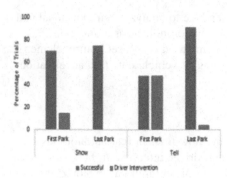

**Fig. 6.** Percentage of trials for which the autoparking was engaged, where the car was able to successfully and fully park itself, and where the driver intervened, split by first and last park, as well as condition.

In a follow-on study, we compared driver intervention rates when either showing the autoparking capabilities to drivers or merely telling them about the features without demonstration. The study showed that the intervention rates when showing the parking features drops significantly compared to when drivers are merely told about the autopark capabilities.

### 3.3 Testbed C: Gaming Research Laboratory at USAFA

#### 3.3.1 AI Teaming Research

One issue that has emerged in HMT research is that human team players communicate less overall with autonomous teammates, which can affect team performance [43, 44]. This may be part of the reason that few AI agents exist that work with humans in a real-world team setting with the ability to communicate with human team members. There is thus a need to evaluate human communication styles with autonomous agents in team environments.

#### 3.3.2 Gaming as a Reasonable and Fun Way to Approximate Teamwork

To study human-AI communication, we have established a video game laboratory to leverage the immersive experiences games afford to many people. Within this game laboratory, we developed a testbed called Cooking with Humans and Autonomy in Overcooked! 2 for studying Performance and Teaming (CHAOPT) [26]. Overcooked 2 is particularly immersive. The game requires coordinated teamwork and good communication to be successful (e.g., earn more points, advance to higher levels, etc.). The game uniquely, cleverly and dynamically manipulates the environment which forces flexible allocation of roles and information sharing (Fig. 7). We have added Wizard of Oz capabilities to observe how human behavior changes when a person believes they are working with an AI agent. This WoZ capability has been valuable because we can examine how humans react differently based on their teammate, with an emphasis on how they communicate, how they evaluate their teammate, and how much they trust their teammates.

#### 3.3.3 Overcooked 2 Game Mechanics

Overcooked 2 is a video game that is teamwork intensive and requires communication to succeed at a high level. Players are given food orders to complete and are required to navigate the kitchen environment, prepare the orders, and deal with distracting features. Once orders are submitted, players are awarded points and tips based on the order correctness and priority. If orders are not fulfilled in time, that order goes away and players lose points and bonuses. Points, bonuses, completion time, and other game-based scores provide handy performance measures for studies. Players have tasks to

 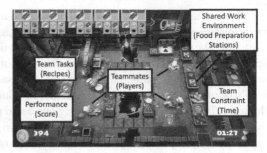

**Fig. 7.** Two cadets playing the Overcooked 2 video game in the Gaming Research Lab along with the game display to show how in-game tasks map to higher-level teaming concepts.

perform such as: collect the required order ingredients, chop food, clean dishes, and cook food to complete orders.

Overcooked stimulates communication through various tasks and levels of difficulty. Different worlds and levels introduce environments where a single player cannot complete the level on their own, which is where the communication aspect of this game is extremely vital. The beginning levels do not require as much communication as the player is still being introduced to new game concepts, but around World 2 Level 1, communication becomes more important to the team's success. The kitchen maps become restrictive and the supply locations become limited to all players. Communication becomes required to fulfill orders and complete the level. The methods of communication that are being measured in this pilot test are push versus pull communication and the amount of communication used. All communication is verbal for this experiment. Push communication refers to one player telling the other player what they want or need. For example, one player that needs cheese to complete an order may tell the other player, "Chop me cheese" or "Throw me cheese". Pull communication refers to one player asking for another player to complete a task. For example, the player may say "Can you throw me cheese?" or "Can you chop cheese for this order?" The communication method may indicate the level of trust the participant has in the human or autonomous agent or which confederate they would prefer to play with.

### 3.4 Testbed D: Social Robotics Laboratory

In future battlefields, some AI systems will likely be embodied in physical robots and not simply exist in virtual environments. Thus, human-robot interaction (HRI) environments must be developed to explore their coordination with humans. Like above, we have developed a testbed to mimic future environments to explore robots in authority, teaming alongside humans, as a moral advisor, and a teammate in a stressful set of tasks.

#### 3.4.1 Warrior-Robot Relations

Using a commercially-available Pepper robot, we created a task to explore empathy towards robots in a human-robot team task under stress. In the transition of robots and other artificial agents from *tools* to *teammates* [45], important questions are raised by forming teammate like relationships with non-human agents, especially in battlefield environments. For instance, in military contexts, there have been powerful stories of military members disusing robots as a result of feeling too much empathy towards them. In 2007, an Army Colonel deeply empathized with an improvised explosive device (IED) detecting robot. The robot was designed to use its many articulating legs to purposefully detonate IEDs, and as a result sacrifice itself in the detonation process. The Army Colonel stopped an exercise demonstrating the robot sacrificing itself to detonate the IEDs, stating that it would be inhumane to continue the exercise [46]. Such inappropriate attributions of empathy towards robots and other artificial agents could prevent humans from making full use of the benefits of autonomous systems deployed in dangerous environments. If a human feels too much empathy, they may sacrifice elements of the mission or task to prevent harm to an autonomous teammate, which could jeopardize the mission or the team.

However, the ability to form empathetic bonds towards others is a major point of emphasis for the training of U.S. military officers especially regarding facilitating respect for the human dignity of others. Empathy may be needed to truly think of and rely upon artificial agents that are specifically designed to be teammates for humans. If a human does not feel enough empathy, the operator may not fully utilize the agent as a teammate or use it in a manner that is unintended or unsustainable. Thus, a balance between too little and too much empathy towards a robot will likely be needed to facilitate good teaming between humans and robots in military contexts. Initial evidence supporting the need for such a balance was provided by a set of studies, using the TEAMMATE simulation, that examined empathy for a robot in the context of a space mining mission [47, 48]. The studies demonstrated that human teammates responded more often and more quickly to a robot request for help when it was portrayed as a helpful teammate companion and when it appeared more damaged.

Additionally, military teams are normally required to operate under high levels of stress and stress is a common way comradery is built between members of the team. Inducing stress can help make people feel stronger perceived understanding of others' emotions and feelings [49]. As a result, interacting with a robot teammate under high stress could cause people to feel more empathetic towards a robot team member than when interacting under low stress.

To induce empathy and stress in a human-robot team task we conducted a study with cadets at USAFA. The task was to interact with Pepper, a humanoid robot developed by Softbank Robotics, as a teammate in an intrinsically incentivized spelling bee game. Participants were tasked with working with Pepper to spell increasingly difficult English words taken from the National Adult Spelling Bee Practice vocabulary list [50]. Participants were responsible for spelling 2/3 of the words, while Pepper was responsible for spelling 1/3 of the words. If participants spelled a word incorrectly, the Pepper would lose 1/8 of its simulated battery health/life. And, Pepper's simulated battery health and system performance would continue to degrade as the participant misspelled words until Pepper ran out of health. Pepper's speech was also slowed as a result of its diminished health. However, the participant could stop the study at any time to prevent Pepper from losing health. Upon doing so, the participant's spelling score would become final. If Pepper's simulated health was completely diminished, participants were told that Pepper lost all memory of the participant and was no longer functional.

As the participant interacted with Pepper, their poor spelling performance harmed Pepper by reducing Pepper's simulated battery life and health by 1/8. Participants' willingness to continue was perceived as less empathy shown towards Pepper. If/when the participant stopped the spelling task to preserve Pepper, it was recorded as an objective measure of empathy shown toward the robot. To induce stress, half of the participants completed the spelling task under time pressure.

Our results suggested an interesting interaction between participant gender and stress on empathy scores, where males and females showed a differential pattern of empathetic behavior toward the Pepper robot under different stress conditions [35]. The data trended such that males were initially more empathetic towards robot partners in unstressed conditions than females but were far more likely to act less empathetic toward robot partners in stressed conditions. Whereas females were far more likely to act more empathetic toward robot partners in stressed conditions. This finding could inform changes in the design of robots intended for stressful scenarios and when working in teams with people. For example, an engineer could design less human-like responses for women operators and more human-like responses for male operators based on the stressful dynamics of the task. However, it is important when designing human-like responses to adhere to intuitive principles of human-likeness [51] and avoid designs that are perceived as uncanny [52]. This work presents some early findings in understanding the nuanced roles that empathy and stress play in teaming with robotic and other artificial agents. This understanding will be important in creating effective human-agent teams for successful deployment in several Air Force contexts.

### 3.4.2   Robots as Moral Advisors

**Fig. 8.** Pepper robot used in the social robotics task

The role of an artificial moral advisor would be to assist people in making decisions that comply with moral standards and values [53–58] and perhaps even serve as a 'cognitive wingman' [59]. Previous research on how people make moral judgments and decisions showed that people in identical moral dilemmas may arrive at diverging decisions depending on various psychological factors, such as gender, time pressure, cognitive load, and language [60–63]. This inconsistency in people's decisions may become a critical problem, especially in contexts where their decisions could bring about significant and irrevocable consequences. One such example would be a military context such as the Air Force, where there is a high demand for morally-laden decisions [64]. To illustrate, imagine that the following ethical dilemma takes place in a military operation: An officer faces a decision of whether to sacrifice one person to save many lives or to take no action and lose many lives. Previous researchers found that, in this situation, the time it took for people to reach the same utilitarian decision of sacrificing one life to save many was approximately 5.8 s under no cognitive load but increased to 6.5 s under cognitive load [61]. This implies that in the military context, the officer's decision may cause strikingly different consequences depending on their cognitive resources available at the time. Given this volatile tendency of people's moral decision-making process, it would be useful to develop an artificial intelligence system that may guide human teammates to follow a systematic and well-informed decision-making process before reaching a moral decision. However, it is critical that such a system has a degree of moral competence [59, 65].

As a first step towards building an artificial moral advisor, we have launched an investigation on how robots can effectively communicate a message that encourages people to make morally right choices. Recently, it was found that a robot's responses to people's request can result in changes in people's judgments of whether a certain behavior is morally permissible or not [66]. If people's moral judgments could be shaped by a robot's response, would people also be receptive to a robot's unsolicited advice on what is morally right or wrong choice in human-robot teams? Drawing from Confucian role-ethics [67], we predict that a piece of moral advice from a robot may influence people's behavior to a varying degree depending on how people relate themselves with the robot. Whereas, in a human-robot team, a human may readily acquire a status of a partner, teammate, or colleague, a robot may not easily be granted with such a status [68]. Therefore, we predict that people would be more willing to follow a robot's moral advice when they perceive the robot as their teammate compared to when they do not perceive the robot as their teammate.

To test this idea, we will program commercial-off-the-shelf (COTS) robots, like pepper (Fig. 8), to interact with human participants. These participants are asked to do a tedious task that will ultimately benefit their team performance, but can stop at any time. In this situation, a moral behavior would be to complete the task. Whenever participants express their intention to stop continuing the task, a robot gives them a response that emphasizes the importance of being a good teammate. We predict that the effect of highlighting the importance of being a good teammate would have a more positive effect on the likelihood of participants' completing the task when they perceived the robot as their teammate compared to when they do not.

Our proposal to seek assistance from an artificial intelligence system in making moral decisions may raise many ethical concerns. Decisions about ethical dilemmas where no absolute answer is available may appear to be outside the purview of any kind of artificial intelligence systems. Perhaps, the value of an artificial moral advisor in human-machine teams may be most evident in situations where morally right or wrong actions are stipulated (e.g., "Do not cheat on tests"). However, even in moral gray areas, we expect that an artificial moral advisor can be useful at various levels of fidelity. First, an artificial moral advisor can quickly and accurately gather and convey to humans information relevant to decisions at hand so that humans can make informed decisions. Next, humans can verify whether their personal moral norms and values are consistent with their team's or the general public's by checking in with an artificial moral advisor. Third, humans can rely on an artificial moral advisor to facilitate clear discussions about moral choices with other human teammates. Finally, an artificial moral advisor can optimize persuasive strategies for encouraging humans to adhere to moral norms by accurately assessing team characteristics.

# 4  Discussion

There are many ways to define fidelity requirements for simulations used in HMT studies. Across the testbeds described above, a wide range of established and novel methods have been used to create presence and appropriate fidelity in research studies. These environments have elicited behaviors that appear natural and appropriate for the task. However, the results of these studies are intended to generalize to high stress environments where real lives are at stake. The primary question is whether we have achieved the appropriate level of fidelity that allows us to make this generalization. In applied studies when technologies are being tested, designing appropriate levels of fidelity to elicit stress and interaction become even more important. For example: How can researchers stimulate the same levels of stress as when a real air threat is detected in an F-35? We discuss these and other considerations in the next sections.

## 4.1  The Need for Appropriate Fidelity in Military Training and Research

The highest level of fidelity a researcher can have is to conduct research in actual operational and naturalistic environments. The use of systems in naturalistic environments can provide helpful data to inform lab-based studies. For example, effectiveness of early human-robot teamwork was demonstrated by operational robot deployment in the 9/11 World Trade Center attack and Fukushima Daiichi disasters [69]. Others have demonstrated the utility of interviewing pilots to discover how they learn their systems in situ. This information can be used in the design and develop automated systems such as the Auto-GCAS [2, 70].

However, conducting research in naturalistic settings poses its own set of challenges. Operational personnel are not always available for research, the occurrence of disasters cannot be predicted in advance, and naturalistic settings and conditions cannot usually not be fully controlled. This reality necessities the use of simulations. By developing high fidelity simulations in military contexts, the gap between developers and users can be bridged prior to production or fielding of new technology. For human-autonomy teaming research there are several unique considerations of fidelity requirements in the creation of such testbeds.

### 4.1.1  Simulating Stress, Risk and Urgency to Study Trust in HMT

The first consideration is the accurate simulation of stress, risk, and urgency. Military members rely on basic and advanced technologies in warfare. Establishing artificial time limits, promoting competition, and using the WoZ techniques may not be enough to create an appropriate sense of urgency. Additionally, IRB regulations require researchers to not put participants at more risk than they are exposed to daily. Researchers have addressed this issue by creating more realistic but still controllable environments. For example, trust in a robot was studied in a building simulated to be on fire with real alarms and simulated smoke [38, 39]. Our studies have used an actual Tesla vehicle with a realistic parking situation [31–33]. All these studies have seemed to elicit genuine behavior from participants making it more likely that constructs such as trust are evaluated more appropriately. However, it is unclear how these approximations differ from real life-or-death domains such as in medicine, aviation, and military operations. It is also unclear how different humans behave in these environments compared to laboratory studies. Further research into HMT fidelity requirements should compare different levels of fidelity to answer this question.

### 4.1.2   Using Wizard of Oz to Simulate Future HMT Capabilities

One difficult challenge in HMT research is the ability to simulate future HMT capabilities such as a fully autonomous agent that can communicate, coordinate and work seamlessly in an HMT. Currently, such an agent does not exist. Therefore, the WoZ method is a useful method to approximate this future capability. Participants usually participate in HRI/HCI studies tabula rasa and believe they are interacting with technology and not a human. Thus, we have found WoZ [28] to be a high-fidelity method to uncover differences in how participants interact with technologies versus humans. Such methods are used in the HRI field to uncover important psychological and human performance issues that might arise assuming the technology is realized. An alternative to this method is the "Oz-of-Wizard" in which the human behavior is simulated or assumed to test how a robot will respond [29]. Eventually, WoZ may become less important in research if autonomous agents become more readily available and configurable. For example, recent work in our laboratory has investigated bonding and trust with Sony's Aibo, a robotic dog that demonstrates a variety of autonomous dog-like behaviors [71, 72].

### 4.2   The Trade-off Space for Fidelity Requirements

Fidelity is multi-faceted and certain tradeoffs should be made while maintaining the integrity of the training exercise, task assessment, or research study, depending on the core reason for using a simulation environment.

The goals an organization has for using a simulator (whether it be as a testbed for training, assessment, or a means of studying trust and teamwork) determine the fidelity requirements for that simulation environment. Therefore, a crucial step in implementing simulation practices is clearly identifying the needs or research focus of that initiative. For example, to get the most realistic results for measuring risk taking and trust in automated tasks, we were able to use the HART. Meanwhile, to understand how pilots may behave with systems that have autonomous capabilities which do not yet exist, in scenarios that are impossible to test in a naturalistic environment, we use the AFT, prioritizing physical and conceptual fidelity over psychological fidelity. Whereas, when testing teamwork and moral decision making, we value cognitive or conceptual fidelity above physical fidelity.

In summary, a brief guideline to assist in selecting the appropriate level of fidelity might be as follows:

1) Clearly identify the research objective for use of simulation
2) Based on the identified research objective, appropriately prioritize level and type of fidelity to accomplish stated goal
3) Both the user and producer should be involved in development process
4) High fidelity does not necessarily mean greater training performance benefits
5) WoZ and other techniques can maintain fidelity while increasing the breadth and depth of HMT testing scenarios, especially for capabilities which are still in development
6) Trust and workload are two important measurements in understanding how HMT can most effectively be implemented in different contexts.

## 5 Conclusion

We have demonstrated through a variety of testbeds the utility of presenting varying levels of appropriate levels of fidelity in human-machine teaming research. Technology available today will change rapidly and levels of fidelity will likely improve accordingly. It will therefore remain a constant challenge to balance the goals of the research with the available technologies to accurately assess human-machine teaming performance for future operations. This paper is a first step to outlining an approach to assessment of appropriate levels of fidelity in human-machine teaming research.

**Acknowledgements.** The authors would like to thank Cadets Jessica Broll and Makenzie Hockensmith for their contributions to this work. The views expressed in this document are the authors and may not reflect the official position of the USAF Academy, USAF, or U.S. Government. The material is based upon work supported by the Air Force Office of Scientific Research under award number 16RT0881.

## References

1. Sheridan, T.B.: Adaptive automation, level of automation, allocation authority, supervisory control, and adaptive control: distinctions and modes of adaptation. IEEE Trans. Syst. Man Cybern.-Part A: Syst. Hum. **41**(4), 662–667 (2011)
2. Lyons, J.B., et al.: Comparing trust in auto-GCAS between experienced and novice air force pilots. Ergon. Des. **25**(4), 4–9 (2017)
3. Ilachinski, A.: Artificial Intelligence and Autonomy: Opportunities and Challenges (No. DIS-2017-U-016388-Final). Center for Naval Analyses, Arlington, United States (2017)
4. Kaber, D.B.: Issues in human–automation interaction modeling: presumptive aspects of frameworks of types and levels of automation. J. Cogn. Eng. Decis. Making **12**(1), 7–24 (2018)
5. Hancock, P.A.: Imposing limits on autonomous systems. Ergonomics **60**(2), 284–291 (2017)
6. Scharre, P.: Army of None: Autonomous Weapons and the Future of War. WW Norton & Company, New York (2018)

7. Kott, A., Alberts, D.S.: How do you command an army of intelligent things? Computer **50**(12), 96–100 (2017)
8. Endsley, M.R.: Autonomous Horizons: System Autonomy in the Air Force-A Path to the Future. United States Air Force Office of the Chief Scientist, AF/ST TR, 15-01 (2015)
9. Parasuraman, R., Sheridan, T.B., Wickens, C.D.: Situation awareness, mental workload, and trust in automation: viable, empirically supported cognitive engineering constructs. J. Cogn. Eng. Decis. Making **2**(2), 140–160 (2008)
10. Miller, C.A., Parasuraman, R.: Designing for flexible interaction between humans and automation: delegation interfaces for supervisory control. Hum. Factors **49**(1), 57–75 (2007)
11. Roscoe, S.N., Williams, A.C.: Aviation psychology (1980)
12. Munshi, F., Lababidi, H., Alyousef, S.: Low-versus high-fidelity simulations in teaching and assessing clinical skills. J. Taibah Univ. Med. Sci. **10**(1), 12–15 (2015)
13. Usoh, M., et al.: Walking> walking-in-place> flying, in virtual environments. In: Proceedings of the 26th Annual Conference on Computer Graphics and Interactive Techniques, pp. 359–364, July 1999
14. Alexander, A.L., Brunyé, T., Sidman, J., Weil, S.A.: From gaming to training: a review of studies on fidelity, immersion, presence, and buy-in and their effects on transfer in PC-based simulations and games. DARWARS Train. Impact Group **5**, 1–14 (2005)
15. Dion, D.P., Smith, B.A., Dismukes, P.: The Cost/Fidelity Balance: Scalable Simulation Technology-A New Approach to High-Fidelity Simulator Training at Lower Cost. MS AND T, 38-45 (1996)
16. Wong, Y.J., Steinfeldt, J.A., LaFollette, J.R., Tsao, S.C.: Men's tears: football players' evaluations of crying behavior. Psychol. Men Masc. **12**(4), 297 (2011)
17. Taylor, H.L., Lintern, G., Koonce, J.M.: Quasi-transfer as a predictor of transfer from simulator to airplane. J. Gen. Psychol. **120**(3), 257–276 (1993)
18. Taylor, H.L., Lintern, G., Koonce, J.M., Kaiser, R.H., Morrison, G.A.: Simulator scene detail and visual augmentation guidance in landing training for beginning pilots. SAE Trans. **100**, 2337–2345 (1991)
19. Flexman, R.E., Stark, E.A.: Training simulators. In: Handbook of Human Factors, vol. 1, pp. 1012–1037 (1987)
20. McClernon, C.K., McCauley, M.E., O'Connor, P.E., Warm, J.S.: Stress training improves performance during a stressful flight. Hum. Factors **53**(3), 207–218 (2011)
21. Lievens, F., Patterson, F.: The validity and incremental validity of knowledge tests, low-fidelity simulations, and high-fidelity simulations for predicting job performance in advanced-level high-stakes selection. J. Appl. Psychol. **96**(5), 927 (2011)
22. Massoth, C., et al.: High-fidelity is not superior to low-fidelity simulation but leads to over-confidence in medical students. BMC Med. Educ. **19**(1), 29 (2019). https://doi.org/10.1186/s12909-019-1464-7
23. Salas, E., Bowers, C.A., Rhodenizer, L.: It is not how much you have but how you use it: toward a rational use of simulation to support aviation training. Int. J. Aviat. Psychol. **8**(3), 197–208 (1998)
24. Choi, W., et al.: Engagement and learning in simulation: recommendations of the Simnovate engaged learning domain group. BMJ Simul. Technol. Enhanc. Learn. **3**(Suppl 1), S23-S32 (2017)
25. Tossell, C., et al.: Human factors capstone research at the united states air force academy. In: Proceedings of the Human Factors and Ergonomics Society Annual Meeting, vol. 63, no. 1, pp. 498–502. SAGE Publications, Los Angeles, November 2019
26. Bishop, J., et al.: CHAOPT: a testbed for evaluating human-autonomy team collaboration using the video game overcooked! 2. In: 2020 Systems and Information Engineering Design Symposium (SIEDS), pp. 1–6. IEEE, April 2020

27. Tanibe, T., Hashimoto, T., Karasawa, K.: We perceive a mind in a robot when we help it. PloS One **12**(7), 1–12 (2017)
28. Bartneck, C., Forlizzi, J.: A design-centred framework for social human-robot interaction. In: Proceedings of the Ro-Man 2004, Kurashiki, pp. 591–594 (2004)
29. Steinfeld, A., Jenkins, O.C., Scassellati, B.: The oz of wizard: simulating the human for interaction research. In: Proceedings of the 4th ACM/IEEE International Conference on Human Robot Interaction, pp. 101–108, March 2009
30. Lorenz, G.T., et al.: Assessing control devices for the supervisory control of autonomous wingmen. In: 2019 Systems and Information Engineering Design Symposium (SIEDS), pp. 1–6. IEEE, April 2019
31. Tomzcak, K., et al.: Let Tesla park your Tesla: driver trust in a semi-automated car. In: 2019 Systems and Information Engineering Design Symposium (SIEDS), pp. 1–6. IEEE, April 2019
32. Tenhundfeld, N.L., de Visser, E.J., Ries, A.J., Finomore, V.S., Tossell, C.C.: Trust and distrust of automated parking in a Tesla model X. Hum. Factors **62**, 194–210 (2019). 0018720819865412
33. Tenhundfeld, N.L., de Visser, E.J., Haring, K.S., Ries, A.J., Finomore, V.S., Tossell, C.C.: Calibrating trust in automation through familiarity with the autoparking feature of a Tesla model X. J. Cogn. Eng. Decis. Making **13**(4), 279–294 (2019)
34. Haring, K., Nye, K., Darby, R., Phillips, E., de Visser, E., Tossell, C.: I'm not playing anymore! A study comparing perceptions of robot and human cheating behavior. In: Salichs, M., et al. (eds.) ICSR 2019. LNCS (LNAI), vol. 11876, pp. 410–419. Springer, Cham (2019). https://doi.org/10.1007/978-3-030-35888-4_38
35. Peterson, J., Cohen, C., Harrison, P., Novak, J., Tossell, C., Phillips, E.: Ideal warrior and robot relations: stress and empathy's role in human-robot teaming. In: 2019 Systems and Information Engineering Design Symposium (SIEDS), Charlottesville, VA, USA, pp. 1–6 (2019)
36. Lee, J.D., See, K.A.: Trust in automation: designing for appropriate reliance. Hum. Factors **46**(1), 50–80 (2004)
37. de Visser, E.J., et al.: Towards a theory of longitudinal trust calibration in human–robot teams. Int. J. Soc. Robot. **12**, 459–478 (2020). https://doi.org/10.1007/s12369-019-00596-x
38. Robinette, P., Li, W., Allen, R., Howard, A.M., Wagner, A.R.: Overtrust of robots in emergency evacuation scenarios. In: 2016 11th ACM/IEEE International Conference on Human-Robot Interaction (HRI), pp. 101–108. IEEE, March 2016
39. Wagner, A.R., Borenstein, J., Howard, A.: Overtrust in the robotic age. Commun. ACM **61**(9), 22–24 (2018)
40. Okamura, K., Yamada, S.: Adaptive trust calibration for supervised autonomous vehicles. In: Adjunct Proceedings of the 10th International Conference on Automotive User Interfaces and Interactive Vehicular Applications, pp. 92–97, September 2018
41. Berka, C., et al.: EEG correlates of task engagement and mental workload in vigilance, learning, and memory tasks. Aviat. Space Environ. Med. **78**(5), B231–B244 (2007)
42. Ekman, P., Friesen, W.V.: Facial Action Coding Systems. Consulting Psychologists Press, Palo Alto (1978)
43. Walliser, J.C., de Visser, E.J., Wiese, E., Shaw, T.H.: Team structure and team building improve human-machine teaming with autonomous agents. J. Cogn. Eng. Decis. Making **13**(4), 258–278 (2019)
44. Demir, M., McNeese, N.J., Cooke, N.J.: Team situation awareness within the context of human-autonomy teaming. Cogn. Syst. Res. **46**, 3–12 (2017)

45. Phillips, E., Ososky, S., Grove, J., Jentsch, F.: From tools to teammates: toward the development of appropriate mental models for intelligent robots. In: Proceedings of the Human Factors and Ergonomics Society Annual Meeting, vol. 55, no. 1, pp. 1491–1495. SAGE Publications, Los Angeles, September 2011

46. Garreau, J.: Bots on the ground. Washington Post 6 (2007)

47. Wen, J., Stewart, A., Billinghurst, M., Dey, A., Tossell, C., Finomore, V.: He who hesitates is lost (… in thoughts over a robot). In: Proceedings of the Technology, Mind, and Society, pp. 1–6 (2018)

48. Wen, J., Stewart, A., Billinghurst, M., Tossell, C.: Band of brothers and bolts: caring about your robot teammate. In: 2018 IEEE/RSJ International Conference on Intelligent Robots and Systems (IROS), pp. 1853–1858. IEEE, October 2018

49. Tomova, L., Majdandžić, J., Hummer, A., Windischberger, C., Heinrichs, M., Lamm, C.: Increased neural responses to empathy for pain might explain how acute stress increases prosociality. Soc. Cogn. Affect. Neurosci. **12**(3), 401–408 (2017)

50. National Adult Spelling Bee Practice. https://www.vocabulary.com/lists/144082. Accessed 23 Feb 2020

51. Phillips, E., Zhao, X., Ullman, D., Malle, B.F.: What is human-like? Decomposing robots' human-like appearance using the Anthropomorphic roBOT (ABOT) Database. In: Proceedings of the 2018 ACM/IEEE International Conference on Human-Robot Interaction, pp. 105–113, February 2018

52. Kim, B., Bruce, M., Brown, L., de Visser, E., Phillips, E.: A comprehensive approach to validating the uncanny valley using the Anthropomorphic RoBOT (ABOT) database. In: 2020 Systems and Information Engineering Design Symposium (SIEDS), pp. 1–6, April 2020

53. Haring, K.S., et al.: Conflict mediation in human-machine teaming: using a virtual agent to support mission planning and debriefing. In: 2019 28th IEEE International Conference on Robot and Human Interactive Communication (RO-MAN), pp. 1–7. IEEE, October 2019

54. Bellas, A., et al.: Rapport building with social robots as a method for improving mission debriefing in human-robot teams. In: 2020 Systems and Information Engineering Design Symposium (SIEDS), pp. 160–163. IEEE, April 2020

55. Haring, K.S., et al.: Robot authority in human-machine teams: effects of human-like appearance on compliance. In: Chen, J., Fragomeni, G. (eds.) HCII 2019. LNCS, vol. 11575, pp. 63–78. Springer, Cham (2019). https://doi.org/10.1007/978-3-030-21565-1_5

56. Giubilini, A., Savulescu, J.: The artificial moral advisor. The "Ideal Observer" meets artificial intelligence. Philos. Technol. **31**(2), 169–188 (2018)

57. Malle, B.F.: Integrating robot ethics and machine morality: the study and design of moral competence in robots. Ethics Inf. Technol. **18**(4), 243–256 (2016). https://doi.org/10.1007/s10676-015-9367-8

58. Savulescu, J., Maslen, H.: Moral enhancement and artificial intelligence: moral AI?. In: Romportl, J., Zackova, E., Kelemen, J. (eds.) Beyond Artificial Intelligence. TIEI, vol. 9, pp. 79–95. Springer, Cham (2015). https://doi.org/10.1007/978-3-319-09668-1_6

59. Coovert, M.D., Arbogast, M.S., de Visser, E.J.: The cognitive Wingman: considerations for trust, humanness, and ethics when developing and applying AI systems. In: McNeese, S., Endsley (eds.) Handbook of Distributed Team Cognition. CRC Press Taylor & Francis, Boca Raton (in press)

60. Costa, A., et al.: Your morals depend on language. PLoS One **9**(4), e94842 (2014)

61. Greene, J.D., Morelli, S.A., Lowenberg, K., Nystrom, L.E., Cohen, J.D.: Cognitive load selectively interferes with utilitarian moral judgment. Cognition **107**(3), 1144–1154 (2008)

62. Sütfeld, L.R., Gast, R., König, P., Pipa, G.: Using virtual reality to assess ethical decisions in road traffic scenarios: applicability of value-of-life-based models and influences of time pressure. Front. Behav. Neurosci. **11**, 122 (2017)

63. Tinghög, G., et al.: Intuition and moral decision-making – the effect of time pressure and cognitive load on moral judgment and altruistic behavior. PLoS One **11**(10), e0164012 (2016)
64. Cook, M.L.: The Moral Warrior: Ethics and Service in the U.S. Military. SUNY Press, Albany (2004)
65. Williams, T., Zhu, Q., Wen, R., de Visser, E.J.: The confucian matador: three defenses against the mechanical bull. In: Companion of the 2020 ACM/IEEE International Conference on Human-Robot Interaction, pp. 25–33, March 2020
66. Jackson, R.B., Williams, T.: Language-capable robots may inadvertently weaken human moral norms. In: 2019 14th ACM/IEEE International Conference on Human-Robot Interaction (HRI), pp. 401–410 (2019)
67. Rosemont Jr, H., Ames, R.T.: Confucian Role Ethics: A Moral Vision for the 21st Century? Vandenhoeck & Ruprecht, Göttingen (2016)
68. Groom, V., Nass, C.: Can robots be teammates?: Benchmarks in human–robot teams. Interact. Stud. **8**(3), 483–500 (2007)
69. Murphy, R.R.: Disaster Robotics. MIT Press, Cambridge (2014)
70. Ho, N.T., Sadler, G.G., Hoffmann, L.C., Lyons, J.B., Johnson, W.W.: Trust of a military automated system in an operational context. Milit. Psychol. **29**(6), 524–541 (2017)
71. Kim, B., et al.: How early task success affects attitudes toward social robots. In: Companion of the 2020 ACM/IEEE International Conference on Human-Robot Interaction, pp. 287–289, March 2020
72. Schellin, H., et al.: Man's new best friend? Strengthening human-robot dog bonding by enhancing the Doglikeness of Sony's Aibo. In: 2020 Systems and Information Engineering Design Symposium (SIEDS), pp. 1–6. IEEE, April 2020

# A Portable Measurement System for Spatially-Varying Reflectance Using Two Handheld Cameras

Zar Zar Tun[1](✉), Seiji Tsunezaki[1], Takashi Komuro[1], Shoji Yamamoto[2], and Norimichi Tsumura[3]

[1] Graduate School of Science and Engineering, Saitama University, 255 Shimo-okubo, Sakura-ku, Saitama 338-8570, Japan
zarzar@is.ics.saitama-u.ac.jp
[2] Tokyo Metropolitan College of Industrial Technology, 8-17-1 Minami-senju, Arakawa-ku, Tokyo 116-8523, Japan
[3] Graduate School of Science and Engineering, Chiba University, 1-33 Yayoi-cho, Inage-ku, Chiba 263-8522, Japan

**Abstract.** In this paper, we propose a system that can measure the spatially-varying reflectance of real materials. Our system uses two hand-held cameras, a small LED light, a turning table, and a chessboard with markers. The two cameras are used as a view and light cameras respectively to acquire incoming and outgoing light directions simultaneously, and the brightness at each position on the target material. The reflectance is approximated by using the Ward BRDF (Bidirectional Reflectance Distribution Function) model. The normal directions and all model parameters at each position on the material are estimated by non-linear optimization. As the result of experiment, the normal directions for all spatial points were properly estimated, and the correct colors of rendered materials were reproduced. Also, highlight changes on the surfaces were observed when we moved the light source or the rendered materials. It was confirmed that our system was easy to use and was able to measure the spatially-varying reflectance of real materials.

**Keywords:** Reflectance measurement · Normal directions · Ward reflectance model

## 1 Introduction

It is effective to measure the spatial reflectance distributions on the surfaces of real materials, such as paper, fabric and metal materials, for the purpose of reproducing realistic computer graphics. The reflectance at a position on a surface can be expressed using the BRDF (Bidirectional Reflectance Distribution Function). The BRDF is a function that defines how light is reflected at an opaque surface, and it is a four-dimensional reflectance function of the incoming direction $(\theta_i, \phi_i)$ and the outgoing direction $(\theta_o, \phi_o)$, where $\theta$ and $\phi$ are the elevation and azimuthal angles, respectively.

© Springer Nature Switzerland AG 2020
C. Stephanidis et al. (Eds.): HCII 2020, LNCS 12428, pp. 266–276, 2020.
https://doi.org/10.1007/978-3-030-59990-4_20

Many researchers developed reflectance measurement systems by setting up existing apparatus in various ways. Some systems use large apparatus, such as a robot-based gonioreflectometer [3,13], and three stepper motors-based gonioreflectometer [9]. They used many motors and controllers, which need to be oriented carefully. Their systems can measure the reflectance of a single point, but the spatially-varying reflectance cannot be measured. This system was extended by additionally using a pan-tilt-roll motor unit [14] to measure the spatial reflectance. An imaging gonioreflectometer was developed using a silicon photodiodes and an incandescent lamp [15]. Some systems were accordingly extended using a half-silvered hemispherical mirror [19], and a curved mirror [5]. Another apparatus like gonioreflectometer was developed using a photometer and a CCD video camera [4]. Their systems can measure the spatially-varying reflectance, but the different combinations of viewing and illumination directions are required. Therefore, they require high cost apparatus and take long capturing time.

On the other hand, some researchers developed their own acquisition apparatus by combining a gantry with robot arms, rotating arms, lenses, cameras, and light arrays in a variety of ways [6–8,10,16–18,20]. Their systems can also measure the spatial reflectance but require many light sources, controllers for moving the devices, and wide space to set up the apparatus. Other systems using portable apparatus consisting of a single light source, mobile phone camera and built-in flashlight were developed [1,2,12]. Their acquisition apparatus can measure the spatially-varying reflectance but has limitations in the movement of cameras, the number of images to be captured, and the type of materials to be measured.

In this study, we propose a system that can measure the spatially-varying reflectance of real materials using two handheld cameras as a light and view cameras. The light camera is used to capture the incoming light directions, and the view camera is used to capture the outgoing directions, and the brightness of each pixel on the surface of the material sample. The light camera is attached to by an LED light and is placed on a stand, and the view camera is moved by hand. The attached LED light is used to illuminate the target material. A chess board with markers is used to acquire the camera poses of two cameras, and additionally the positions of all pixels on the surface. A turning table is used to acquire the different views of the material. The Ward BRDF model is used to approximate the reflectance. The spherical polar coordinate system is used to calculate the distributions of normal directions across the surface. All model parameters as well as the normal directions are estimated by non-linear optimization. Our system can be used to easily measure the spatially-varying reflectance of real materials.

## 2   Related Work

Measuring the reflectance on the surface highly depends on the configuration of apparatus, and many measurement systems were built by researchers. We categorized those systems into three groups.

**Measurement Systems Using Gonioreflectometer.** Baribeau et al. [3] configured a robot-based gonioreflectometer in which a five-axis robot was used to tilt the material sample, and a rotation stage was used to hold the light source. An array spectroradiometer was used to detect the reflected light from the sample and also detect the direct light from the light source. Their system can measure the reflected spectral radiance but requires to synchronically orient the sample and light source. Lyngby et al. [13] set up a gonioreflectometer using a six-axis industrial robotic arm, an arc-shaped light source used as a static light array, and an RGB camera. The sample was placed on a stand under the static light array. The robotic arm was used to position and orient a camera to capture images of the material sample. Their system can reduce the number of controllers to one, but their systems can only measure the reflectance at one point.

Murray-Coleman et al. [15] developed an imaging gonioreflectometer by using silicon photodiodes as two photodetectors, and an incandescent lamp. All components and the sample were positioned by stepper motors. Their system can measure the spatial reflectance, but the configuration is complex and takes long measurement time. Ward [19] extended it by using a half-silvered hemispherical mirror instead of using silicon photodiodes, a quartz-halogen lamp, and a CCD camera with a fish-eye lens. The sample was moved by hand and illuminated by a quartz-halogen lamp. The light source was controlled by a controller to move along the hemisphere. His system can cover the entire hemisphere of reflected light directions simultaneously, but the configuration is still complex and expensive.

Dana [4] did not use the hemispherical mirror arrangement, and developed an apparatus like gonioreflectometer, in which a robot, a photometer, a CCD video camera, a halogen bulb with a Fresnel lens, and a personal computer were used. The camera was mounted on a tripod, and the texture sample and halogen bulb were oriented by the robot arm. A few years later, Dana et al. [5] used a curved mirror instead of using the hemisphere and proposed another type of gonioreflectometer including a beam splitter, a concave parabolic mirror, a CCD camera, a collimated beam of light, and translation stages. The stages were used for automatic material scanning by placing the illumination aperture on an X–Z stage and the mirror on an X–Y–Z stage. Their systems can obtain multiple views of the same surface points simultaneously, but many devices are required. Mcallister et al. [14] designed a spatial gonioreflectometer based on the system [9] using a pan-tilt-roll motor unit, which held the surface sample with fiducial markers, a calibrated light mounted on a motorized rail, and a CCD camera. All components were mounted on an optical bench and controlled by four motors separately. Their system can acquire the spatial reflectance distribution, but many devices and controllers are required.

**Measurement Systems Using Light Array.** Filip et al. [8] proposed an acquisition setup by using a mechanical gantry, a consumer digital camera, and two LED lights. One of the two arms held two LEDs and another arm held the camera. These two arms were synchronically controlled by the gantry to capture images of the material sample placed below the setup. Their system can capture sparse texture data but limits the size of the material sample. Tunwattanapong et al. [17] used a hemispherical gantry, in which a motion control motor was used to rotate a 1 m diameter semi-circular arc attached with 105 white LEDs series. A sample was placed on a platform that was rotated to eight positions by a motor. An array of five machine vision cameras was configured in the form of a mathematical plus sign, and it was placed in front of the sample. Their system can acquire the geometry and spatially-varying reflectance, but many light sources and cameras were used.

Gardner et al. [10] used a translation gantry, a white neon tube as a linear light source, and a digital camera. The gantry is used to move the light source horizontally across the surface of a planar material. The camera was fixed on a tripod at approximately a 55-degree incidence angle to the material. Their system can estimate the spatial reflectance by a single pass of a linear light source, but it limits the materials, such as printed paper and fabrics. Wang et al. [18] adapted the system in [10] by replacing the white neon tube with a linear array of 40 LEDs and using a color checker pattern for camera calibration. A stepping motor was used to control the light array to move horizontally over the material sample. A static camera was placed about 1.0 m away and 0.6 m above the center of the sample surface. Their system can capture images of the material under different illuminations from one view, but motor and many light sources are still required.

**Measurement Systems Using Portable Apparatus.** Aittala et al. [1] developed a system using an LCD display (laptop screen) and a single-lens reflex camera. The material sample was illuminated by the LCD screen by placing it in a static position below the screen, and the camera took the reflected light from the sample. Their system is portable and can measure the spatial reflectance. Aittala et al. [2] also used only a mobile phone, that was hold in parallel with a material sample and captured a flash-no-flash image pair under a point light source (built-in flash-light). Li et al. [12] also used only one mobile camera set-up that took a pair of images under the ambient light, and a point light source (built-in flash-light). Their systems are light-weight and can measure the spatially-varying reflectance but has limitations in the movement of cameras, the number of images to be captured, and the type of materials to be measured.

# 3  Measuring the Spatially-Varying Reflectance

## 3.1  Measurement Apparatus

The measurement setup is shown in Fig. 1. The system consists of a PC, two handheld cameras, a turning table, a small LED light, and a ChArUco board [11]. The two handheld cameras are used as a light camera and a view camera, respectively. The light camera is mounted with a small LED light, and is placed on a stand. The LED light is used to illuminate the target material placed on the

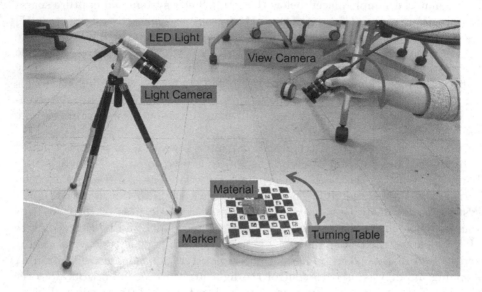

**Fig. 1.** Measurement setup.

turning table. The view camera is moved by hand from 0 to 90° of polar angles, and 0 to 360° of azimuthal angles of the outgoing light directions. The ChArUco board, a chessboard with markers, is placed under the material on the turning table. While conducting the experiment, the room was set dark except the small LED light to prevent entering light from the environment. Using this apparatus, dynamically changing incoming and outgoing light angles and different views of each spatial point can be acquired.

The color images were captured by the two handheld cameras at the frame rate of 15 frames per second. The cameras were 0.5 m away from the material, and 50 frames were captured for each material. The resolution of color images was 1920 × 1024 pixels. The specifications of the PC were Intel Core $i7 - 7700$ CPU, 16 GB memory and NVIDIA GeForce GTX 1070 6 GB PC for GPU.

## 3.2   Measurement Method

Our system measures the incoming and outgoing light directions, $\omega_i(t)$ and $\omega_o(t)$ using the chessboard with markers, and the intensity $I(x, t)$ on a point $x$ at time $t$. Then, the system estimates the reflectance at several points on the material according to the BRDF model as shown in Fig. 2. The model parameters, the diffuse albedo $\rho_d$, the specular albedo $\rho_s$, and the roughness $\alpha$, as well as the normal direction $N$ are simultaneously estimated using the Levenberg–Marquardt optimization method, which finds the parameters that minimizes the square difference between the observed intensity $I(x, t)$ and the estimated luminance on a pixel $x$.

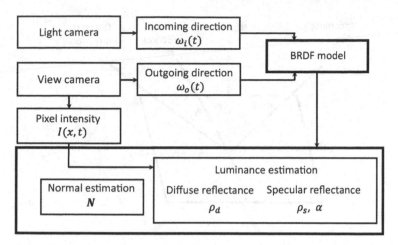

**Fig. 2.** Measurement method: The system measures the incoming and outgoing light directions $\omega_i(t)$ and $\omega_o(t)$, and the intensity $I(x,t)$ on a point $x$ at time $t$. Then, the model parameters as well as the normal direction are estimated by non-linear optimization.

The system represents the reflectance using the Ward BRDF model [19], which determines how much light is reflected from the incoming direction $i$ to the outgoing direction $o$ at a surface point $x$ as follows:

$$f_r(\theta_i, \phi_i, \theta_o, \phi_o) = \frac{\rho_d}{\pi} + \frac{\rho_s}{4\pi\alpha^2\sqrt{\cos\theta_i \cos\theta_o}}e^{-\frac{\tan^2\theta_h}{\alpha^2}} \tag{1}$$

where $\theta_i$ and $\theta_o$ are the incoming and outgoing light angles, $\theta_h$ is the halfway angle between the normal and the halfway vector, respectively.

The ChArUco marker board is used to acquire the calibration parameters, which is used to estimate the poses of the two cameras. Moreover, it is used to know the pixel positions of the material sample.

As the view camera is moved by hand while capturing the color images, some of markers can be missed out. Therefore, more than four detected markers in each frame are used to interpolate the marker corners, otherwise the frames that do not have enough number of markers are skipped. In this way, the calibration parameters for each camera in the world coordinate system defined by the markers are acquired. Then, the camera pose $\mathbf{x}_c$ for each 3D position $\mathbf{p}$ is estimated as follows:

$$\mathbf{x}_c = R^{-1}\mathbf{p} - R^{-1}\mathbf{t} \tag{2}$$

where $R$ is the rotation matrix, and $\mathbf{t}$ is the translation vector of each camera.

**Estimation of the BRDF Parameters.** The variables used in the BRDF model are shown in Fig. 3, where $i$ and $o$ represent the incoming and outgoing

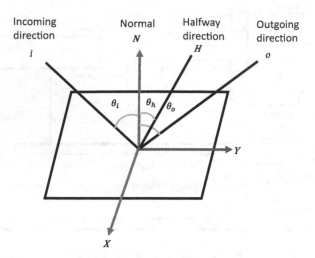

**Fig. 3.** Variables used in the Ward BRDF model.

light directions on a surface point, $H$ is the halfway vector between $i$ and $o$, $N$ is the normal direction, $\theta_i$ is the angle between the normal and the incident light direction, $\theta_o$ is the angle between the normal and the outgoing light direction, and $\theta_h$ is the angle between $N$ and $H$. The reflectance parameters are estimated using the Levenberg–Marquardt optimization. It is a non-linear optimization method and accordingly it is required to give the initial values.

As the object is rotating and the view camera is moving, the intensity on the vertex is changing from frame to frame. To obtain the diffuse and specular albedos, the correspondence between the object vertex and the color pixel is calculated according to the projection of the vertex to the image plane. The initial values for the diffuse and specular albedos are initialized by the average of the color values at the reprojected points over all frames. The roughness parameter is initialized from 0 to 1, and the best-fitted one is chosen.

**Normal Estimation.** The distribution of normal directions for all points across the material surface is estimated using the spherical polar coordinate system as shown in Fig. 4, where $N(\theta_n, \phi_n)$ is the estimated normal direction at a pixel position, which is described by two angles, the polar angle $\theta_n$ and the azimuthal angle $\phi_n$, respectively. The normal direction for each pixel position is estimated by giving the initial values to the two angles. The initial values range from 0 to 180° for $\theta_n$, and from 0 to 360° for $\phi_n$, respectively, and the best-fitted pair is chosen.

## 4    Experimental Result

In our experiment, real materials, such as metallic-colored paper and paper with golden texture were used as samples. The colored frames were captured about

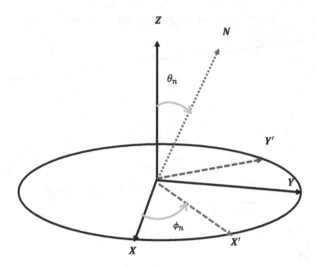

**Fig. 4.** Estimation of normal direction using the spherical polar coordinate system: $N(\theta_n, \phi_n)$ is the estimated normal direction, which is described by two angles, the polar angle $\theta_n$ and the azimuthal angle $\phi_n$, respectively.

50 frames for each sample, and the reflectance is approximated using the Ward BRDF model. The reflectance of totally 62901 points were estimated for each material. It took totally about 12.5 min of capturing time, and about 9 min of computation time for each material. Therefore, our system took less than 22 min of total measuring time for each material.

**Fitting Result.** The fitting result for the estimated reflectance, which is expressed using the parameters of the Ward BRDF model, over observed brightness of RGB components with regard to frame numbers at a spatial point, is shown in Fig. 5. As the model is well fitted with observed brightness of each component, we can confirm that the estimated BRDF parameters are optimized correctly.

**Rendering Result.** Figure 6 shows the rendering results for metallic-colored paper and paper with golden texture. Figure 6 (a) shows the appearance of the original materials, Fig. 6 (b) shows the estimated normal maps, and Fig. 6 (c)(d) shows the rendered materials in different views. We can see the correct colors of rendered materials that correspond to those of the original materials. Additionally, highlight changes on the surfaces were observed when we moved the light source or the rendered materials.

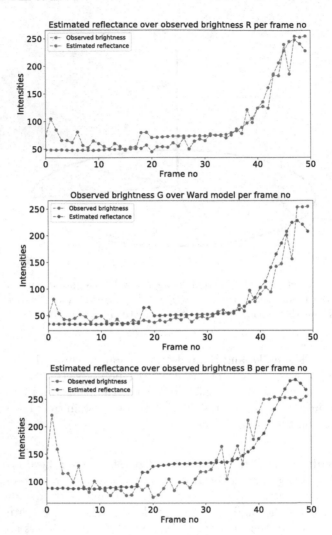

**Fig. 5.** Fitting result for the estimated reflectance over observed brightness of RGB components with regard to frame numbers at a spatial point of metallic-colored paper.

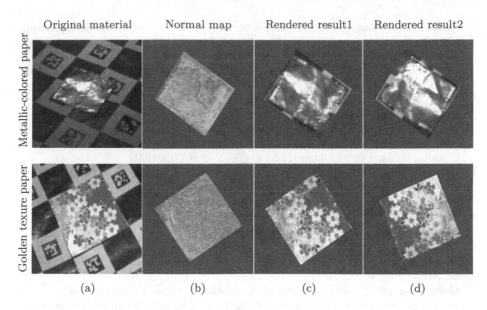

**Fig. 6.** The rendering results for metallic-colored paper and paper with golden texture:
(a) Original materials (b) Estimated normal maps (c–d) Rendered materials in different
views. (Color figure online)

## 5   Conclusion

We proposed a portable system that can easily measure the spatially-varying
reflectance of real materials. This system requires only a small space compared
to the apparatus using gonioreflectometer and gantry, and can move the cameras
freely by hand to capture images. We conducted the experiment using only plan-
ner materials; however, this configuration could be applied to the measurement
of non-planner materials as well. In the future, we will extend our method to
measure more complex reflectance such as the anisotropic reflectance.

## References

1. Aittala, M., Weyrich, T., Lehtinen, J.: Practical SVBRDF capture in the frequency
   domain. ACM Trans. Graph. **32**(4) (2013). https://doi.org/10.1145/2461912.
   2461978
2. Aittala, M., Weyrich, T., Lehtinen, J.: Two-shot SVBRDF capture for stationary
   materials. ACM Trans. Graph. **34**(4) (2015). https://doi.org/10.1145/2766967
3. Baribeau, R., Neil, W.S., Côté, E.: Development of a robot-based gonioreflectometer for spectral BRDF measurement. J. Mod. Opt. **56**(13), 1497–1503 (2009).
   https://doi.org/10.1080/09500340903045702
4. Dana, K.J., van Ginneken, B., Nayar, S.K., Koenderink, J.J.: Reflectance and
   texture of real-world surfaces. ACM Trans. Graph. **18**(1), 1–34 (1999). https://
   doi.org/10.1145/300776.300778

5. Dana, K.J., Wang, J.: Device for convenient measurement of spatially varying bidirectional reflectance. J. Opt. Soc. Am. A **21**(1), 1–12 (2004). https://doi.org/10.1364/JOSAA.21.000001. http://josaa.osa.org/abstract.cfm?URI=josaa-21-1-1
6. Dong, Y., et al.: Manifold bootstrapping for SVBRDF capture. ACM Trans. Graph. **29**(4) (2010). https://doi.org/10.1145/1778765.1778835
7. Fichet, A., Sato, I., Holzschuch, N.: Capturing spatially varying anisotropic reflectance parameters using Fourier analysis. In: Proceedings of the 42nd Graphics Interface Conference, GI 2016, pp. 65–73. Canadian Human-Computer Communications Society, School of Computer Science, University of Waterloo, Waterloo, Ontario, Canada (2016). https://doi.org/10.20380/GI2016.09
8. Filip, J., Vávra, R., Krupička, M.: Rapid material appearance acquisition using consumer hardware. Sensors (Basel, Switzerland) **14**(10), 19785–19805 (2014). https://doi.org/10.3390/s141019785. https://pubmed.ncbi.nlm.nih.gov/25340451
9. Foo, S.C.: A gonioreflectometer for measuring the bidirectional reflectance of material for use in illumination computation (1997)
10. Gardner, A., Tchou, C., Hawkins, T., Debevec, P.: Linear light source reflectometry. ACM Trans. Graph. **22**(3), 749–758 (2003). https://doi.org/10.1145/882262.882342
11. Garrido-Jurado, S., Muñoz Salinas, R., Madrid-Cuevas, F., Marín-Jiménez, M.: Automatic generation and detection of highly reliable fiducial markers under occlusion. Pattern Recogn. **47**(6), 2280–2292 (2014). https://doi.org/10.1016/j.patcog.2014.01.005
12. Li, B., Feng, J., Zhou, B.: A SVBRDF modeling pipeline using pixel clustering (2019)
13. Lyngby, R.A., Matthiassen, J.B., Frisvad, J.R., Dahl, A.B., Aanæs, H.: Using a robotic arm for measuring BRDFs. In: Felsberg, M., Forssén, P.-E., Sintorn, I.-M., Unger, J. (eds.) SCIA 2019. LNCS, vol. 11482, pp. 184–196. Springer, Cham (2019). https://doi.org/10.1007/978-3-030-20205-7_16
14. Mcallister, D.K., Lastra, A.: A generalized surface appearance representation for computer graphics. Ph.D. thesis (2002). aAI3061704
15. Murray-Coleman, J., Smith, A.: The automated measurement of BRDFS and their application to luminaire modeling. J. Illum. Eng. Soc. **19**(1), 87–99 (1990). https://doi.org/10.1080/00994480.1990.10747944
16. Rump, M., Mueller, G., Sarlette, R., Koch, D., Klein, R.: Photo-realistic rendering of metallic car paint from image-based measurements. Comput. Graphics Forum **27**(2), 527–536 (2008). https://doi.org/10.1111/j.1467-8659.2008.01150.x
17. Tunwattanapong, B., et al.: Acquiring reflectance and shape from continuous spherical harmonic illumination. ACM Trans. Graph. **32**(4) (2013). https://doi.org/10.1145/2461912.2461944
18. Wang, J., Zhao, S., Tong, X., Snyder, J., Guo, B.: Modeling anisotropic surface reflectance with example-based microfacet synthesis. ACM Trans. Graph. **27**(3), 1–9 (2008). https://doi.org/10.1145/1360612.1360640
19. Ward, G.J.: Measuring and modeling anisotropic reflection. SIGGRAPH Comput. Graph. **26**(2), 265–272 (1992). https://doi.org/10.1145/142920.134078
20. Yu, J., Xu, Z., Mannino, M., Jensen, H.W., Ramamoorthi, R.: Sparse sampling for image-based SVBRDF acquisition. In: Klein, R., Rushmeier, H. (eds.) Workshop on Material Appearance Modeling. The Eurographics Association (2016). https://doi.org/10.2312/mam.20161251

# Influence of Visual Gap of Avatar Joint Angle on Sense of Embodiment in VR Space Adjusted via C/D Ratio

Takehiko Yamaguchi[1]([✉]), Hiroaki Tama[1], Yuya Ota[1], Yukiko Watabe[2], Sakae Yamamoto[3], and Tetsuya Harada[3]

[1] Suwa University of Science, Toyohira, Chino, Nagano 50001, Japan
tk-ymgch@rs.sus.ac.jp
[2] Yamanashi Prefectural Board of Education, Marunouchi, Kofu, Yamanashi 400-8501, Japan
[3] Tokyo University of Science, 6-3-1 Niijuku, Katsushika, Tokyo, Japan

**Abstract.** The movement of an avatar in the VR space is often handled in complete synchronization with the body of the operator, both spatially and temporally. However, if the operator uses an interaction device with a narrow work area, the range of motion of the avatar may be adjusted by tuning the control/display (C/D) ratio. This tuning of the C/D ratio is a technique often used in mouse interaction; however, upon applying the concept of the C/D ratio on an avatar that is connected with the body of an operator, the sense of embodiment felt by the operator of the avatar is affected. In this study, we investigate the subjective effects of the sense of embodiment, presence, and mental workload while performing a point-to-point task between two points, by using the effects of the avatar appearance and the C/D ratio as independent variables.

**Keywords:** Sense of embodiment · Immersive virtual reality (VR) · Proteus effect

## 1 Introduction

### 1.1 General VR Interaction in Immersive Environments

In immersive virtual reality (VR), the positional relationship between the virtual body (avatar) in the VR space and the body in real space is perfectly synchronized spatiotemporally such that the sense of self-location perceived through the avatar and that of the self-body completely match with each other. However, in this situation, the movable range of the avatar is restricted by the physical movable range of its body.

The hand of an avatar in the VR space acts as a pointer, and the user interacts with an object in the VR space via the hand of the avatar. However, depending on the type of interaction device employed by the user to control the hand of the avatar, we must adjust the operation amount of the device and the movement range of the hand of the avatar.

For example, many haptic devices are designed such that the range of the workspace is narrower than that of the movement of the actual hand owing to the structure [1–4].

© Springer Nature Switzerland AG 2020
C. Stephanidis et al. (Eds.): HCII 2020, LNCS 12428, pp. 277–287, 2020.
https://doi.org/10.1007/978-3-030-59990-4_21

In such a case, we must set a ratio to ensure that the movement amount of the hand of the avatar becomes larger than the operation amount of the device. This ratio, called the control/display (C/D) ratio, is a commonly used tool for mouse interaction [5–8].

## 1.2 Full-Body VR Interaction via the C/D Ratio

To date, most avatars used in immersive VR were displayed with a pointer above their wrists. In recent years, however, the improvements in the motion-sensor technology have enabled to display, in real time, an avatar that is synchronized with the entire body movement of a user.

Even when an avatar is full-bodied, its hand is a pointer in the VR space. That is, even upon using a full-body avatar, when the movement amount of the avatar is adjusted, its movable range can be adjusted by tuning the C/D ratio. However, because the hand of a full-body avatar is connected to its arm, we must limit the C/D ratio to be within the range of movement of the joint angle of the arm.

## 1.3 Persistence of the Sense of Embodiment

Upon adjusting the C/D ratio for a full-bodied avatar, the sense of proprioception might be uncomfortable, even if the C/D ratio is adjusted within the range of movement of the joint angle of the arm. For example, as depicted in Fig. 1, the actual hand joints are curved, whereas the joints of the avatar in the VR space are completely extended. Conversely, a user visually feels that the joint is completely extended; however, in the sense of proprioception, he or she feels that the joint is bent.

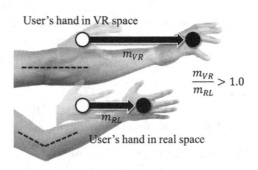

User's hand in VR space

User feel that his/her hand is bent because of the proprioception, but the hand is visually not bent in VR space so that the user might feel a sense of incongruity.

$m_{VR}$

$\dfrac{m_{VR}}{m_{RL}} > 1.0$

$m_{RL}$

User's hand in real space

**Fig. 1.** Task environments in real and VR spaces.

The effect of the C/D ratio has been discussed from the viewpoint of operability when an avatar is displayed only in the pointer part [9–12]. However, when a body, such as an arm, is connected to the pointer in the manner previously described, the sense of embodiment associated with the avatar, as well as its operability, might be affected. The sense of embodiment is the sensation through which an individual misperceives another body (not his or her own body) as his or her own body. It comprises three elements: sense of agency, sense of body ownership, and sense of self-location [13]. The C/D ratio might affect both the sense of agency, which is the sense that one's body is moving, and sense of self-location, which is the sense of one's body position.

This study aims to investigate how the avatar appearance and C/D ratio affect the sense of embodiment, especially in terms of the sense of agency and sense of self-location.

## 2  Methods

### 2.1  Participants

A total of 17 students were recruited from Suwa University of Science. They were males with the mean age of 21 years (SD = 1.32). They had minimal or no experience with VR, as well as mixed reality, and related applications.

### 2.2  Apparatus

**Visual Display/Rendering.** A head-mounted display (HMD) (manufacturing details: HTC VIVE) was used for the experimental task. The headset covered a nominal field-of-view, approximately 110°, using two 2,160 × 1,200 pixel displays, which were updated at 90 Hz. For visual rendering, Unity3D was employed, rendering the graphics at 60 Hz.

**3D Motion Tracker.** A leap-motion sensor was used to track the 3D position of the hand of the user, as well as the orientation at 120 Hz.

**Task Environment.** A customized experimental environment was developed for this study. It had three blocks on a table both in real and VR spaces. These blocks were spatially synchronized both in real and VR spaces such that a user could visually recognize them in the VR space, as well as haptically recognize them in the real space (see Fig. 2).

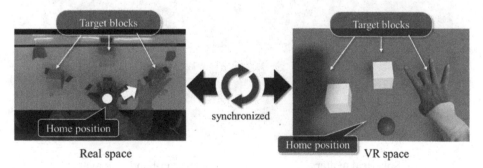

**Fig. 2.** Task environments in real and VR spaces.

## 2.3  Experimental Task

The protocol for the experimental task was the same for all the conditions. The participants blindfolded themselves before entering the experimental room so that they could not recognize the task conditions. After the experimenter provided them with the instructions of the experiment, each participant was made to wear the HMD and then calibrate his position in a virtual environment that was displayed through the HMD. In the experimental task, a target block was randomly displayed from three directions. Each participant was required to set his hand on the home position when the task began. Upon displaying the target block, each participant was required to touch it using his virtual avatar. After touching the block, he had to return to the home position, following which the next target block was displayed. This trial was conducted 30 times for one experimental condition.

## 2.4  Experimental Design and Independent Variables

To examine the effect of the C/D ratio and external appearance of the virtual avatar, a "within-subjects factorial experiment" design was used, resulting in six experimental tasks.

**External Appearance of Virtual Avatar.** In this experiment, two types of avatars were created, as depicted in Fig. 3. The spherical avatar in (a) was synchronously controlled using the position of the palm of the participant. The human-hand-like avatar depicted in (b) was controlled by synchronizing the joint positions and angles of the fingers and arms of the hand of the participant.

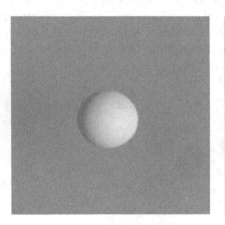        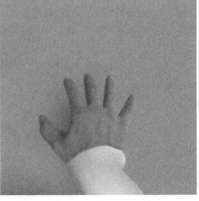

      (a) Spherical avatar                (b) Human-hand-like avatar

**Fig. 3.** Two types of avatar for the task.

**C/D Ratio in the VR Space.** The C/D ratio is the ratio of the amplitude of the hand movement of a real to that of his or her virtual hand. In this study, the C/D ratio was changed thrice (0.9, 1.0, and 1.3), as depicted in Fig. 4. When the C/D ratio was 0.9, the position of the avatar shifted to the front of the position of the hand of the actual user. When it was 1.0, the position of the hand of the actual user exactly matched that of the avatar. When the C/D ratio was 1.3, the position of the avatar shifted to behind the hand of the actual user.

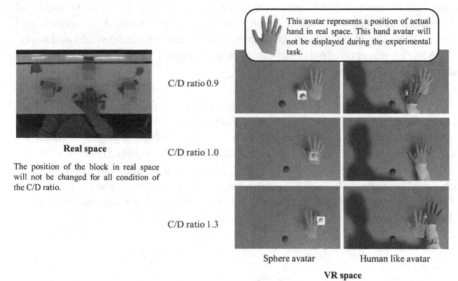

**Fig. 4.** Concept of the C/D ratio in the experimental task.

### 2.5 Dependent Variables

To systematically investigate the effect of both the external appearance of virtual avatar and the C/D ratio in customized task environments, we utilized several dependent measures, which can be categorized into two types of variables. The first one is task performance, which includes the velocity changes of the hand motion. The second one is subjective performance, which includes the sense-of-embodiment questionnaire (SoEQ), presence questionnaire, and NASA-TLX.

**Subjective Performance.** It includes three types of questionnaires through which we asked the participants regarding their feelings associated with the sense of embodiment, presence, and mental workload.

### 2.6 Procedure

The participants were required to read and sign an informed consent. After completing the paperwork, a short training was conducted to familiarize them to interact in the customized experimental environment. The experimental tasks began after the training session. Each participant completed 30 trials for each of the six tasks in a randomized order. After completing each task, the participants answered the questionnaires.

# 3 Results and Discussions

## 3.1 Subjective Performance

*Presence Questionnaire.* It is a common measure of presence in immersive VR environments. It comprises seven subscales: 1) realism, 2) possibility to act, 3) quality of interface, 4) possibility to examine, 5) self-evaluation of performance, 6) sounds, and 7) haptics. However, in our experimental task, because there were no sound and haptic effects, ANOVA was applied only to five subscales, excluding the "sounds" and "haptics" subscales.

1) *Realism.* A two-factor ANOVA, as the within-subject factors, revealed the main effect on C/D ratio factor, $F(1, 16) = 5.554, p < .01, \eta^2 = .067$. A post-hoc comparison (performed using Bonferroni's method, where $\alpha = .05$) indicated a significant difference between the 0.9 condition ($M = 4.441; SD = 1.156$) and 1.0 condition ($M = 5.1389; SD = 1.034$). However, no significant difference was observed between the 0.9 and 1.3 conditions, and between the 1.0 and 1.3 conditions. Table 1 presents the results of ANOVA for "realism." Figure 5 depicts the result of the average plot.

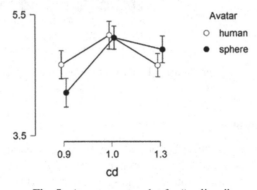

**Fig. 5.** Average score plot for "realism."

**Table 1.** Result of ANOVA for "realism."

|  | Sum of Squares | df | Mean Square | F | p | $\eta^2$ |
|---|---|---|---|---|---|---|
| Avatar | 0.168 | 1 | 0.168 | 0.170 | 0.686 | 0.001 |
| Residual | 15.835 | 16 | 0.990 |  |  |  |
| cd | 8.271 | 2 | 4.135 | 5.554 | 0.008 | 0.067 |
| Residual | 23.825 | 32 | 0.745 |  |  |  |
| Avatar * cd | 2.239ª | 2ª | 1.120ª | 1.447ª | 0.250ª | 0.018 |
| Residual | 24.768 | 32 | 0.774 |  |  |  |

*Although realism* was considered susceptible to visual effects, no visual main effects were observed. However, the main effect of the C/D ratio was different, indicating that the subjective view of realism is influenced more by the restriction of movement than the appearance. In particular, when the C/D ratio was 0.9, the realism was low.

2) *Possibility to act.* A two-factor ANOVA, as the within-subject factors, revealed the main effect on C/D ratio factor, $F(2, 32) = 7.526, p < .01, \eta^2 = .089$. A post-hoc comparison (performed using Bonferroni's method, where $\alpha = .01$) demonstrated a significant difference between the 0.9 condition ($M = 4.919; SD = 1.036$) and 1.0 condition ($M = 5.566; SD = 0.835$) and between the 0.9 condition ($M = 4.919; SD = 1.036$) and 1.3 condition ($M = 5.441; SD = 0.839$). However, no significant difference was noticed between the 1.0 and 1.3 conditions. Table 2 presents the results of ANOVA for "possibility to act." Figure 6 depicts the result of the average plot.

**Table 2.** Result of ANOVA for "possibility to act."

|  | Sum of Squares | df | Mean Square | F | p | η² |
|---|---|---|---|---|---|---|
| Avatar | 0.885 | 1 | 0.885 | 1.048 | 0.321 | 0.010 |
| Residual | 13.511 | 16 | 0.844 | | | |
| cd | 8.011 | 2 | 4.006 | 7.526 | 0.002 | 0.089 |
| Residual | 17.031 | 32 | 0.532 | | | |
| Avatar ∗ cd | 0.413 | 2 | 0.206 | 0.410 | 0.667 | 0.005 |
| Residual | 16.129 | 32 | 0.504 | | | |

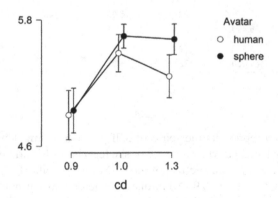

**Fig. 6.** Average score plot for "possibility to act."

Because *possibility to act* is a measure related to the easiness of behavior, it might become low when the leaching action is restricted. Consequently, it was observed that *possibility to act* was low when the C/D ratio was 0.9.

4) *Quality of interface, 5) Possibility to examine, and 6) Self-evaluation of performance.* The two-factor ANOVA, as the within-subject factors, revealed no main effect on both the factors, namely, the external appearance of the virtual avatar and the C/D ratio.

***Sense of Embodiment Questionnaire.*** It is a standard measure of the feeling of sense of embodiment in immersive VR environments. It comprises six subscales: 1) location, 2) agency 3) ownership, 4) tactile sensations, 5) appearance, and 6) response. In our experimental task, because there was no tactile effect, ANOVA was applied only to five subscales, excluding the "tactile sensations" subscale.

1) *Location.* A two-factor ANOVA, as the within-subject factors, revealed the main effect on the C/D ratio factor, $F(2, 32) = 7.743, p < .01, \eta^2 = .039$. A post hoc comparison (performed using Bonferroni's method, where $\alpha = .01$) demonstrated a significant difference between the 0.9 condition ($M = -0.588, SD = 3.000$) and 1.0 condition ($M = 0.588, SD = 2.697$) and between the 0.9 condition ($M = -0.588; SD = 3.000$) and 1.3 condition ($M = 0.618; SD = 2.606$). However, no significant difference was noticed between the 1.0 and 1.3 conditions. Table 3 presents the results of ANOVA for "location." Figure 7 depicts the result of the average plot.

**Table 3.** Result of ANOVA for "location."

|  | Sum of Squares | df | Mean Square | F | p | $\eta^2$ |
|---|---|---|---|---|---|---|
| Avatar | 44.010 | 1 | 44.010 | 3.938 | 0.065 | 0.053 |
| Residual | 178.824 | 16 | 11.176 |  |  |  |
| CD | 32.176 | 2 | 16.088 | 7.743 | 0.002 | 0.039 |
| Residual | 66.490 | 32 | 2.078 |  |  |  |
| Avatar * CD | 5.667 | 2 | 2.833 | 0.563 | 0.575 | 0.007 |
| Residual | 161.000 | 32 | 5.031 |  |  |  |

Because *location* represents the sense of self-position, it might be affected if the avatar of a user is not aligned with where he or she feels his or her arm is. Consequently, there was a difference in the main effect of the C/D ratio. Specifically, it is significantly affected when the C/D ratio is 0.9. Additionally, when the avatar appears similar to a human, a visual gap occurs in the joint position; thus, a synergistic effect with the main effect of the appearance was expected; however, no significant difference was observed in the interaction effect.

2) *Agency.* A two-factor ANOVA, as the within-subject factors, revealed the main effect on the external appearance of the virtual avatar factor, $F(1, 16) = 7.136, p <$

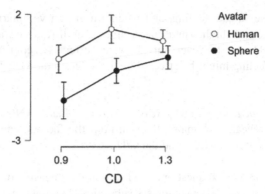

**Fig. 7.** Average score plot for "location."

.01, $\eta^2$ = .152. The result demonstrated a significant difference between the human-like-avatar condition ($M$ = 4.3373; $SD$ = 3.068) and spherical-avatar condition ($M$ = 0.902, $SD$ = 4.933). Table 4 presents the results of ANOVA for "agency." Figure 8 depicts the result of the average plot.

**Table 4.** Result of ANOVA for "agency."

|  | Sum of Squares | df | Mean Square | F | p | $\eta^2$ |
|---|---|---|---|---|---|---|
| Avatar | 307.147 | 1 | 307.147 | 7.136 | 0.017 | 0.152 |
| Residual | 688.686 | 16 | 43.043 |  |  |  |
| CD | 28.608 | 2 | 14.304 | 2.262 | 0.121 | 0.014 |
| Residual | 202.392 | 32 | 6.325 |  |  |  |
| Avatar * CD | 4.294 | 2 | 2.147 | 0.363 | 0.699 | 0.002 |
| Residual | 189.373 | 32 | 5.918 |  |  |  |

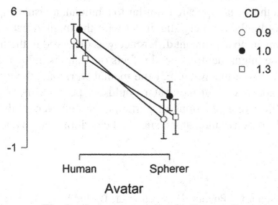

**Fig. 8.** Average score plot for "agency."

*Agency* is an exercise subject feeling, and when a user can voluntarily control his or her avatar, it might become high. Therefore, a significant difference was expected in the main effect of the C/D ratio; however, only the visual effect was confirmed. Conversely, the motion subject feeling might be affected by a change in the visual joint position of the avatar.

3) *Ownership, 4) Appearance, and 5) Response.* A two-factor ANOVA, as the within-subject factors, revealed no main effect on both the factors, namely, the external appearance of the virtual avatar and the C/D ratio.

*NASA-TLX.* The NASA-TLX questionnaire is a standard measure of the mental work-load in a task environment. It comprises six subscales: 1) mental demand, 2) physical demand 3) temporal demand, 4) performance, 5) effort, 6) and frustration. In this study, we calculated the weight workload (WWL) on the basis of these six subscales to assess the mental work load; thus, ANOVA was applied only to WWL.

1) *Weighted Workload.* A two-factor ANOVA, as the within-subject factors, revealed no main effect on both the factors, namely, the external appearance of the virtual avatar and the C/D ratio.

## 4 Conclusion

In this study, we investigated the effects of the differences in the visual joint angles of avatars and the actual joint angles fed back from participants' sense of proprioception on realistic sensation, somatization sensation, and mental workload. Specifically, we investigated the effects of repeated simple point-to-point movements by using different avatar C/D ratios and appearances. Consequently, the main effect of the C/D ratio was recognized in the subscale of *realism* in the presence, and it was proved that *realism* decreased as the constraint of the movement increased. In SoE, the strongest effect of the C/D ratio was observed on the *location* subscale when the C/D ratio was 0.9. Additionally, when the avatar appeared similar to a human, a visual gap occurred in the joint position; accordingly, a synergistic effect with the main effect of the appearance was expected; however, no significant difference was observed in the interaction effect. The *agency* subscale might identify visual effects only and is influenced by changes in the visual joint position of the avatar. For the mental workload, the main effects of the C/D ratio and the appearance of the avatar could not be confirmed. In this study, we focused on the subjective evaluations of participants; however, in the future, we will compare them with objective indicators, such as behavioral and physiological data.

## References

1. Birglen, L., Gosselin, C., Pouliot, N., Monsarrat, B., Laliberte, T.: SHaDe, a new 3-DOF haptic device. IEEE Trans. Robot. Autom. **18**, 166–175 (2002)

2. Silva, A.J., Ramirez, O.A.D., Vega, V.P., Oliver, J.P.O.: PHANToM OMNI haptic device: kinematic and manipulability. In: 2009 Electronics, Robotics and Automotive Mechanics Conference (CERMA), pp. 193–198 (2009)
3. Martin, S., Hillier, N.: Characterisation of the Novint Falcon haptic device for application as a robot manipulator. In: Australasian Conference on Robotics and Automation (2009)
4. Najdovski, Z., Nahavandi, S.: Extending haptic device capability for 3D virtual grasping. In: Ferre, M. (ed.) EuroHaptics 2008. LNCS, vol. 5024, pp. 494–503. Springer, Heidelberg (2008). https://doi.org/10.1007/978-3-540-69057-3_63
5. Blanch, R., Guiard, Y., Beaudouin-Lafon, M.: Semantic pointing: improving target acquisition with control-display ratio adaptation. In: Proceedings of the 2004 Conference on Human Factors in Computing Systems - CHI 2004, pp. 519–526. ACM Press, Vienna (2004)
6. Dominjon, L., Lecuyer, A., Burkhardt, J.-M., Richard, P., Richir, S.: Influence of control/display ratio on the perception of mass of manipulated objects in virtual environments. In: IEEE Proceedings, VR 2005, Virtual Reality, pp. 19–25 (2005)
7. Casiez, G., Vogel, D., Balakrishnan, R., Cockburn, A.: The impact of control-display gain on user performance in pointing tasks. Hum.-Comput. Interact. 23, 215–250 (2008)
8. Yamaguchi, T., Richard, P., Veaux, F., Dinomais, M., Nguyen, S.: Upper-body interactive rehabilitation system for children with cerebral palsy: the effect of control/display ratios. Virtual Reality 6 (2012)
9. Hinckley, K., Tullio, J., Pausch, R., Proffitt, D., Kassell, N.: Usability analysis of 3D rotation techniques. In: Proceedings of the 10th Annual ACM Symposium on User interface Software and Technology - UIST 1997, pp. 1–10. ACM Press, Banff (1997)
10. Kwon, S., Choi, E., Chung, M.K.: Effect of control-to-display gain and movement direction of information spaces on the usability of navigation on small touch-screen interfaces using tap-n-drag. Int. J. Ind. Ergon. 41, 322–330 (2011)
11. van Mensvoort, K., Hermes, D.J., van Montfort, M.: Usability of optically simulated haptic feedback. Int. J. Hum.-Comput. Stud. 66, 438–451 (2008)
12. Riyal, R., Patel, A.D., Murugesan, T., Turini, G., Young, J.G.: Effect of control-display transfer function on pointing performance for a hand/finger based touchless gestural controls: a preliminary investigation. Proc. Hum. Factors Ergon. Soc. Ann. Meet. 59, 1085–1089 (2015)
13. Kilteni, K., Groten, R., Slater, M.: The sense of embodiment in virtual reality. Presence: Teleoper. Virtual Environ. 21, 373–387 (2012)

# User Experience in Virtual, Augmented and Mixed Reality

# Analysis of Differences in the Manner to Move Object in Real Space and Virtual Space Using Haptic Device for Two Fingers and HMD

Yuki Aoki[1], Yuki Tasaka[1], Junji Odaka[1($\boxtimes$)], Sakae Yamamoto[1], Makoto Sato[2], Takehiko Yamaguchi[3], and Tetsuya Harada[1]

[1] Tokyo University of Science, 6-3-1 Niijuku, Katsushika-ku, Tokyo, Japan
8119518@ed.tus.ac.jp, odaka.junji.hrlb@gmail.com
[2] Tokyo Institute of Technology, 4259 Nagatsuta-cho, Midori-ku, Yokohama, Kanagawa, Japan
[3] Suwa University of Science, Toyohira, Chino-City, Nagano 5000-1, Japan

**Abstract.** One of the elements that make up VR is "self-projection". This is to create a state where a person can enter the virtual space and experience it as a first person. If not only HMD but also haptic devices can be used to touch and move objects in the virtual space with the haptic sensation, it is thought that self-projection performance will improve. Therefore, we developed a haptic device with two rings (two-point control type SPIDAR-GCC) so that the user can perform knob operation in the same way as in reality. And we constructed workspaces in virtual and real space with highly similarity to investigate the difference between the work performed in virtual space and in real space. Using this device and HMD, the experimental participants were asked to accomplish a pick-and-place task which is pinching the peg with two fingers and insert it into the hole in the pegboard. In this experiment, the difference in the manner to move objects between in real space and in virtual space was observed. It was mainly due to the errors of measuring the displacements of fingers in the haptic device.

**Keywords:** Haptic device · Force feedback · Work analysis

## 1 Introduction

### 1.1 Background and Purpose

The three elements required for VR are three-dimensional spatiality, real-time interaction, and self-projection [1]. We have made experiments to clarify the self-projection property so called Sense of Agency (SOA) in VR using a desktop haptic device SPIDAR-GCC [2]. This device has a spherical grip with buttons as an end effector, and the user can pick and move a virtual object by grasping the grip and pressing one of the buttons. However, this manner of picking and grasping an object is different from the manner to do them in real world. The user always needs to grasp the end effector regardless of

© Springer Nature Switzerland AG 2020
C. Stephanidis et al. (Eds.): HCII 2020, LNCS 12428, pp. 291–301, 2020.
https://doi.org/10.1007/978-3-030-59990-4_22

whether holding the virtual object or not and needs to push the button to hold it. In this way, it is thought that the self-projection property cannot be produced enough on such a device.

One of the two purposes of this research is to improve the self-projection performance by developing a two-point control type SPIDAR-GCC that can pinch virtual objects by the action of picking up an object with two fingers actually. Hereinafter, two-point control type SPIDAR-GCC is referred as SPIDAR-GCC-TP. The other one is to clarify the self-projection property in the virtual environment using it.

### 1.2  Outline of the Research

In this experiment, using the VIVE PRO [3] and the SPIDAR-GCC-TP as HMD and haptic device respectively, the experimental participants pinched, moved, and fitted the peg into the hole of the peg board with thumb and index finger in virtual space. The same work was performed in the real space, and the difference in the distance and trajectory of the object was analyzed and evaluated.

## 2  Proposed Device and Application Configuration

### 2.1  SPIDAR-GCC-TP

In this experiment, the SPIDAR-GCC-TP shown in Fig. 1 was developed and used. This is an improvement of the space interface device SPIDAR-GCC for presenting force sense developed by Sato. Table 1 shows the specifications of the SPIDAR-GCC-TP. The original SPIDAR-GCC is a device that can grasp and release objects in virtual space by pressing or releasing a button with a finger, and is connected to an end effector equipped with a button. This is a device that can be operated in 6 degrees of freedom, by controlling eight strings with motors. In this study, we improved this, divided eight threads into two four threads groups, attached a ring end effector to each, and made it possible to operate

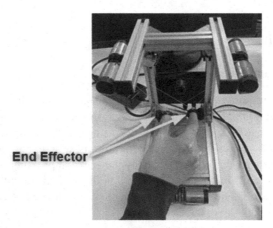

**Fig. 1.**  SPIDAR-GCC-TP

with two fingers through the end effectors. With this improvement, each end effector has lost the degree of freedom of rotation, however, the operation of pinching with two fingers has become available.

**Table 1.** The specifications of SPIDAR-GCC [2]

| | |
|---|---|
| Update rate [Hz] | 500 |
| Minimum tension [N] | 0.45 |
| Maximum tension [N] | 2.0 |
| Motor terminal voltage [V] | 12 |
| Motor terminal resistance [ohm] | 7.84 |
| Motor torque constant [Nm/A] | 0.0092 |
| Motor speed constant [rpm/V] | 900 |
| No load current [A] | 0.06 |
| Maximum duration of force presentation [s] | 60 |
| Grip radius [m] | 0.0565 |
| Connection | USB 2.0 |

## 2.2  Vive Pro

In this study, we used VIVE PRO as a visual presentation device. By using two SteamVR Base Station 2.0, this device is capable of accurate tracking in millimeters or less on a wide room scale of 5 m × 5 m. Figure 2 shows the VIVE PRO, and Table 2 shows the specifications of it.

**Fig. 2.**  VIVE PRO [3]

**Table 2.**  the specifications of VIVE PRO [3]

| Screen | Dual AMOLED 3.5 inch (diagonal) |
|---|---|
| Resolution | 2880 × 1600 |
| Refresh rate [Hz] | 70 |
| Viewing angle [°] | 110 |
| Connection | Bluetooth, USB-C port |
| Sensor | SteamVR tracking, G sensor, gyroscope, proximity sensor, IPD sensor |

### 2.3  Application Configuration

In this research, using a SPIDAR-GCC-TP and HMD, an application of Pick and Place Task in which user holds a small peg with two fingers and inserts it into a hole of the same dimensions. Figure 3 shows the virtual space created. This application created using Unity 3D [4].

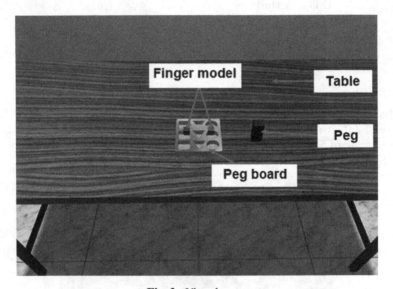

**Fig. 3.**  Virtual space

As shown in Fig. 4, in the virtual space, when the experiment cooperator sits with the body facing the desk, the Left and right direction is the x, the vertical direction is the y, and the depth direction is the z-axis. All the objects in the virtual space were set to the same dimension, shape, and weight as the real ones.

When the user puts the thumb and the index finger into the rings, which are the end effectors of the SPIDAR-GCC-TP, and moves them, the finger model in the virtual space moves similarly. However, the displacement amplification ratio of the finger model was set to three times of the finger displacement due to the dimensional limitation of

**Fig. 4.** Coordinate axes in virtual space

the device workspace. There are three types of pegs: cylindrical, triangular prism, and quadrangular prism, and in this experiment, only quadrangular prism peg was used for its easy hold ability in the virtual space. Therefore, only square holes on the pegboard were used. There was some difficulty in the construction of the physical model for the haptics of small objects. In addition, for the simplify of the physical calculations, the degree of freedom of the object was limited to 3 of translation. Table 3 shows the specifications for the peg and the pegboard.

**Table 3.** Specifications of the peg and the pegboard

|  | Peg board | Peg |
| --- | --- | --- |
| Weight | 98 g | 26 g |
| Size | 12.85 cm × 12.85 cm × 1.7 cm | 2.7 cm × 2.7 cm × 6 cm |
| Hole size | 3.25 cm × 3.25 cm × 1.4 cm | – |

The texture of the wood-grained desk top used in the experiment in the real space was photographed and mapped on the desk top in the virtual space. In order to analyze the behaviors in the virtual workspace, it was necessary to record them. Therefore, instead of capturing movie, we implemented a function that records the coordinates of the objects in a csv file and replay them based on the data.

## 3 Evaluation Experiment of the Behavior During the Task

### 3.1 The Method of Experiment in Real and Virtual Space

**Experiment in Real Space.** The number of participants in the experiment was 13. Figure 5 shows the working space in the real space. As shown in Fig. 6, in order to analyze the behavior in the real space using movies, three video cameras A, B and C were installed on right front, right beside, and right above.

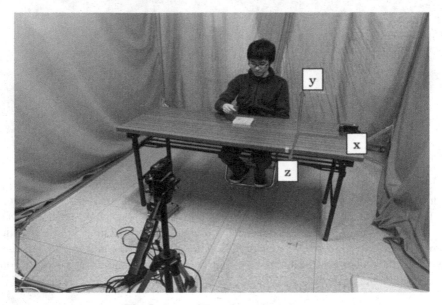

**Fig. 5.** State of Experiment in real space

The participants seated in two positions, one was right in front of the pegboard, and the other was slightly on the left of the pegboard where they felt most comfortable in working. Hereinafter, the experiments performed while sitting at each position are referred to as "real space front task" and "real space left task". The experimental procedures for the two tasks were the same. Table 4 shows them.

**Experiment in Virtual Space.** Figure 7 shows the experiment in the virtual space. Since the work on the pegboard is the main focus, the SPIDAR-GCC-TP was installed at the position where the pegboard in the real space was located, and the position of the pegboard in the virtual space was also the same as in the real space.

The participants seated in two positions: one was right in front of the pegboard, and the other was slightly on the left of the pegboard where they felt most comfortable in working. Hereinafter, the tasks performed while sitting at each position are referred to

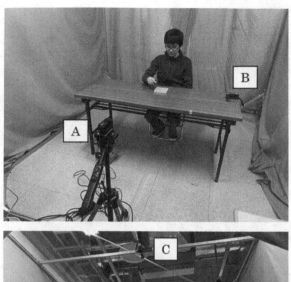

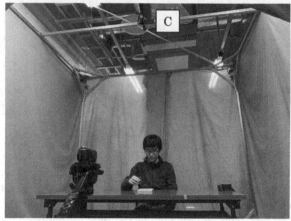

**Fig. 6.** Position of the three video cameras

**Table 4.** Experimental procedures in real space

| Order | Procedure |
|---|---|
| 1 | Waits while sitting on the chair and places both hands on laps |
| 2 | With the start signal of an experimenter, grasps the peg with the right index finger and thumb and inserts it into the square hole of the pegboard |
| 3 | After putting the peg into the hole, releases fingers from the peg |
| 4 | Picks the peg out of the hole with the same fingers, and puts it at the position originally located |
| 5 | Places both hands on laps |
| 6 | Repeats 2 to 5 nine more times without start signal |

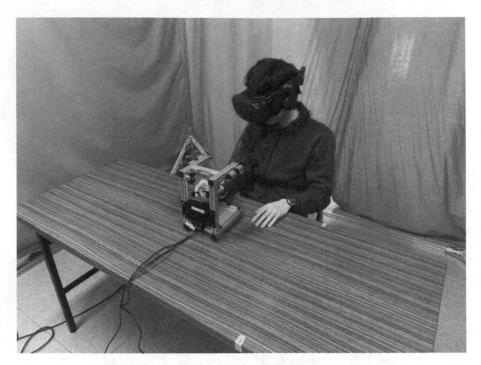

**Fig. 7.** Experiment in virtual space

as "virtual space front task" and "virtual space left task". The device was placed in the place where the experiment cooperator was most comfortable to operate in the virtual space front task, and in the place where the pegboard was placed in the real space task in the virtual space left task. The experimental procedures for the two tasks are the same. Table 5 shows the experimental procedure in the virtual space.

**Table 5.** Experimental procedure in virtual space

| Order | Procedure |
|---|---|
| 1 | Sits on the chair and puts on the end effectors and locates them at the center of the workspace of SPIDAR-GCC-TP for the calibration |
| 2 | With the start signal of an experimenter, grasps the peg with the right index finger and thumb and inserts it into the square hole of the pegboard |
| 3 | After putting the peg into the hole, releases fingers from the peg |
| 4 | Picks the peg out of the hole with the same fingers, and puts it at the position originally located |
| 5 | Place the fingers at the initial position where the calibration was performed |
| 6 | Repeats 2 to 5 nine more times without start signal |

## 3.2  Results and Discussion

**Working Time.**  Figure 8 shows the average of the working time in the real space and in the virtual space for all the participants. The working time in the virtual space was about 2 s longer than the working time in the real space. This suggests that participants tended to perform the task in the virtual space more carefully than the task in real space. The reason for this is that the SPIDAR-GCC-TP had errors in the measurement of the displacement of the end effectors.

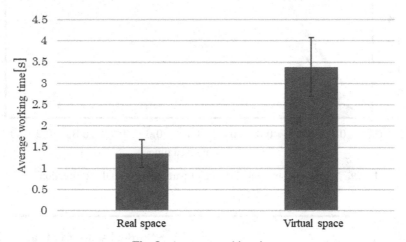

**Fig. 8.** Average working time

**Trajectory of Peg Movement.**  The video analysis software Kinovea [5] was used to analyze the work in the real space. This is a software that can track the movement of objects and humans in moving images, display their trajectories, and record their coordinates. Using it, the pegs in the video were tracked and the coordinates were recorded. The tracking was performed from the moment when the peg was pinched, until the moment when it was completely stand still in the hole on the pegboard. In the analysis of the work in the virtual space, the coordinate records of the objects described in Sect. 2.3 were used.

The task time of the trajectories was normalized and divided into ten equal parts. And the graphs of normalized time change of the normalized movement distance in x and y-axis direction were created for each participant. Since most of the participants had similar tendencies, an example of a typical one is shown in Fig. 9 and 10. No clear and distinctive difference was found in the trajectories between in the real space and in the virtual space. Also, from Fig. 8, 9 and 10, since the graphs of the real task and the virtual task have almost the same shape, it is considered that in virtual space, the overall operation is slower rather than the specific operation.

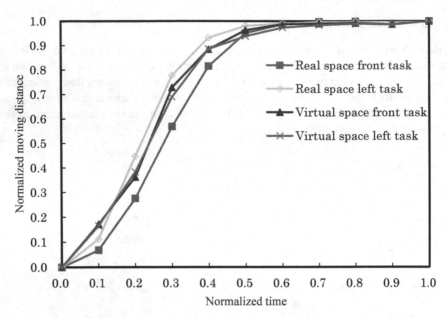

**Fig. 9.** Normalized moving distance of participant A in the x direction

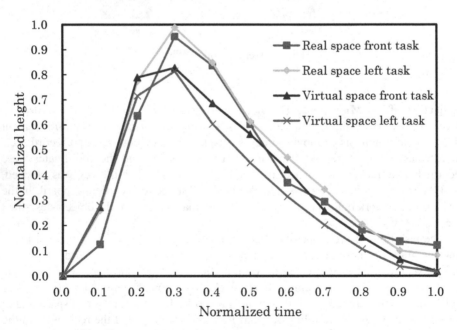

**Fig. 10.** Normalized height of participant A in the y direction

# 4   Conclusions and Prospects

In this study, we have improved the 6 degree of freedom desktop haptic device SPIDAR-GCC and developed SPIDAR-GCC-TP that can be operated with two fingers so that Pick and Place task can be performed. In the evaluation experiment, the working time in the virtual space was about 2 s longer than the working time in the real space. The reason for this is that the SPIDAR-GCC-TP had errors in the measurement of the displacement of the end effectors. In future research, we will improve above problem and make evaluation experiments more precisely.

**Acknowledgments.** We would like to thank all the research participants. This work was supported by JSPS KAKENHI Grant Number JP17H01782.

# References

1. Tachi, S., Satou, M., Hirose, M.: Virtual Reality Science. Virtual Reality Society of Japan (2011)
2. Tasaka, Y., Ichimaru, H., Yamamoto, S., Sato, M., Yamaguchi, T., Harada, T.: Analysis of differences in the manner to move objects in a real and virtual space. In: Yamamoto, S., Mori, H. (eds.) HCII 2019. LNCS, vol. 11570, pp. 58–69. Springer, Cham (2019). https://doi.org/10.1007/978-3-030-22649-7_6
3. VIVE Pro|VIVE Enterprise. https://enterprise.vive.com/us/product/vive-pro/. Accessed 12 June 2020
4. Unity Technologies. https://unity3d.com/. Accessed 12 June 2020
5. Kinovea. https://www.kinovea.org/. Accessed 12 June 2020

# A Study of Size Effects of Overview Interfaces on User Performance in Virtual Environments

Meng-Xi Chen[✉] [iD] and Chien-Hsiung Chen

National Taiwan University of Science and Technology, Taipei 10607, Taiwan
cmx12677@gmail.com

**Abstract.** Many virtual environment applications use an overview interface showing a survey of the entire space. Little research has been conducted on the size effects of overview interfaces on users' performance and experiences in virtual environments. The experiment is two (overview interface size) × two (familiarity with mobile devices) between-subject design. Participants completed three tasks on a mobile device and filled out the NASA task load index (TLX) questionnaire as a measure of mental workload. Thirty-two participants were invited to take part in the experiment based on convenient sampling method. The results are as follows: (1) Participants using the smaller overview interface performed significantly better than those using the larger overview interface in the most difficult task. (2) Participants who were more familiar with mobile devices performed significantly better than those who were less familiar with mobile devices in their first visit to the unfamiliar virtual environment. (3) The larger overview interface required significantly more mental workload than the smaller overview interface in terms of the sum score and performance score on NASA TLX. (4) The impacts of overview interface size on users' performance in the most difficult task and the effort they felt appear to be different for participants with different levels of familiarity with mobile devices. Thus the design of overview interface should consider users' familiarity with mobile devices.

**Keywords:** Virtual environment · Interface design · Overview · Size

## 1 Introduction

The virtual environments have gradually been used universally. It is hard to find directions towards a destination in an unfamiliar virtual environment. There is a growing research demand on visual aids to ensure the users' engagement in virtual environments. Many virtual environment applications use an overview interface showing a survey of the entire space to preserve users' sense of position and context. A larger overview interface on a limited screen means a smaller detail interface displaying various details of the information space [2]. The size of the overview interface in virtual environments has not been researched as yet.

Studies have indicated that differences in gender, experience, age and other traits affect users' performance and subjective experience in virtual environments [4, 10, 13]. Lee et al. [9] suggested that individuals who were familiar with mobile devices performed

© Springer Nature Switzerland AG 2020
C. Stephanidis et al. (Eds.): HCII 2020, LNCS 12428, pp. 302–313, 2020.
https://doi.org/10.1007/978-3-030-59990-4_23

better in wayfinding with interactive maps. New information related to the usage of mobile devices that stored in users' long-term memory is easy to understand. Designers of overview interfaces in virtual environments should pay attention to the issue of whether users are familiar with information technology devices.

The purpose of the study is to examine the use of overview interfaces by people with different levels of familiarity with mobile devices, specifically investigating how the size of the overview interface affects users' performance and subjective experience in a virtual environment.

## 2  Related Work

Virtual reality technology is used to create virtual environments simulating different kinds of space, such as an architecture, a tourist attraction and a game space. Designers propose solving problems in virtual environments by a map, a grid, nodes, paths, audio cues, and other assistive aids [5, 12]. Overview and detail interfaces can display the entirety and details of an information space [11]. A detail interface is always used to control position at a short distance and undertake diagnostic tasks, while monitoring tasks might benefit from an overview interface [8]. Empirical studies have demonstrated that the overview interface helps increase user efficiency, satisfaction, preference and the sense of position and control [8]. Nevertheless, the authors note that integration of overview and detail interfaces increases the mental and motor efforts required from a user.

Visual variables including size, value, grain, color, orientation, and shape [1] are used to explicitly represent spatial relationships on an overview interface. Earlier studies have evaluated the effectiveness and efficiency of some visual variables of overview interfaces [4, 6]. In some cases, the size of an overview interface can be changed by clicking on a button on the screen. To enlarge the size of the overview interface can attract users' attention and vary the amount of information that can be displayed at one point in time. However, there have not been much studies on the important factor that may affect users' performance and mental workload in virtual environments.

The typical mobile devices, such as smartphone and tablet computer, provide the opportunity to move virtual environments from fixed places to more convenient locations. Mobile virtual environment platforms now are equipped with high display resolution, sufficient computing and networking capability. When the technology devices are unfamiliar to the users, it is prone to cause operation errors. With the accumulation of hands-on experience using the devices, there is less demand for the description of system operation and users might feel less pressure to meet performance expectations. Users expect to use familiar interaction techniques instead of unfamiliar ones in 3D interfaces [3]. Effects of users' experience of using mobile devices on the usage of overview interfaces on a mobile virtual environment platform are still to be confirmed.

## 3  Methods

### 3.1  Participants

A total of thirty-two participants (9 men and 23 women) were invited to take part in a wayfinding experiment. Half of the participants used mobile devices for an average of less than six hours (1–6 h) a day. The other half used mobile devices for an average of more than or equal to six hours (6–12 h) a day. All participants used the device of experiment with no problem in basic operation.

All participants were university students, 30 undergraduates and 2 graduate students, between 18 and 24 years old. Seven people used overview interfaces in virtual environments once a week (21.88%). Participants who used overview interfaces in virtual environments less than once a week were 14 people (43.75%). Eleven people did not have experience of using overview interfaces in virtual environments (34.38%).

### 3.2 Materials and Apparatus

A simple virtual exhibition was created with 3DS Max software by reference to the memorial museum of Liu Zigu, who is a famous Chinese artist (see Fig. 1). The overview interface was created with Photoshop software (see Fig. 2). The experiment operation was configured with Unity 3D game engine. This experiment was conducted on an iPad Air 2 tablet computer with screen size of 9.7 in., resolution of 2048 × 1536 pixels, and iOS 9.7 operating system.

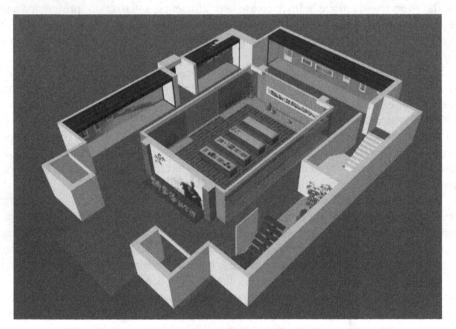

**Fig. 1.** View from the top to see the whole virtual model set up for this study

### 3.3 Experimental Design

This experiment adopted a two (overview interface size) x two (familiarity with mobile devices) between-subjects design. There were two types of overview interfaces adopted

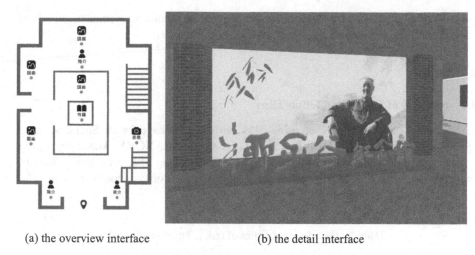

(a) the overview interface                    (b) the detail interface

**Fig. 2.** The overview and detail interfaces used in this study

in this study: the larger overview interface (1024 × 1536 pixels) whose size is close to the detail interface, and the smaller overview interface (384 × 512 pixels) whose size is significantly smaller than the detail interface. Participant were randomly assigned to either of two experimental conditions described above. People who spend less than six hours a day and who spend more than or equal to six hours a day using mobile devices were compared experimentally in terms of their performance and experiences. Participants were divided into four groups. Three research questions were addressed:

- Whether the size of the overview interface can affect users' performance and subjective experience in a virtual environment?
- Whether high familiarity with mobile devices can improve users' performance in a virtual environment?
- What is the interaction effect between overview interface size and users' familiarity with mobile devices?

### 3.4 Procedure

Each participant was asked to complete three tasks with increasing complexities to get them interacting with the virtual exhibition. The time of completing each task was recorded. We asked participants to find virtual objects matching certain criteria (e.g., shape, color, distance, orientation). In the more difficult tasks, participants needed to memorize and compare the information of the space. Participants had to accomplish these search and comparison tasks without help from the experimenter.

After completing all the tasks, participants were required to fill out a questionnaire to gather data regarding their subjective experience in the virtual environment. Mental workload was gathered by NASA task load index (TLX) [7], since subjective experience is crucial to what extent an overview interface will be used. NASA TLX consists of six subscales: mental demand, physical demand, temporal demand, performance, effort and frustration. Lower is better for all measures (1-best; 7-worst).

# 4    Results

## 4.1    Analysis of Task Completion Time

The collected data were analyzed using IBM Statistical Package for the Social Sciences (SPSS) software in terms of two-way analysis of variance (ANOVA). Basic descriptive statistics including the means and standard errors regarding task completion time are presented in Table 1. Table 2 presents the ANOVA results of task completion time for each independent variable level.

**Table 1.** Descriptive statistics of task completion time (s)

|  |  | Task 1 | | Task 2 | | Task 3 | | N |
|---|---|---|---|---|---|---|---|---|
|  |  | M | SD | M | SD | M | SD | |
| Size | Large | 47.274 | 39.485 | 15.908 | 7.447 | 33.956 | 20.593 | 16 |
|  | Small | 28.999 | 35.676 | 17.531 | 18.548 | 21.253 | 11.634 | 16 |
| Familiarity | <6 h/day | 52.614 | 46.251 | 14.881 | 6.121 | 24.174 | 13.364 | 16 |
|  | ≥6 h/day | 23.659 | 20.583 | 18.558 | 18.874 | 31.036 | 21.019 | 16 |

**Table 2.** Two-way ANOVA of task completion time

| Source | | SS | df | MS | F | P |
|---|---|---|---|---|---|---|
| Task 1 | Size | 2671.988 | 1 | 2671.988 | 2.178 | .151 |
|  | Familiarity | 6706.847 | 1 | 6706.847 | 5.467 | .027* |
|  | Size * Familiarity | 1418.181 | 1 | 1418.181 | 1.156 | .291 |
| Task 2 | Size | 21.076 | 1 | 21.076 | .106 | .747 |
|  | Familiarity | 108.155 | 1 | 108.155 | .545 | .466 |
|  | Size * Familiarity | 331.982 | 1 | 331.982 | 1.674 | .206 |
| Task 3 | Size | 1290.955 | 1 | 1290.955 | 5.358 | .028* |
|  | Familiarity | 376.683 | 1 | 376.683 | 1.563 | .222 |
|  | Size * Familiarity | 1268.694 | 1 | 1268.694 | 5.266 | .029* |

$\alpha = 0.05$, *$p < 0.05$.

The first task was the easiest task, which is to search for a specific object. Table 2 reveals no significant main effect of overview interface size regarding the Task 1 completion time ($F = 2.178$, $p = 0.151 > 0.05$). There was a significant main effect of users' familiarity with mobile devices on the Task 1 completion time ($F = 5.467$, $p = 0.027$

< 0.05). The results suggest that participants who spend less than six hours a day using mobile devices (M = 52.614, Sd = 46.251) took significantly more time to complete the first task than those who spend more than or equal to six hours a day (M = 23.659, Sd = 20.583). There was no significant interaction effect between the variables of overview interface size and users' familiarity with mobile devices on the Task 1 completion time (F = 1.156, p = 0.291 > 0.05).

The second task was a more difficult task in which participants needed to compare information about distance and look up relevant details. As is shown in Table 2, the main effect of overview interface size on the Task 2 completion time was not significant (F = 0.106, p = 0.747 > 0.05). The main effect of users' familiarity with mobile devices on the Task 2 completion time was also not significant (F = 0.545, p = 0.466 > 0.05). There were no significant main and interaction effects (F = 1.674, p = 0.206 > 0.05) in terms of the Task 2 completion time. It indicates that in the second task, overview interface size and users' familiarity with mobile devices did not significantly affect user performance, as participants might already be familiar with the virtual environment.

The third task was the most difficult task in which participants needed to compare information about orientation and search for a specific object. A significant main effect of overview interface size on the Task 3 completion time was detected (F = 5.358, p = 0.028 < 0.05). The results suggest that the Task 3 completion time for the larger overview interface (M = 33.956, Sd = 20.593) was significantly longer than that for the smaller overview interface (M = 22.253, Sd = 11.634). Table 2 shows that the main effect of users' familiarity with mobile devices on the Task 3 completion time was not significant (F = 1.563, p = 0.222 > 0.05). An interaction effect of overview interface size and users' familiarity with mobile devices was found for the Task 3 completion time (F = 5.266, p = 0.029 < 0.05).

According to Fig. 3, when using the larger overview interface, participants who spend more than or equal to six hours a day using mobile devices took more time (M = 43.684, Sd = 19.619) to complete the third task than those who spend less than six hours a day (M = 24.229, Sd = 17.537). Whereas, when using the smaller overview interface, participants who spend more than or equal to six hours a day using mobile devices took less time (M = 18.387, Sd = 14.005) to complete the third task than those who spend less than six hours a day (M = 24.119, Sd = 8.668). Participants who spend less than six hours a day using mobile devices took quite similar time to complete the third task using both overview interfaces.

## 4.2  Analysis of NASA TLX

The descriptive statistics and two-way ANOVA of NASA TLX questionnaire are shown in Table 3 and Table 4. We analyzed the mental workload on six dimensions with each overview interface and found no significant difference between our tested conditions on the sections of mental demand, physical demand, temporal demand and frustration.

**Mental Demand.** Participants who spend more than or equal to six hours a day using mobile devices gave the lowest mental demand scores (M = 3.625, Sd = 3.335). Participants who spend less than six hours a day using mobile devices gave the highest mental demand scores (M = 6.000, Sd = 4.556). Participants supporting by the larger overview

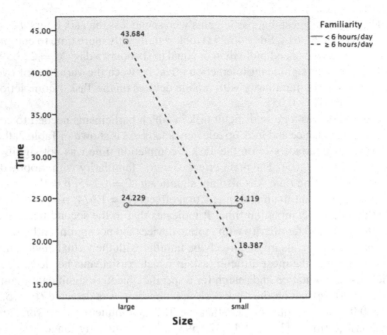

**Fig. 3.** The interaction diagram of overview interface size and users' familiarity with mobile devices in terms of the Task 3 completion time

interface (M = 5.250, Sd = 4.468) and the smaller overview interface (M = 4.375, Sd = 3.822) reported similar mental demand scores.

**Physical Demand.** All experimental groups' reports on physical demanding were rather low. Participants who spend more than or equal to six hours a day using mobile devices gave the lowest physical demand scores (M = 0.125, Sd = 0.363). Participants who spend less than six hours a day using mobile devices gave the highest physical demand scores (M = 0.938, Sd = 2.059). Participants using the larger overview interface (M = 0.813, Sd = 2.062) reported higher physical demand scores than those using the smaller overview interface (M = 0.250, Sd = 0.552).

**Temporal Demand.** The smaller overview interface leaded to the lowest temporal demand (M = 5.875, Sd = 4.778). The other conditions were considered to a higher temporal demand.

**Performance.** A significant main effect of overview interface size was detected in terms of the performance score on NASA TLX (F = 5.313, p = 0.029 < 0.05). Participants using the smaller overview interface reported the lowest performance scores (M = 2.084, Sd = 3.663). Participants supporting by the larger overview interface reported the highest performance scores (M = 6.104, Sd = 5.893).

**Effort.** An interaction effect of overview interface size and users' familiarity with mobile devices was found on effort score (F = 5.363, p = 0.028 < 0.05). According to Fig. 4, when using the smaller overview interface, participants who spend more

**Table 3.** Descriptive statistics of NASA TLX

|  |  | Size | | Familiarity | |
| --- | --- | --- | --- | --- | --- |
|  |  | Large | Small | <6 h/day | ≥6 h/day |
| Mental demand | M | 5.250 | 4.375 | 6.000 | 3.625 |
|  | SD | 4.468 | 3.822 | 4.566 | 3.335 |
|  | N | 16 | 16 | 16 | 16 |
| Physical demand | M | .813 | .250 | .938 | .125 |
|  | SD | 2.062 | .552 | 2.059 | .363 |
|  | N | 16 | 16 | 16 | 16 |
| Temporal demand | M | 7.750 | 5.875 | 6.625 | 7.000 |
|  | SD | 5.492 | 4.778 | 5.466 | 4.992 |
|  | N | 16 | 16 | 16 | 16 |
| Performance | M | 6.104 | 2.084 | 5.209 | 2.979 |
|  | SD | 5.893 | 3.663 | 6.255 | 3.871 |
|  | N | 16 | 16 | 16 | 16 |
| Effort | M | 5.250 | 4.458 | 5.145 | 4.563 |
|  | SD | 5.113 | 4.080 | 4.291 | 4.953 |
|  | N | 16 | 16 | 16 | 16 |
| Frustration | M | 2.333 | .896 | 2.146 | 1.083 |
|  | SD | 3.445 | 1.298 | 3.101 | 2.106 |
|  | N | 16 | 16 | 16 | 16 |
| Sum | M | 27.500 | 17.937 | 26.062 | 19.375 |
|  | SD | 10.642 | 12.383 | 13.804 | 10.048 |
|  | N | 16 | 16 | 16 | 16 |

than or equal to six hours a day using mobile devices gave significantly lower effort scores (M = 2.375, Sd = 2.292) than those who spend less than six hours a day (M = 6.542, Sd = 4.528). Whereas, when using the larger overview interface, participants who spend more than or equal to six hours a day using mobile devices gave higher effort scores (M = 6.750, Sd = 6.031) than those who spend less than six hours a day (M = 3.749, Sd = 3.808).

**Frustration.** Participants reported low frustration scores in all conditions. Participants using the smaller overview interface gave the lowest frustration scores (M = 0.896, Sd = 1.298). Participants using the larger overview interface reported the highest frustration scores (M = 2.333, Sd = 3.445).

**Table 4.** Two-way ANOVA of NASA TLX

| Source | | SS | df | MS | F | P |
|---|---|---|---|---|---|---|
| Mental demand | Size | 6.125 | 1 | 6.125 | .385 | .540 |
| | Familiarity | 45.125 | 1 | 45.125 | 2.838 | .103 |
| | Size * Familiarity | 28.128 | 1 | 28.128 | 1.769 | .194 |
| Physical demand | Size | 2.529 | 1 | 2.529 | 1.186 | .285 |
| | Familiarity | 5.284 | 1 | 5.284 | 2.478 | .127 |
| | Size * Familiarity | 3.335 | 1 | 3.335 | 1.564 | .221 |
| Temporal demand | Size | 28.125 | 1 | 28.125 | 1.011 | .323 |
| | Familiarity | 1.123 | 1 | 1.123 | .040 | .842 |
| | Size * Familiarity | 15.125 | 1 | 15.125 | .544 | .467 |
| Performance | Size | 129.323 | 1 | 129.323 | 5.313 | .029* |
| | Familiarity | 39.761 | 1 | 39.761 | 1.634 | .212 |
| | Size * Familiarity | .780 | 1 | .780 | .032 | .859 |
| Effort | Size | 5.009 | 1 | 5.009 | .261 | .613 |
| | Familiarity | 2.718 | 1 | 2.718 | .142 | .709 |
| | Size * Familiarity | 102.746 | 1 | 102.746 | 5.363 | .028* |
| Frustration | Size | 16.531 | 1 | 16.531 | 2.473 | .127 |
| | Familiarity | 9.031 | 1 | 9.031 | 1.351 | .255 |
| | Size * Familiarity | 7.031 | 1 | 7.031 | 1.052 | .314 |
| Sum | Size | 731.563 | 1 | 731.563 | 5.643 | .025* |
| | Familiarity | 357.759 | 1 | 357.759 | 2.760 | .108 |
| | Size * Familiarity | 11.277 | 1 | 11.277 | .087 | .770 |

$\alpha = 0.05$, *$p < 0.05$.

**Sum.** We also analyzed data of the sum score on NASA TLX. Physical demand and frustration were less weighted contributing factors than the others. There was a significant difference between the larger overview interface and the smaller overview interface ($F = 5.643$, $p = 0.025 < 0.05$). As is shown in Table 3, participants using the larger overview interface reported the highest overall workload (M = 27.50, SD = 10.64). Users of the smaller overview interface reported the least overall workload (M = 17.94, SD = 12.383).

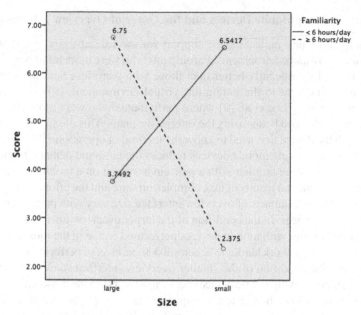

**Fig. 4.** The interaction diagram of overview interface size and users' familiarity with mobile devices in terms of the effort score on NASA TLX

## 5  Discussion

### 5.1  Size Effects of Overview Interfaces

The results of task completion time indicate that the size of the overview interface can affect user performance in a virtual environment. Participants using the smaller overview interface performed significantly better in the most difficult search and comparison task. This might be because the smaller overview interface, by displaying more virtual objects, can help participants acquire more information and then reduce time spent at the task of finding and comparing information. This may indicate that participants did not take much time to integrate information on the overview and detail interfaces.

The results from NASA TLX suggest that the size of the overview interface has a significant influence on subjective experience in a virtual environment. Users with the smaller overview interface experienced significantly less overall workload and were more satisfied with their performance. For each measurement of NASA TLX, participants using the larger overview interface gave higher scores. Users' subjective experience with the overview interface are consistent with the results generated from their performance in the most difficult task. Most probably, the smaller overview interface means a larger detail interface providing more information to make users feel easier to know the environment well.

Overall, measured by task completion time and NASA TLX, the smaller overview interface is a better visual aid than the larger overview interface in virtual environments.

## 5.2   Familiarity with Mobile Devices and the Usage of Overview Interfaces

The results of this study indicate some support for the role of users' familiarity with mobile devices in virtual environments. Participants who were more familiar with mobile devices performed significantly better than those who were less familiar with mobile devices in their first visit to the unfamiliar virtual environment. It is consistent with the results presented by Lee et al. [9] where participants who were more familiar with mobile devices performed better using the interactive maps. This might be because users unfamiliar with the device first need to know the method of operation. This may indicate that the experience of using mobile devices reduces the temporal demand to understand system usage and become familiar with a new environment on a mobile device.

From the experimental results of task completion time and the effort score on NASA TLX, we found that the impacts of overview interface size vary with participants' familiarity with mobile devices. In the condition of the larger overview interface, participants who were more familiar with mobile devices performed worse in the most difficult task and felt like they have to work harder to accomplish their level of performance. The opposite results from the condition of the smaller overview interface show that participants who were less familiar with mobile devices performed worse in the most difficult task and felt like they have to work harder to accomplish their level of performance. Subjective experience of participants who were more familiar with mobile devices were consistent with the results generated from their performance in the most difficult task. One possible explanation is that although the smaller overview interface means a larger detail interface displaying more detailed information, the overview interface showing a survey of the entire virtual environment can provide participants with relative position and distance information to help form cognitive map for decision making and preserve users' sense of position and context. This may indicate that the requirement of overview interface as an aid might decrease as users' familiarity with mobile devices increases. Users who were more familiar with mobile devices would like to pay more attention to the detail interface.

Concerning subjective assessments of mental workload, the low physical demand scores and frustration scores reflected the comfortable and effective interactions with overview interfaces. Although participants who were more familiar with mobile devices gave lower scores than those who were less familiar with mobile devices on most dimensions except temporal demand, the NASA TLX results show no significant main effect of users' familiarity with mobile devices. It is perhaps due to the limited size and complexity of the virtual exhibition used in this study.

All of the participants in this study were university students and relatively young. Differences in education backgrounds and age may affect user performance in virtual environments. These limitations provide the focus for future work.

## 6   Conclusion

In this study, the size of overview interfaces and users' familiarity with mobile devices were manipulated to assess possible effects on users' performance and experiences in virtual environments. Our results show that the smaller overview interface can improve users' performance in the most difficult task, reduce users' overall mental workload and enhance their satisfaction with performance in virtual environments. High familiarity with mobile devices can improve users' performance in their first visit to the unfamiliar virtual environment. The size effects of overview interfaces vary with users' familiarity with mobile devices. Designers of virtual environment applications might be well advised to use the smaller overview interfaces and consider users' familiarity with mobile devices.

# References

1. Bertin, J.: Semiology of Graphics: Diagrams, Networks, Maps. The University of Wisconsin Press, Madison (1983)
2. Büring, T., Gerken, J., Reiterer, H.: Usability of overview-supported zooming on small screens with regard to individual differences in spatial ability. In: Proceedings of the Working Conference on Advanced Visual Interfaces, AVI 2006, pp. 233–240. Association for Computing Machinery, New York (2006)
3. Bowman, D.A., et al.: 3D user interfaces: new directions and perspectives. IEEE Comput. Graph. Appl. **28**(6), 20–36 (2008)
4. Chen, M.-X., Chen, C.-H.: User experience and map design for wayfinding in a virtual environment. In: Yamamoto, S., Mori, H. (eds.) HCII 2019. LNCS, vol. 11570, pp. 117–126. Springer, Cham (2019). https://doi.org/10.1007/978-3-030-22649-7_10
5. Darken, R.P., Sibert, J.L.: Wayfinding strategies and behaviors in large virtual worlds. In: Proceedings of the SIGCHI Conference on Human Factors in Computing Systems, Vancouver, BC, pp. 142–149 (1996)
6. Devlin, A.S., Bernstein, J.: Interactive wayfinding: map style and effectiveness. J. Environ. Psychol. **17**(2), 99–110 (1997)
7. Hart, S.G., Staveland, L.E.: Development of NASA-TLX (task load index): results of empirical and theoretical research. In: Hancock, P.A., Meshkati, M. (eds.) Human Mental Workload, pp. 139–183. North Holland, Amsterdam (1988)
8. Hornbæk, K., Bederson, B.B., Plaisant, C.: Navigation patterns and usability of zoomable user interfaces with and without an overview. ACM Trans. Comput.-Hum. Interact. **9**(4), 362–389 (2002)
9. Lee, S., Kim, E.Y., Platosh, P.: Indoor wayfinding using interactive map. Int. J. Eng. Technol. **7**(1), 75–80 (2015)
10. Moffat, S.D., Zonderman, A.B., Resnick, S.M.: Age differences in spatial memory in a virtual environment navigation task. Neurobiol. Aging **22**, 787–796 (2001)
11. Plaisant, C., Carr, D., Shneiderman, B.: Image browsers: taxonomy, guidelines, and informal specifications. IEEE Softw. **12**(2), 21–32 (1995)
12. Ramloll, R., Mowat, D.: Wayfinding in virtual environments using an interactive spatial cognitive map. In: Proceedings of the 5th International Conference on Information Visualization, London, England, UK, pp. 574–583 (2001)
13. Tlauka, M., Brolese, A., Pomeroy, D., Hobbs, W.: Gender differences in spatial knowledge acquired through simulated exploration of a virtual shopping centre. J. Environ. Psychol. **25**(1), 111–118 (2005)

# Text Input in Virtual Reality Using a Tracked Drawing Tablet

Seyed Amir Ahmad Didehkhorshid, Siju Philip, Elaheh Samimi,
and Robert J. Teather$^{(\boxtimes)}$

Carleton University, 1125 Colonel by Dr, Ottawa, ON K1S 5B6, Canada
SijuPhilip@cmail.carleton.ca, rob.teather@carleton.ca

**Abstract.** We present an experiment evaluating the effectiveness of a tracked
drawing tablet for use in virtual reality (VR) text input. Participants first completed a text input pre-test, entering several phrases using a physical keyboard.
Participants then entered text in VR using an HTC Vive, with a tracker mounted
on a drawing tablet with a QWERTY soft keyboard overlaid on the virtual tablet.
This was similar to text input using stylus-supported mobile devices. Our results
indicate that not only did participants prefer the Vive controller, it also offered
superior entry speed (16.31 wpm vs. 12.79 wpm with the tablet and stylus) and
error rates (4.1% vs. 6.4%). Pre-test scores were also correlated to measured entry
speeds, and reveal that user typing speed on physical keyboards provides a modest
predictor of VR text input speed (R2 of 0.6 for the Vive controller, 0.45 for the
tablet).

**Keywords:** Text input · Virtual reality · Drawing tablet

## 1 Introduction

Virtual reality (VR) devices are becoming increasingly affordable and performant.
Between the falling prices, and the recent emergence of wireless head-mounted displays
(HMDs), VR is also becoming more accessible. Despite these recent advancements in
interaction technologies for VR systems, an ongoing problem is symbolic and text input
in VR. Text input has traditionally received somewhat less attention from the research
community than other interaction tasks in VR [3]. This is likely because substantial text
input has been a more "niche" task than selection, manipulation, or navigation. Though
the exact reasons for this are unclear, it may be related to lower quality VR systems of
the past that were uncomfortable to use for lengthy composition. In the past few years,
however, there have been several studies on different text entry methods in VR [11, 12,
34, 41], suggesting increasing application demand.

Replacing standard, physical keyboards for heavy text-entry tasks such as writing a
paper or coding would be difficult. The absence of physical embodiment and difficulty
using keyboards in midair have given rise to alternative techniques to address the many
use cases of short, yet arbitrary, text input. Consider, for example sending quick SMS-
like messages to another user, annotating parts of a virtual environment during a design

© Springer Nature Switzerland AG 2020
C. Stephanidis et al. (Eds.): HCII 2020, LNCS 12428, pp. 314–329, 2020.
https://doi.org/10.1007/978-3-030-59990-4_24

review in VR/AR, or calling up a webpage by typing a URL. In each situation, it is undesirable to break flow by switching context from VR to using a physical keyboard in order to perform such tasks, especially when using HMDs. We thus explore alternative approaches that can be used for VR text entry that are easier to use for a novice user, and which offer acceptable performance levels. In this study, we compared the most common method of text input in VR – using a 3D tracked controller employing ray-casting – to text input using a tracked tablet with a stylus.

We present an experiment evaluating text entry using a tracked tablet and stylus. Such devices support other tasks in VR and may thus integrate into the user's workflow better than keyboards or other tracked controllers. Notably, previous work has shown keyboards offer efficient and precise text entry in VR [11], but they can be used for just that – text entry. A tablet and stylus, on the other hand, offers various functions including selection, menu navigation, swiping, drawing, etc. People are also already familiar with using styluses as text input devices due to their similarity to pencils/pens.

Using a pen and stylus with an on-screen QWERTY keyboard is common in the mobile computing domain. Many modern smartphones and tablets include a stylus (e.g., Samsung's Galaxy Note line), which among other operations (e.g., drawing) can be used with onscreen keyboards to support text input. Using the stylus has the potential for better performance than fingers due to the "fat finger" problem [37]. This style of interaction is naturally familiar from writing with a pen and paper, and it has been used in VR before [4, 10, 32]. We propose to leverage this familiarity using modern VR hardware by adding a tracker to a digital drawing tablet while using the tablet's digitizer to detect the stylus contact point. Unlike using a ray to select keys from a virtual keyboard using a tracked controller, a tablet/stylus also offers tactile feedback.

Previous work has taken a similar approach, using a 3D tracked physical pen and tablet metaphor [4], or using a wooden tablet and pen [10]. Our approach is based on the observation that simply tracking a drawing tablet yields higher precision on the contact point than using a secondary tracker on the stylus while also keeping the stylus unencumbered. Moreover, the tablet (and Vive tracker) are relatively inexpensive compared to light-weight optical trackers (e.g., Vicon) that could be used to track a stylus. The result is similar to text input on mobile devices; the main difference between our scenario and mobile devices is that the user is sitting in a virtual environment instead of the real one. In our experiment, participants used the tracked tablet in VR and entered text using a stylus by selecting characters on a virtual QWERTY keyboard displayed on the tablet screen in VR. See (Fig. 1).

However, the text entry potential of tablets and styluses in VR has not been well studied, especially in comparison to common approaches used with recent commercial VR devices. A motivating hypothesis of our work is that using a stylus and tablet may yield comparable text input performance compared to a ray/controller method. The main contribution of our work is an experiment showing that, although the tablet underperforms relative to the controller, the performance difference is small, and can be potentially improved.

**Fig. 1.** (Left) The VR view, showing the QWERTY soft keyboard. The "H" key is highlighted to reflect that the stylus is currently hovering over it. (Right) Participant performing the experiment with the Tablet and Stylus.

## 2  Related Work

Using a tablet and stylus for text entry has been studied extensively in the HCI community in non-VR scenarios [6, 9, 51, 54]. Entry rates using a stylus for input with a QWERTY soft keyboard range between 8.9 wpm and 30.1 wpm, according to Soukoreff et al. [49]. Here we focus on previous research using tablet devices and different text entry techniques for VR.

### 2.1  Tablets and Mobile Devices in VR

One of the earliest examples of using a tablet-like device in VR was the HARP system developed by Lindeman et al. [22]. Their results showed that a 2D stylus and tablet metaphor used in VR provided better support for precise selection actions. They argued this was a direct result of providing a tactile surface via the virtual tablet. The benefits of so-called "passive" haptic feedback are well-known in the VR research community [1, 5, 8, 15, 27]. They also showed that using a hand-held device is preferable to fixed-position devices, as they provide freedom for working effectively in IVEs.

Poupyrev et al. developed one of the earliest text entry methods in VR, the Virtual Notepad. It used a spatially tracked tablet and stylus system for taking notes in VR [32]. Using the Virtual Notepad users could take notes, modify them, add or remove pages, and manipulate the documents within the VE. The system employed character recognition to detect and issue handwritten commands in the form of individual letters. Medeiros et al. proposed using mobile devices for interaction in virtual environments [36]. They concluded that user familiarity with these mobile devices reduces their resistance to immersive virtual environments (IVEs). Other researchers have explored the use of mobiles as input devices in VR, for example, to select and manipulate objects [17].

Several other studies have explored the use of mobile devices and tablets in VR contexts [7, 14, 42, 47]. Perhaps the most similar work to ours, Kim and Kim's proposed method, using a smartphone and its hovering function for text entry in VR [18]. HoVR used a smartphone's hovering function for text entry in VR. However, they did not report text entry speed or error rate and reported only task completion time, and also did not use commonly used stimulus text [24], which makes comparing results difficult.

## 2.2 Text Entry Methods in VR

Compared to interaction tasks like selection and navigation, there are comparatively few studies on text input in VR. (Table 1) presents a summary of several key studies.

**Table 1.** Several text entry studies in immersive virtual reality environments and their performance results.

| 1st author | Text entry method | Entry rate (wpm) | Error rate | Notes |
|---|---|---|---|---|
| Poupyrev [32] | Tablet & stylus with digital ink | Not reported | Not reported | Switched between showing hands and the stylus and used character recognition for text input |
| Bowman [4] | QWERTY keyboard with pen and tablet metaphor | 10$^*$ | 7.14 errors per subject | Original result were reported in character per minute (cpm). *Note: we converted to wpm by dividing the original results by five |
| | Pinch keyboard | 5$^*$ | 43.17 error per subject | |
| González [10] | Pen based QWERTY keyboard | 7$^*$ | 7% character error rate | Used tablet and pen made out of wood, tracked via sensor. Users could see the pen. * Note: Entry rate originally reported in CPM, converted to wpm |
| | Pen based disk keyboard | 4$^*$ | 2% character error rate | |
| Grubert [11, 12] | QWERTY desktop keyboard | 26 | 2.1% character error rate | Different hand representations were used. i.e. full hand vs. fingers only. Also looked into repositioning the keyboard in VR |
| | QWERTY touchscreen keyboard | 11 | 2.7% character error rate | |
| Yu [55] | Head pointed & Gesture based | 10 to 19 | 1.23% to 3.08% corrected error rate | Investigated TapType, DwellType, and GestureType techniques |
| Yu [56] | Dual joystick controller | 7 to 15 | 1.57% to 1.59 uncorrected error rate | Used a circular keyboard layout |

*(continued)*

**Table 1.** (*continued*)

| 1st author | Text entry method | Entry rate (wpm) | Error rate | Notes |
|---|---|---|---|---|
| Kim [18] | QWERTY touchscreen keyboard | Not reported | Not reported | Used smartphone's hovering function for finger tracking. Reported task completion time |
| Kuester [21] | Wearable glove | Not reported | Not reported | Used the concept of column and rows found in traditional keyboards |
| Rajanna [34] | Gaze typing | 6 to 9 | .02% to .08% rate of back space | Sitting and biking were conditions in the experiments |
| Xu [53] | Head motions | 8 to 12 | 2.25% to 2.46% uncorrected error rate | Dwell and hands free interaction method, used a circular keyboard layout |
| Prätorius [33] | Thumb to finger taps | Not reported | Not reported for text entry. | Reported keystroke per character |
| Gugenheimer [13] | Split QWERTY touchscreen keyboard | 10 | Not reported | Used displayed fixed UIs, users wore a touch sensor on the HMD |
| Speicher [41] | QWERTY keyboard with Controller Pointing | 15 | 0.97% Corrected error rate | Also looked into the physical demand required and cyber sickness in different text entry methods in VR |
| | QWERTY keyboard with Controller Tapping | 12 | 1.94% Corrected error rate | |

Bowman et al. empirically compared four different text entry techniques including a pen and tablet, voice recognition, a one-hand chord keyboard and a method using pinch gloves [4]. Speech recognition was fastest, at around 14 words per minute (wpm). Their pen and tablet metaphor offered entry rates of up 12 wpm. Unlike our study, which employs a tablet digitizer to "track" the stylus, their study used a 3D tracked stylus to touch letters, indicating selection by pressing a stylus button. Another example of using a pen-based approach, González et al. tracked a wooden tablet and stylus with a sensor, although the type of sensor is not reported. In a series of text entry experiments, they report entry rates between 7 and 8 wpm [10] with the stylus and tablet.

Speech recognition is a potentially attractive method of VR text input. The SWIFTER speech recognition system improved on Bowman's speech-based approach [4], achieving average entry speeds of up to 23 WPM [31]. However, although speech to text-based techniques are fast, they are not well-suited to loud environments, or in situations requiring discretion (e.g., users having private conversations, or entering password securely).

Also, editing is challenging with speech-based methods; for example, cursor positioning is problematic. Moreover, speech can interfere with the cognitive process required to enter text [38, 43].

Several studies have explored the use of game controllers as text input mechanisms both in VR, and in other similar use scenarios, such as games [20, 29, 50, 56]. Isokoski et al. used a controller and a tablet with a stylus in their experiments [16]. They reported entry rates ranging from 6 to 8 wpm using the Quickwriting method [16, 30]. Entry speed with conventional game controllers tends to range between 6 to 15 wpm [50, 56].

Other researchers explored the use of head motion and gaze direction for typing in VR [34, 53, 55]. Yu et al. explored three head-based techniques for text entry in VR [55]. They reported average entry speed of 24.73 WPM with their GestureType technique after one hour of training. Gugenheimer et al. introduced FaceTouch, which employed display-fixed UIs [13]. Their approach employed touchpads mounted on the front face of the HMD, which users touched to enter text. In an informal study on text entry using a split QWERTY keyboard, the authors reported average text entry speed of approximately 10 wpm. Rajanna et al. focused on investigating how keyboard design, selection method, and motion in the field of view impact typing performance and user experience [34]. They concluded that VR gaze typing is viable, if somewhat unnatural.

Several recent VR text input studies have investigated the use of physical keyboards with various hand visualizations [2, 11, 12, 44, 45]. Knierim et al. evaluated the effects of virtual hand representation and hand transparency on typing performance of experienced and inexperienced typists in VR [19]. Their results suggest that experienced users (e.g., touch typists) performance is not significantly affected by missing hands or different hand visualization. However, inexperienced users are impacted by these factors. Similar work by Grubert et al. also investigated methods for virtual hand representation with minimalistic fingertip visualization [11, 12]. Specifically, their minimalistic visualization showed only dots at the fingertips rather than an entire hand visualization. They report that even with minimalistic fingertip representations, entry speeds ranging from 34 to 38 wpm are possible, depending on hand and finger representation.

Other VR text input methods employing 3D tracked controllers employing direct touch, or ray-casting, to select keys from a virtual keyboard. Entry rate ranged from 12 to 15 wpm [41]. Several other studies employed gloves and hand gestures both in VR and non-VR context [21, 26, 28, 33, 35, 46, 48]. Yi et al. reported entry rates of up to 29 wpm using their hand based method in a non-VR setting [52]. As suggested by Grubert et al., methods that use a controller for text entry tend to have a higher learning curves and require more training than keyboards, and can also cause user fatigue [11].

## 3 Methodology

### 3.1 Participants

We recruited 28 participants from our local community but ended up removing four of them. Two were extreme outliers (entry speed scores more than 3 SDs from the mean), and there were logging errors with the other two. This left us with 24 participants upon which our analysis is based. There were 9 female and 15 male participants, aged between 18 and 54 (M = 26.21, SD = 8.19). Eighteen participants reported that they had not played 3D games, or only played them infrequently. Twenty-one participants had very little or no prior experience with VR. Seven participants reported regularly using a pen or stylus for typing on their tablet or smartphone. One participant indicated that they did not text at all, while nine texted frequently during a typical day. The rest of them reported moderate texting. All participants had normal or corrected-to-normal stereo vision, assessed by having participants correctly determine the depth of two spheres presented in the scene.

### 3.2  Apparatus

**Hardware.** We used a PC with an Intel Core i7 CPU with an NVIDIA Geforce GTX 1080 graphics card. We used the HTC Vive VR platform, which includes an HMD with 1080 × 1200 per eye resolution, 90 Hz refresh rate, and a 110° field of view. The tablet was an XP-PEN STAR 06 wireless drawing tablet. Its dimensions were 354 mm × 220 mm × 9.9 mm with a 254 mm × 152.4 mm active area, and a 5080 LPI resolution. The tablet included a stylus with a barrel button and a tip switch to support activation upon pressing it against the tablet surface. The 2D location of the stylus is tracked along its surface by the built-in electromagnetic digitizer. We affixed a Vive tracker to the top-right corner of the tablet using velcro tape. See (Fig. 2).

**Fig. 2.** Tablet and stylus with Vive tracker and Vive HMD along with the physical keyboard used in pre-tests.

**Software.** We used MacKenzie's "Typing Text Experiment"[1] software for the pre-test. The pre-test consisted of 30 randomly determined phrases the participants entered using a real keyboard.

The main experiment used our VR software developed in Unity3D. We developed a custom library to get stylus input from the XP-PEN STAR tablet into Unity. The tablet

---

[1] Available at http://www.yorku.ca/mack/HCIbook/.

is ordinarily seen as fitting the human interaction device (HID) profile by Windows 10, which, by default would map it to the mouse cursor, which was not desired in VR. To avoid this, we installed a custom LibUSB driver that allowed direct access to the raw data from the tablet. The library provided data such as the coordinate position of the stylus on the tablet surface, the amount of pressure applied by on the stylus tip switch, and whether the stylus was touching the tablet surface or hovering above it within approximately 2 cm.

The software polled the Vive tracker to map a virtual model of the tablet to the physical tablet, co-locating the two. The tablet stylus was used to interact with the tablet. The stylus itself was not tracked, hence tracking was limited to the tip of the stylus in a close range to the tablet surface (about 2 cm). Due to this limitation, we did not render a model of the stylus or hands. However, when the stylus was in the range of the tablet, we displayed a cursor at the stylus tip. Notably, this is how such graphics tablets are typically used, as the display is not collocated, which has the advantage of not covering part of the drawing with the hand. By applying pressure on the stylus tip switch by pressing it against the tablet surface, input events were detected for corresponding keys on the tablet and the corresponding character was entered.

Participants sat in the virtual room seen in (Fig. 3). A QWERTY soft keyboard was displayed on the virtual tablet. The keys were 1.5 cm × 1.5 cm in size. Participants sat between 30 to 40 cm from the keyboard, depending on their position. Participants were always presented with a simulated QWERTY soft keyboard displayed centered on the tablet (see Fig. 3). The tablet was positioned on a table as seen in (Fig. 3). The current target phrase appeared near the virtual keyboard to reduce the need for glancing during entry. As participants entered the phrase, each keystroke was presented, giving them immediate feedback. While hovering on keys, the letters changed color to indicate which would be selected if the tip switch was pressed. Upon pressing a key in this fashion, an auditory "click" sound was played and the key letters change colour to yellow. The SPACE bar and the ENTER key each had distinct button press sounds.

**Fig. 3.** View of the virtual room seen by participants. Red were spheres used for stereo viewing test (Color figure online)

**Fig. 4.** Vive Controller with soft keyboard.

Figure 4 depicts the other text entry method, the Vive controller. The QWERTY soft keyboard used with the Vive controller had keys sized 10 cm × 10 cm with a 0.5 cm gap between each adjacent key. This keyboard faced the participant, and was slightly tilted towards the participant (by 10°) and was positioned about 110–120 cm from the participant's seated position. With this text entry method, participants pointed a ray from the Vive controller at the desired key and pressed the trigger button to select. Upon being intersected by the selection ray, a key changed colour from blue to yellow. Upon pressing the trigger, the selected key would turn grey. The sound effects were the same as the tablet condition.

### 3.3 Procedure

Before they began, participants first read and signed consent forms and completed a pre-questionnaire to gather demographic data such as age and gaming/VR experience, stylus usage, and mobile text input habits. The experimenter then explained the procedure. Correction (i.e., backspace) was disabled with both the real keyboard during the pre-test as well as the virtual keyboard during the experiment. Auto-correction or similar features were not implemented. Participants were instructed to enter each phrase as quickly and accurately as possible, ignoring any mistakes and pressing the ENTER key to end each trial. Timing started as soon as the participant entered the first character and stopped as soon as they hit the ENTER key.

Upon starting the experiment, participants first entered 30 phrases using MacKenzie's "Typing Text Experiment" software with real keyboard as a pre-test. After the pre-test, the experimenter demonstrated how to use the Vive controller and the tablet. The participants then entered VR, and were screened for stereo viewing prior to continuing the experiment. They were presented with two red spheres at different depths (Fig. 3). They were instructed to reach out and intersect the Vive controller with the spheres to make them disappear. All participants were able to reliably detect the depth of the spheres and hence passed this screening.

The task then started without any training with the interaction devices. Participants were instructed to enter a random pool of phrases from a phrase set commonly used in text entry experiments [24]. Participants were presented with one line from the phrase set at a time, entering 30 phrases with each text input method (both in pre-test and in VR). After completing all 30 phrases, participants completed a questionnaire related to their experience of that method. In the end, they completed a post-questionnaire about their preferred text input method and any comments or suggestions they had.

### 3.4 Design

Our experiment employed a within-subjects design with two independent variables:

> ***Text Input Method***:     Tablet and stylus, Vive Controller
> ***Trial***:                         1, 2, 3, …, 30

Text input method order was counterbalanced by having half the participants use the Vive Controller then Tablet and stylus, and the other half in the reverse order. Entry speed, in words per minute (wpm), was calculated as:

$$wpm = \frac{|s|}{T} \times 60 \times \frac{1}{5}$$

where T is the text entry time in seconds, and |s| is the input string length (in characters). We used every five characters (including spaces) as a single word, consistent with the text input literature [25].

Error rates were calculated using character error rate (CER), calculated as:

$$CER\% = \frac{MSD(stimulus\,text,\ response\,text)}{CharLenght\,(stimulus\,text)} \times 100$$

CER is the minimum number of character-level insertion, deletion, and substitution operations required to transform the response text into the stimulus text, i.e., the minimum string distance (MSD) between the two, divided by the number of characters in the stimulus text [23, 39]. This metric better represents errors. For example, a single character insertion early in a phrase yields a single error, rather than a "cascade" of mismatched characters [40]. CER is expressed as a percentage of errors.

## 4   Results

Since participants completed only 30 trials with each text input method, we did not evaluate the effect of trial, and averaged all trials together. We compared performance using a t-test. The assumption of sphericity was met in all cases based on Mauchly's test. We were also interested in determining if the participant's touch-typing speed predicts their VR text entry speed, so used linear regression to see if they were correlated.

### 4.1   Performance

Figure 5 depicts average entry speed for each text entry method. The tablet and stylus had a mean entry speed of 12.79 WPM ($SE = .71$). In contrast, the Vive controller offered faster entry speed at 16.31 WPM ($SE = .44$). A t-test revealed a significant main effect of text entry method for entry speed ($t(29) = -32.3, p < .001$).

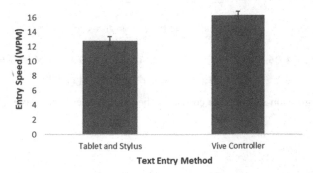

**Fig. 5.** Average text entry speed for each VR input method. Error bars show 95% confidence interval.

Figure 6 depicts the average error rate for each text entry method. The tablet and stylus technique had a mean error rate of 6.42% ($SE = 1.66$), and the Vive controller had a mean error rate of 4.14% ($SE = .95$). A t-test revealed that the main effect of text entry method on error rate was significant ($t(29) = 3.74, p < .001$).

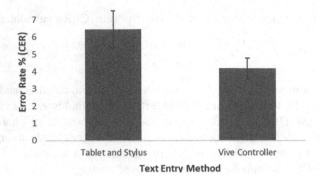

**Fig. 6.** Average error rate for each VR input method. Error bars show 95% confidence intervals.

Figure 7 depicts a regression analysis between pretest entry speeds and measured VR entry speed for both text input methods. As seen in the figure, there is a modest relationship between pre-test text input speed (i.e., on a standard keyboard) and VR text entry speed, for both the Vive controller and the tablet and stylus. Indeed, faster typists on a typical desktop setup had better entry speed with both VR text input methods.

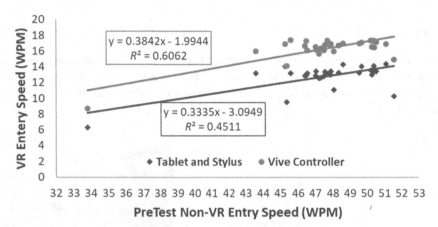

**Fig. 7.** Linear regression showing the correlation between pre-text entry speeds and VR entry speeds.

### 4.2 User Experience

All participants except one preferred the Vive controller text entry method. A few reported having trouble reading the 3D text on the tablet noting that "the text gets quite blurry." It is possible larger keys might resolve this issue. Another issue was the weight of the HMD; several participants noted it bothered them while using the tablet, which necessitated looking down. Participants also reported confusion with the sensitivity of the pen and tablet, they were not sure about the amount of tip switch pressure required to register input. Most of them found it hard without knowing where the cursor and the stylus were when tracking was lost and reported they wanted to see the stylus at all times. One participant stated that the tablet was "good to use when you get the hang of it, but the distance and specific angle you have to hold the stylus at is a little difficult to get used to". One of them also suggested that implementing two styluses could have improved the speed and efficiency of typing as one could at least be used for spacing.

## 5 Discussion

Our performance results (16.32 vs. 12.80 wpm) indicate that our proposed method performed worse than the Vive controller. Notably, the tablet and stylus performance is comparable to several previous text input methods seen in Table 1. In particular, its performance is comparable to Bowman's tracked stylus and tablet [4], several game controller methods [56], head motion [53] and touchscreen typing on a soft keyboard [12]. It performs better than several previous techniques, including an alternative tracked tablet and stylus [10] and gaze typing [34].

Despite these results, we believe with improvements, there is still potential for the tablet and stylus to be considered a viable text entry method in VR. In particular, participants noted several issues that could improve the design of tablet-based text input schemes, including larger keys and more ergonomic positioning of the tablet, perhaps letting participants hold the tablet or move it where comfortable. Participant comments also suggest that adding the stylus and hand visualization and improving the tip switch sensitivity could further improve the tablet and stylus performance.

Some participants leaned forward and moved very close to the tablet with their head down to tap on the tablet; this could have caused the HMD to move slightly on their head and cause the blurry effect they reported while using the tablet. Some participants also reported noticing the weight of the HMD only while using the tablet. This also might have to do with the fact that they were hunching over the tablet. In contrast, with the Vive controller condition, they held their heads up to see the soft keyboard.

Participants also reported their arms getting tired because of tapping, and that it also caused finger fatigue. This was mainly because more hand/arm movement was required with the tablet keyboard. While using the Vive controller with ray-casting, participants could easily move the cursor across the keyboard via small wrist movements. We believe this could be improved by implementing a swiping text entry method for the tablet, or by using a different soft keyboard layout. Notably, the Vive controller can also be used bimanually, which could potentially have increased typing performance relative to the one-handed operation with the stylus. We opted to give them two controllers to more accurately simulate the real experience of text input on commercial VR systems. Participants noted that they liked the tactile feedback provided by the tablet, in line with previous work on haptics in VR [5, 15, 27].

## 6 Conclusion

In this paper, we evaluated the effectiveness of using a tablet and stylus for VR text input. We designed and developed a text entry method using a QWERTY virtual keyboard on a tablet in VR. We found that the Vive controller performed faster, had lower error rates, and was preferred by participants. We also found that VR text input speed can be – to a modest extent – predicted by touch typing entry speed with a conventional keyboard. We believe that by improving some limitations of our experiment, we can improve the tablet and stylus efficiency might be better. Also, if users are already using a tablet for other functions in VR, like a VR design session, using the tablet instead of a controller for text entry may be inherently preferable and feel more natural.

## 7    Limitations and Future Work

There are a few limitations with the current experiment. First, tablet position was always fixed. We had chosen this option, rather than allowing the user to hold the tablet, as we suspected the alternative would increase arm fatigue. However, because of this decision, the tablet's virtual keyboard and the target phrases were not in the same location, necessitating that participants look up and down to see the sentence first, and then type it with the keyboard. In contrast, this head motion was not required with the Vive controller condition, which may partly explain the difference observed. A future implementation would better visually co-locate the keyboard and stimulus phrases.

A second limitation is that the stylus itself was not tracked. Several participants commented on this. This was a limitation of the type of tablet we used (i.e., no in-air tracking) and could be addressed by using an optical tracker to maintain a lightweight tracker on the stylus. New devices like Logitech's VR Ink stylus may address this.

Several other comparisons are possible, but were excluded from the current experiment to keep it simple. Future studies might, for example, focus on stylus vs. finger input, holding the tablet in 3D space vs. resting on a surface, and the effect of scaling stylus motion when entering text (i.e., applying CD gain to the stylus). Other options including using different keyboard layouts and swipe-based text input. We are also considering using a new HMDs with higher resolutions, lighter designs, and a wider field of view to address the ergonomic concerns mentioned by participants of this study.

**Acknowledgements.** We would like to thank Kyle Johnsen and A. J. Tuttle for sharing their tablet driver source code. We also thank our participants. This research was supported by NSERC.

## References

1. Aguerreche, L., Duval, T., Lécuyer, A.: Reconfigurable tangible devices for 3D virtual object manipulation by single or multiple users. In: Proceedings of the 17th ACM Symposium on Virtual Reality Software and Technology, pp. 227–230 (2010)
2. Bovet, S., et al.: Using traditional keyboards in VR: steam VR developer kit and pilot game user study. In: IEEE Games, Entertainment, Media Conference, pp. 132–135 (2018)
3. Bowman, D.A., Kruijff, E., LaViola, J.J., Poupyrev, I.: 3D User Interfaces: Theory and Practice. Addison Wesley Longman Publishing Co., INC (2004)
4. Bowman, D.A., Rhoton, C.J., Pinho, M.S.: Text input techniques for immersive virtual environments: an empirical comparison. Proc. Human Factors Ergon. Soc. Ann. Meeting **46**(26), 2154–2158 (2002)

5. Brooks, J.F.P., Insko, B.E.: Passive Haptics Significantly Enhances Virtual Environments. The University of North Carolina at Chapel Hill
6. Cechanowicz, J., Dawson, S., Victor, M., Subramanian, S.: Stylus based text input using expanding CIRRIN. Adv. Vis. Interfaces, pp. 163–166 (2006)
7. Chen, T.-T., Hsu, C.-H., Chung, C.-H., Wang, Y.-S., Babu, S.V.: iVRNote: design, creation and evaluation of an interactive note-taking interface for study and reflection in VR learning environments. In: IEEE Conference on Virtual Reality and 3D User Interfaces, 172–180 (2019)
8. Franzluebbers, A., Johnsen, K.: Performance benefits of high-fidelity passive haptic feedback in virtual reality training. In: ACM Symposium on Spatial User Interaction, pp. 16–24 (2018)
9. Goldberg, D., Richardson, C.: Touch-typing with a stylus. In: Proceedings of the INTERACT 1993 and Conference on Human Factors in Computing Systems, pp. 80–87 (1993)
10. González, G., Molina, J.P., García, A.S., Martínez, D., González, P.: Evaluation of text input techniques in immersive virtual environments. New Trends on Human-Computer Interaction: Research, Development, New Tools and Methods. Springer. 109–118
11. Grubert, J., Witzani, L., Ofek, E., Pahud, M., Kranz, M., Kristensson, P.O.: Effects of hand representations for typing in virtual reality. In: IEEE Conference on Virtual Reality and 3D User Interfaces (VR), pp. 151–158 (2018)
12. Grubert, J., Witzani, L., Ofek, E., Pahud, M., Kranz, M., Kristensson, P.O.: Text entry in immersive head-mounted display-based virtual reality using standard keyboards. In: IEEE Conference on Virtual Reality and 3D User Interfaces (VR), pp. 159–166 (2018)
13. Gugenheimer, J., Dobbelstein, D., Winkler, C., Haas, G., Rukzio, E.: FaceTouch: enabling touch interaction in display fixed uis for mobile virtual reality. In: Proceedings of the 29th Annual Symposium on User Interface Software and Technology, pp. 49–60 (2016)
14. Henrikson, R., Araujo, B., Chevalier, F., Singh, K., Balakrishnan, R.: Multi-device storyboards for cinematic narratives in VR. In: Proceedings of the 29th Annual Symposium on User Interface Software and Technology, pp. 787–796 (2016)
15. Hoffman, H.G.: Physically touching virtual objects using tactile augmentation enhances the realism of virtual environments. In: Proceedings of the Virtual Reality Annual International Symposium, pp. 59–63 (1998)
16. Isokoski, P., Raisamo, R.: Quikwriting as a multi-device text entry method. In: Proceedings of the Third Nordic Conference on Human-Computer Interaction, pp. 105–108 (2004)
17. Katzakis, N., Teather, R.J., Kiyokawa, K., Takemura, H.: INSPECT: extending plane-casting for 6-DOF control. In: Proceedings IEEE Symposium on 3D User Interfaces, pp. 165–166 (2015)
18. Kim, Y.R., Kim, G.J.: HoVR-Type: smartphone as a typing interface in VR using hovering. In: IEEE International Conference on Consumer Electronics (ICCE), pp. 200–203 (2017)
19. Knierim, P., Schwind, V., Feit, A.M., Nieuwenhuizen, F., Henze, N.: Physical keyboards in virtual reality: analysis of typing performance and effects of avatar hands. In: Proceedings of the Conference on Human Factors in Computing Systems, pp. 1–9 (2018)
20. Költringer, T., Isokoski, P., Grechenig, T.: TwoStick: writing with a game controller. In: Proceedings of Graphics Interface, 103–110 (2007)
21. Kuester, F., Chen, M., Phair, M.E., Mehring, C.: Towards keyboard independent touch typing in VR, p. 86, February 2006
22. Lindeman, R.R.W., Sibert, J.J.L., Hahn, J.K.J.J.K.: Towards usable VR: an empirical study of user interfaces for immersive virtual environments. In: Proceedings of the SIGCHI conference on Human Factors in Computing Systems, pp. 64–71 (1999)
23. MacKenzie, I.S., Soukoreff, R.W.: A character-level error analysis technique for evaluating text entry methods. In: Proceedings of the Second Nordic Conference on Human-Computer Interaction, pp. 243–246 (2004)

24. MacKenzie, I.S., Soukoreff, R.W.: Phrase sets for evaluating text entry techniques. In: Extended Abstracts on Human Factors in Computing Systems, pp. 754–755 (2003)
25. MacKenzie, I.S., Soukoreff, R.W.: Text entry for mobile computing: models and methods, theory and practice. Hum.-Comput. Interact. **17**(2–3), 147–198 (2002)
26. Markussen, A., Jakobsen, M.R., Hornbæk, K., Markussen, A., Jakobsen, M.R., Hornbæk, K.: Vulture: a mid-air word-gesture keyboard. In: Proceedings of the 32nd Annual ACM Conference on Human Factors in Computing Systems - CHI 2014, pp. 1073–1082 (2014)
27. McClelland, J.C., Teather, R.J., Girouard, A.: Haptobend: shape-changing passive haptic feedback in virtual reality. In: Proceedings of the 5th Symposium on Spatial User Interaction, pp. 82–90 (2017)
28. Mehring, C., Kuester, F., Singh, K.D., Chen, M.: KITTY: keyboard independent touch typing in VR. In: Virtual Reality Annual International Symposium, pp. 243–244 (2004)
29. Natapov, D., Castellucci, S.J., MacKenzie, I.S.: ISO 9241-9 evaluation of video game controllers. In: Proceedings of Graphics Interface, pp. 223–230 (2009)
30. Perlin, K.: Quikwriting: continuous stylus-based text entry. In: Proceedings of the 11th Annual ACM Symposium on User Interface Software and Technology, pp. 215–216 (1998)
31. Pick, S., Puika, A.S., Kuhlen, T.W.: SWIFTER: design and evaluation of a speech-based text input metaphor for immersive virtual environments. In: IEEE Symposium on 3D User Interfaces (3DUI), pp. 109–112 (2016)
32. Poupyrev, I., Tomokazu, N., Weghorst, S.: Virtual notepad: handwriting in immersive VR. In: Proceedings IEEE Virtual Reality Annual International Symposium, pp. 126–132 (1998)
33. Prätorius, M., Burgbacher, U., Valkov, D., Hinrichs, K.: Sensing thumb-to-finger taps for symbolic input in VR/AR environments. IEEE Comput. Graph. Appl. **35**(5), 42–54 (2015). https://doi.org/10.1109/MCG.2015.106
34. Rajanna, V., Paulin Hansen, J.: Gaze typing in virtual reality: impact of keyboard design, selection method, and motion. Proc. ACM Sympos. Eye Track. Res. Appl. **15**(1–15), 10 (2018)
35. Reyal, S., Zhai, S., Kristensson, P.O.: Performance and User Experience of Touchscreen and Gesture Keyboards in a Lab Setting and in the Wild, pp. 679–688, April 2015
36. De Sa Medeiros, D.P., De Carvalho, F.G., Raposo, A.B., Dos Santos, I.H.F.: An interaction tool for immersive environments using mobile devices. In: XV Symposium on Virtual and Augmented Reality, pp. 90–96 (2013)
37. Shibata, T., Afergan, D., Kong, D., Yuksel, B.F., MacKenzie, S., Jacob, R.J.K.: Text entry for ultra-small touchscreens using a fixed cursor and movable keyboard. In: Proceedings of the Conference Extended Abstracts on Human Factors in Computing Systems, pp. 3770–3773 (2016)
38. Shneiderman, B.: The limits of speech recognition. Commun. ACM, **2000**, 319–365 (2000)
39. Soukoreff, R.W., MacKenzie, I.S.: Measuring errors in text entry tasks: an application of the Levenshtein string distance statistic. Extended Abstracts Hum. Factors Comput. Syst., pp. 319–320 (2001)
40. Soukoreff, R.W., MacKenzie, I.S.: Metrics for text entry research: an evaluation of MSD and KSPC, and a new unified error metric. Proceedings of the ACM Conference on Human Factors in Computing Systems – CHI **2003**, 113–120 (2003)
41. Speicher, M., Feit, A.M., Ziegler, P., Krüger, A.: Selection-based text entry in virtual reality. In: Proceedings of the Conference on Human Factors in Computing Systems, pp. 1–13 (2018)
42. Tuttle, A.J., Savadatti, S., Johnsen, K.: Facilitating collaborative engineering analysis problem solving in immersive virtual reality. In: American Society for Engineering Education (2019)
43. Vertanen, K.: Efficient Correction Interfaces for Speech Recognition. University of Cambridge (2009)

44. Walker, J., Kuhl, S., Vertanen, K.: Decoder-assisted typing using an HMD and a physical keyboard. In: Extended Abstracts of the ACM Conference on Human Factors in Computing Systems, p. 16 (2016)
45. Walker, J., Li, B., Vertanen, K., Kuhl, S.: Efficient typing on a visually occluded physical keyboard. In: Proceedings of the Conference on Human Factors in Computing Systems, pp. 5457–5461 (2017)
46. Wang, C.-Y., Chu, W.-C., Chiu, P.-T., Hsiu, M.-C., Chiang, Y.-H., Chen, M.Y.: PalmType: using palms as keyboards for smart glasses. In: Proceedings of the 17th International Conference on Human-Computer Interaction with Mobile Devices and Services - MobileHCI 2015 (2015)
47. Wang, J., Lindeman, R.W.: Object impersonation: towards effective interaction in tablet- and HMD-based hybrid virtual environments. In: IEEE Virtual Reality (VR), pp. 111–118 (2015)
48. Whitmire, E., et al.: DigiTouch: reconfigurable thumb-to-finger input and text entry on head-mounted displays. Displays. PACM Interact. Mob. Wearable Ubiquitous Technol. **1**, 113 (2017)
49. William Soukoreff, R., Scott Mackenzie, I.: Theoretical upper and lower bounds on typing speed using a stylus and a soft keyboard. Behav. Inf. Technol. **14**(6), 370–379 (1995)
50. Wilson, A.D., Agrawala, M.: Text entry using a dual joystick game controller. In: ACM Conference on Human Factors in Computing Systems, 475–478 (2006)
51. Wobbrock, J.O., Myers, B.A., Kembel, J.A.: EdgeWrite: a stylus-based text entry method designed for high accuracy and stability of motion. In: ACM Symposium on User Interface Software and Technology, pp. 61–70 (2003)
52. Xin, Y., Chun, Y., Mingrui, Z., Sida, G., Ke Sun, Y.S.: ATK: enabling ten-finger freehand typing in air based on 3D hand tracking data. In: ACM Symposium on User Interface Software & Technology, pp. 539–548 (2015)
53. Xu, W., Liang, H.N., Zhao, Y., Zhang, T., Yu, D., Monteiro, D.: RingText: dwell-free and hands-free text entry for mobile head-mounted displays using head motions. IEEE Trans. Visual. Comput. Graph. **25**(5), 1991–2001 (2019)
54. Yatani, K., Truong, K.N.: An evaluation of stylus-based text entry methods on handheld devices studied in different user mobility states. Pervasive Mobile Comput. **5**(5), 496–508 (2009)
55. Yu, C., Gu, Y., Yang, Z., Yi, X., Luo, H., Shi, Y.: Tap, Dwell or Gesture?: exploring head-based text entry techniques for HMDs. In: Proceedings of the Conference on Human Factors in Computing Systems, pp. 4479–4488 (2017)
56. Yu, D., Fan, K., Zhang, H., Monteiro, D., Xu, W., Liang, H.N.: PizzaText: text entry for virtual reality systems using dual thumbsticks. IEEE Trans. Visual. Comput. Graph. **24**(11), 2927–2935 (2018)

# Behavioral Indicators of Interactions Between Humans, Virtual Agent Characters and Virtual Avatars

Tamara S. Griffith[1]([⊠]), Cali Fidopiastis[2]([⊠]), Patricia Bockelman-Morrow[3]([⊠]), and Joan Johnston[1]([⊠])

[1] U.S. Army Combat Capabilities Development Command, Soldier Center,
Simulation and Training Technology Center, Orlando, FL, USA
{Tamara.s.griffith.civ,Joan.h.johnston.civ}@mail.mil
[2] Design Interactive, Orlando, FL, USA
cfidopia@gmail.com
[3] University of Central Florida, Institute for Simulation and Training (UCF IST), Orlando,
FL, USA
pbockelm@ist.ucf.edu

**Abstract.** Simulations and games allow us to experience events as if they were really happening in a way that is safer and less expensive. Despite improvements in realism in these types of environments, one area that still presents a challenge is interpersonal interactions. The subtleties of what makes an interaction rich are difficult to define. As such, there is value in building on existing research into how individuals react to virtual characters to inform future investments.

Ultimately, the goal is to understand what might cause people to engage or disengage with virtual characters. To answer that question, it is important to establish metrics that would indicate when people believe their interaction partner is real, or has agency. This paper describes behavioral metrics explored as part of this research. The results provide valuable feedback on how users need to see and be seen by their interaction partner to ensure non-verbal cues provide context and additional meaning to the dialog. This study provides insight into areas of future research, offering a foundation of knowledge for further exploration and lessons learned.

This was a field study incorporating a novel approach to a real-life experience, a dialog with another individual. Two metrics are explored in this paper; gestural data and open-ended questions, which together provided insight into the information humans rely on and apply in these types of interactions to understand and be understood.

**Keywords:** Simulation · Game · Virtual character · Avatar · Agent · Real-Time character control · Puppeteering · Virtual environment

## 1 Introduction

Unlike previous generations, modern character interactions present people with the opportunity to interact with characters who may provide extemporaneous exchanges

C. Stephanidis et al. (Eds.): HCII 2020, LNCS 12428, pp. 330–342, 2020.
https://doi.org/10.1007/978-3-030-59990-4_25

driven by either (1) other humans or (2) artificial intelligence of some sort. The first driver category, humans driving computer entities, often works much as classic puppeteering would, with the avatar acting and responding at the human driver's will. The second category includes a spectrum of technologies that may loosely be defined as "Artificial Intelligence" (AI) and may be unsophisticated linked scripts with logic cues or leverage highly advanced neural networked evolutionary language learning models to generate entirely novel yet consistently context-sensible responses.

If people sense the character "driver" (regardless of whether they are aware that it is an AI or human driver), their interaction behaviors may alter. To measure this, we have constructed a study to examine human-to-human, human-to-avatar, and human-to-agent interactions. The results of the study help to identify strategies to measure correlates associated with human-computer affective interaction and provide direction for future research efforts in the area of representative machine agency.

Our approach is informed by a variety of disciplinary traditions, bringing insight from work in the realms of computer science, behavioral psychology, cognition, and learning. This work encompasses simulated experiences that are effectively engaging and immersive and that may involve the user so completely that (s)he will experience stress, fear, excitement or anger as a result of the unfolding events [1]. The goal is to create an experience with minimal physiological, emotional and physical difference between the immersive environment and the real world [2]. Keeping the user engaged in the environment is important to reach a state of "flow". Flow is a mental state where a person feels so completely immersed in their experiences to the point of actually gaining energy and losing track of time [3]. This theory is a cornerstone to current-day game design and is applied in learning research [4–6].

Vast improvements in realism in simulated and game environments can support the goal of immersing the user in the environment and establishing an experience that brings about a sense of flow. However, one area still presents a challenge; simulated interpersonal interactions. Many games use cinematics to give the sense of realism to interactions, but this strategy can minimize the sense of control the user experiences. While cinematics is a cost-effective strategy to push the story line forward in the commercial game world, it can change the player to a passive viewer as the interchange unfolds. The challenge is in maintaining user engagement while involving the human dimension of Modeling & Simulation (M&S) experiences.

Consider a U.S Army course on Advanced Situational Awareness (ASA). During this course, soldiers perform surveillance on a village from a distance to look for patterns and changes over time. They may also enter the village and talk with key leaders. During face-to-face interactions soldiers need to be aware of subtle physiological changes that may indicate stress, fear or dishonesty. The U.S. Army currently hires live actors to play the role of villagers within a mockup of a village. They demonstrate normal patterns of life, react to soldier's presence and effect complex social interactions in face-to-face interchanges. This training is critical but carries a high manpower cost. One way to maintain the training value while driving down cost is to conduct the training in an M&S environment. AI agents can demonstrate patterns of life and even disruptions in those patterns. Detailed interpersonal tasks can be performed by human-controlled characters, or avatars. Actors control the avatar's facial expressions, eye movement and gestures in real-time as easily as they manage their own natural body movement. A small number of actors control multiple avatars by jumping from one character to the next as the scenario requires. This functionality provides an equivalent level of training at a fraction of the price. Live training may still be desired after the virtual training experience, but the duration and cost could be significantly reduced. The research described in this study is intended to inform both real-time avatar control as well as AI to support training in the future.

A task was created to establish a dialog between two people in order to study interpersonal interactions. The dialog topic selected for this research was driven by the AI functionality available at the time this research was proposed. Specifically, the realistic interactive dialog of the Digital Survivor of Sexual Assault (DS2A) application developed by the University of Southern California (USC) Institute for Creative Technology (ICT). DS2A is an application that introduces participants to Specialist Jarett Wright. He is an actual individual who survived sexual assault that occurred while he was in the Army. Jarett is a video representation that responds to various questions via voice parsing and complex branching. The development of the application involved the video recording of Jarett providing hours of responses to a wide array of questions along with idle footage that makes the interchange seem like a dialog. A still image depicting a representation from the dialog interaction with Jarett is shown in Fig. 1.

**Fig. 1.** Image of Specialist Jarrett Wright from the Digital Survivors of Sexual Assault (DS2A) Application

An actor represented Specialist Wright in the face-to-face, video, text and human-controlled avatar conditions. He used information from Specialist Jarett Wright's depiction in the D2SA application to shape the story he shared as if it had happened to him. An image of each condition is shown in Figs. 2, 3, 4, 5 and Fig. 6. Each participant provided demographic information and was randomly assigned to one condition then participated in the 8–10 min long dialog session about the sexual assault incident driven by questions chosen and asked by the participant through natural dialog. The participants and the actor, as appropriate, wore an EEG sensor array and Electrodermal Activation monitor and were video recorded during the session. After the interaction participants completed survey questionnaires.

**Fig. 2.** Condition 1 - Participant interacting with a live actor face-to-face

**Fig. 3.** Condition 2 - Participant interacting with an actor via video teleconferencing

The scenario was intended to evoke an emotional response and empathy/sympathy with the character during the interaction. A virtual character that exceeded the threshold necessary to convince an interaction partner that they were real could support various types of purposes in a virtual environment, in addition to that described for ASA. For

**Fig. 4.** Condition 3 - Participant interacting with human-controlled avatar

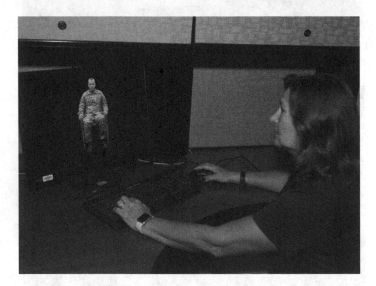

**Fig. 5.** Condition 4 - Participant interacting with computer-controlled agent (AI)

example, the Defense Equal Opportunity Management Institute (DEOMI) is exploring this technology to train human resource personnel to recognize indications of sexual harassment, Post-Traumatic Stress Disorder (PTSD) and depression. In addition, the U.S. Army models realistic, sympathetic characters in military simulations for stress inoculation, to improve rapport-building skills and to depict body posture and movements with the long-term goal of portraying characteristics such as dishonesty, fear, friendliness and distrust.

**Fig. 6.** Condition 5 - Participant interacting with human via text

## 2  Background

While the capabilities of agents (driven by artificial intelligence) and avatars (driven by humans) have improved in recent years, there is still a significant gap in the realism between interactions with humans and interactions with an agent. There has been a great deal of research into this phenomenon [7–10], but it is still unclear what it is about computer agents that might cause us to disengage during the interaction. This research expands our understanding of the differences between various modalities of interaction to inform the existing body of literature on the subject.

The goal is to understand what makes virtual characters appear to have agency. A character's agency is described as the sense that the character can think, has a history, can plan and act, and has opinions [11]. An additional factor, experience, has also been explored, which is the character's ability to sense and feel emotion [12]. To answer this question, it was important to establish metrics that would indicate when people believe their interaction partner had agency. This paper describes the behavioral measures applied during the research as well as open-ended survey responses.

Conversational gestures might function as an indicator that a conversational partner has met a threshold of realism. Conversational hand gestures are "movements of the hands that co-occur with speech but do not appear to be consciously produced by the speaker [13]. For the purpose of this paper, distinct facial expressions were considered in the same light. Hand gestures fall into four categories: representational gestures that can be iconic, metaphoric or spatial; botanic or rhythmic gestures; emblematic, which can replace words in language; and interactive gestures [14]. For the purpose of this paper, interactive gestures are of greatest interest.

Multiple researchers have explored how the number of gestures varies based on the visibility of the listener [14–17]. Jacobs and Garnham [13] designed research that strongly supports the primarily communicative function of gestures. Their research indicated that, "More gestures were produced when the listener appeared attentive than when the listener appeared inattentive." And that "speakers adapt their gesture usage to the perceived requirements of the listener." Their research demonstrated that gestures occur to benefit the listener. If gesturing is a tool used to bolster communication, it can provide insight into variations in communication strategies and communication partners as is described in this work.

Jacobs and Garnham [13] explored two different strategies to measure frequency of gestures. One was gestures per minute of speech and another was gestures per 100 words. Each strategy correlated closely, but the rate per 100 words accommodates changes in speech rate. Given that, gestures per 100 words was used in this research.

## 3   Method

This was a field study which took a novel approach to determine effective strategies to measure whether an individual believes their interactive partner has agency. It was an attempt to move studies on agency into real-life situations while assessing various measurement strategies to inform future research.

Twelve participants (7 males and 5 females) from a local population participated in the study, their mean age was 39.6 years and the standard deviation was 15.18 years. Participants were not compensated for participation. Target age of sample population was military age, or eighteen years or older. Seventeen percent of the participants were black and 83% were white. 8% of the participants had Doctorate Degrees, 25% had Master's Degrees, 17% had Bachelor's Degrees, 17% had either Associates Degrees or Professional Degrees and 33% had some college. 50% of the participants were single, 33% were married and 17% were divorced. 50% of the participants worked full time, 17% were both student and military, 8% worked full time with multiple jobs, 8% worked part time, 8% were homemakers, and 8% were retired. 58% of the participants never played video games, 17% play frequently each week, 17% play rarely and 8% play 1–2 times per month. 33% of the participants had been sexually assaulted in the past, while 67% had not. None of the participants had worked in the mental health profession.

Participants arrived and were provided a briefing on the study. They were warned that the subject matter included topics related to sexual abuse. The Informed Consent was provided and briefed. They were reminded that they could abandon the experiment at any time and were asked to sign the Informed Consent form. Each participant was then asked to complete the Demographic Questionnaire, the Social Phobia Inventory (SPIN) and an EEG Questionnaire which requested data on the use of stimulants, such as coffee, sleep and medication that might affect EEG data.

The study design applied a standardized script from a US Army application DS2A in the order driven by the questions of the participants. The content was based on the story provided by PFC Jarett White in the ICT DS2A application (functioning as the AI condition). The actor was able to apply this same basic script to respond to questions the same way in all the other conditions maintaining a consistent story. The participant was told by the proctor that they would meet an individual who had experienced sexual abuse or assault. Their task was to ask the person for information about the event, giving their interaction partner the opportunity to share what happened and their feelings about it so that the participant could respond to questions about the event and the emotional state of the victim after the meeting. This was intended to motivate the participant to pay attention and engage the actor or agent. However, the participant was able to choose the line of questioning, taking the conversation as deep or as safe as they chose. Each dialog began when the proctor introduced the participant to the actor or the DS2A agent. A timer was set to allow the discussion to continue for 8–10 min but was visible only to the proctor. This allowed the proctor to stop the discussions at a natural stopping point and avoid interrupting either speaker. Each participant experienced one dialog experience which was randomly assigned. After the dialog the equipment was removed, the videos were turned off and the participants were given a rapport, presence and interaction questionnaire, the latter of which contained open-ended questions.

Each dialog was transcribed, providing the number of words per session. Behavioral indicators, such as hand gestures and nods were tabulated during the dialogs. Exaggerated facial expressions were included as gestures. If a gesture continued for more five second it was tabulated as an additional gesture. The number of gestures was calculated for each participant (and in some cases the actor) by dividing the number of gestures by the number of words divided by 100. These data were collected to determine if behavioral indicators, such as gestures provided a meaning measure of the sense of agency in a dialog partner.

While actor data were collected in the first two conditions, it was not included in the chart for conditions three through five. In several places video footage of the actor was unavailable. The video footage for the participant in one session was also unavailable due to equipment failure. Condition 4 did not include actor data since it made use of AI.

# 4  Results

Setting aside individual differences, it was evident that gesturing behavior was more prevalent in the face-to-face condition (condition 1) and in one session in the avatar condition (condition 3). Gesturing was nearly non-existent when dialoging with the AI character and while texting. The results are shown in Fig. 6. While research shows that some people will gesture even when they are on a telephone call [14], gesturing during a dialog might have indicated that the dialog partner had passed the threshold of realism and appeared to have agency.

The open-ended responses to the interaction questionnaire provided insightful feedback that supported the gesture data. In the interest of space this will be summarized and described further in the next section (Fig. 7).

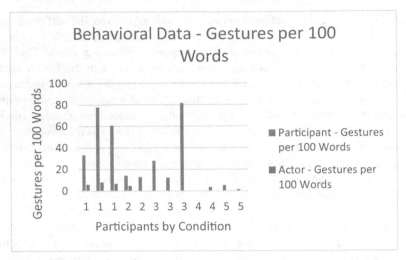

**Fig. 7.** Gestures per 100 Words for Each Session by Condition

## 5 Discussion

It is important to understand that this study was more of a field study than a controlled lab study. It removed many of the constraints and controls a lab study provides which can lead to a wide range of variability. As such, there seemed to be a great deal of individual differences. Despite that, there was a clear difference in the amount of gestures per 100 words in the face-to-face condition as compared to the others. However, there was one session in the avatar condition that had an even higher relative number of gestures than the face-to-face condition. That participant was very conscious of how (s)he was seen and how to modify his/her own gestures to regulate interaction with the dialog partner. It was unclear to what extent this affected the actor, since there wasn't any video footage of the actor's gestures during that particular interchange. Other people seemed to use their gestures to encourage the speaker or to indicate that they were along with him on his story-telling journey by nodding. The gestural data between the actor and the participant do have a directly proportional relationship to one another in interchanges where there is data for both interaction partners.

Of course, if an individual knew that they were communicating with a non-human character, it would have been unlikely that they would gesture. However, if they believe their dialog partner was real and that the partner could see them, it is theorized that they would be more likely that they would augment their dialog with gestures. This seems to be a rich area for further research.

Questionnaires were included in the research, and one in particular included open-ended questions. The feedback from this questionnaire was nicely correlated with the gesture data and indicated that across all conditions, participants were most focused on the content of the dialog with very little focus on the interface strategy or condition. Insights from the survey are shown in Table 1.

**Table 1.** Insights into Open-Ended Responses to Interaction Survey

| Face-to-Face | Focus on facial expressions, eye contact, hands shaking, voice modulations |
|---|---|
| Video | Focus on eyes, not body movement |
| Avatar | - Difficult to determine attitude and emotions by looking at facial expressions<br>- Didn't know how I appeared to avatar<br>- Visual and vocal indicators were contrary – laughed but didn't see that on his face<br>- Arms were straight and not moving |
| Agent | - Delayed response time/fading 'between scenes'<br>- Didn't answer questions or answered same question<br>- Distracted by body language, fidgeting, shaky voice, hand and leg movements<br>- Couldn't 'break the ice'<br>- Robotic<br>- One participant was falling asleep during interchange |
| Text | - Time issues with getting to the point<br>- Miscommunication<br>- Slow responses |

It isn't surprising that in the face-to-face condition participants perceived gestural information from the actor's entire body and that in the video condition the gestural information was now limited to the face and eyes since that is the focus in a video interchange. The more interesting feedback came in the avatar condition and the agent conditions. In the avatar condition a human is able to control the avatar's body and facial expression; however, there is still a good amount of information that is lost when transferred to the avatar. While technology allows for more realistic avatars that can be controlled in real-time, the cost for creating and controlling them still prevents widespread availability. High-cost Hollywood movies have made it difficult to differentiate a character that is real or computer generated. However, the level of realism necessary to help an interaction partner read into your expression is simply not available at a reasonable price-point (under $10 K). Without being able to model facial subtleties, some level of communication is lost, as was pointed out by the participants' responses. This was not indicated in any other data in this study, but provides valuable feedback on where investments might bring about payoffs in this technology.

One participant expressed concern with how he looked to the character controlling the avatar. The participant stated that had (s)he had a sense of how (s)he looked, (s)he might have made adjustments in body language to better aid in the dialog. This was an interesting point that was expressed multiple times by this participant. It was even part of his/her dialog with Jarett, as the participant tried to establish what Jarett could see of him/her. This establishes that it isn't just important during a dialog that humans see their dialog partner and have the ability to read subtle expressions to gain understanding, but that the dialog partner also needs to see the speaker (in this case the participant) in order to get their subtle meaning of gestures. This participant made great use of his/her eyebrows and other facial expressions during the dialog to express surprise, shock and warmth. This feedback could be very valuable in ensuring future AI has the capacity to pick up these details and respond to them.

Another concern was that "visual and vocal indicators were often contradictory, such as hearing the actor laugh but only seeing a faint smile with 'cold eyes' on the avatar." This response re-emphasizes the importance for an avatar to mirror more subtle facial expressions. Detailed expressions can be costly, however, losing a learning partner is costly in different ways. A cost benefit analysis should be used to determine the level of fidelity needed for specific training tasks.

Participants also noted that the avatars posture was not natural. The actor had to take special precautions to ensure that his movements did not cause the avatar's body to behave in awkward ways, so he kept his arms straight. The facial mapping was not perfect and subtle movements like crinkling eyes and moving cheeks were not achieved for this study. This is something that would likely be worth the investment to improve on, depending on the tasks to be achieved within the simulated environment.

In the agent condition the insights had much to do with technology. While the AI branching and voice recognition used by the ICT DS2A were probably sufficiently advanced if an individual knew they are speaking to AI, these participants were not given that insight. They tried to speak with the AI agent as if he were a person, and this led to a good deal of frustration. It also did not appear that the branching program tracked if the question had been asked and answered previously causing disconcerting repetition. Additionally, the voice recognition and branching used in the ICT DS2A are somewhat dated. It is likely that conversational branching has improved since its development. Another interesting factor was that when Jarett responded to detailed questions, his responses were often several minutes long. Since an agent does not have the capacity to assess his listener's continued interest, it might be helpful to break monologues into shorter segments. A human might wait for a nod or 'uh-huh' from their listening partner as an indicator of interest and gentle prodding to go on. Without this conversational tool, other strategies might be necessary when interacting with an agent to indicate further interest.

One participant was uncomfortable with missing 'niceties' that occur in conversation between humans. Basic introductory phrases, or two-way information gathering, might be useful in building rapport. The ICT DS2A was not designed to have a two-way dialog, but rather to respond to questions. Noting the importance of this to a dialog partner helps to define requirements for future AI.

One participant stated that the interaction felt a bit robotic or like an animation. There are moments between videos as the AI interprets the question asked by the participant where a listening video is shown. Before a response video is started there is a pause. This sequence might make the interaction seem animated or robotic, even though the video is of a real person. Participants felt the pauses made the interaction feel less real.

This being a first foray into this type of research, there were several lessons learned. One lesson from the study was the necessity to synchronize measurement strategies and ensuring video equipment was functioning during each session. There were several times that the video equipment began recording but stopped only a couple of minutes into the session. Another challenge was that there were a limited number of participants in the study. This was partially due to set up time associated with the other equipment, the availability of the necessary individuals (proctor, actor and EEG technician) and limited time available due to IRB and organizational limitations. Future research should use a larger number of participants and streamlined setup to reduce the overall time for each participant.

# 6 Conclusion

The purpose of this study was focused on understanding what might cause people to engage or disengage with virtual characters. To answer that question, it was important to assess what measurements might indicate that people believe their interaction partner has agency. The results can be applied to development strategies to improve interactions with virtual characters by informing future requirements.

Behavioral metrics did appear to carry meaning in that the number of gestures per 100 words showed that the face-to-face condition was demonstrated to be the "gold standard." The number of gestures were higher in the face-to-face condition with some interesting peaks in the avatar condition. The agent and text conditions had a negligible amount of gestural behavior. Despite studies that show that people gesture even while they are on the phone, the gesturing data collected during the dialog is clearly higher when the interaction partner seemed real and able to see the speaker. This indicated that gestures might be a meaningful measure to indicate a belief that an interaction partner is an actual person. Future research could explore this premise further.

The self-report data proved to contain some very valuable information. There were some meaningful insights that can be applied to virtual characters to improve their sense of realism. Participants pointed out the importance of subtle behavioral cues that they expect from an interactional partner with agency. Shaky voice and hands, eyes that flit about, and moving about in their chairs were some behavioral signs that participants noticed. Additionally, participants noticed that the avatar, which was controlled by a human's movements, did not mirror the actual crinkling of the corner of the eyes as was indicated by the actor's voice. This discontinuity immediately had an effect on the participants. One participant had trouble picking up cues about the interaction partner's attitude toward the participant because of missing cues in the avatar's face.

People who interacted with the AI noted time delays as the program interpreted the spoken question into text and selected the appropriate video footage to play. They also noted the jumps from one animation scene to the next and found it jarring. Insight into what did not give the sense of agency helps create an experience that does give the sense of agency in future requirements.

Future research could help understand what specific stimulus affects an individual's perception of agency by synchronizing sensors more tightly and possibly using eye-tracking. It would be interesting to assess if a speaker uses more gestures with their dialog partner, would the dialog partner also use more gestures. It would also be interesting to see if that affects engagement. Finally, feedback from the open-ended questions should be implemented to agent and avatar functionality. The quality of the agent and avatar software might influence the outcome of future research.

# References

1. Maxwell, D.B., Griffith, T.S., Finkelstein, N.: The use of virtual worlds in the military services as part of a blended learning strategy. In: Handbook of Virtual Environments: Design, Implementation, and Applications, Second Edition, Orlando, CRC Press, pp. 959–999 (2014)
2. Blascovich, J., Loomis, J., Beall, A., Swinth, K., Hoyt, C., Bailenson, J.N.: Immersive virtual environment technology as a methodological tool for social psychology. Psychol. Inquiry, 13, 103–124 (2002)
3. Csikszentmihalyi, M.: Finding Flow: The Psychology of Engagement with Everyday Life. Basic Books, New Hork (1998)
4. Pearce, J.: Engaging the learner: how can the flow experience support e-learning?," In: World Conference on E-Learning in Corporate, Government, Healthcare, and Higher Education (2005)
5. Shernoff, D.J., Csikszentmihalyi, M., Schneider, B., Shernoff, E.S.: Student engagement in high school classrooms from the perspective of flow theory. In: Application of Flow in Human Development and Education, Dordrecht (2014)
6. Chang, C.-C., Liang, C., Chou, P.-N., Lin, G.-Y.: Is game-based learning better in flow experience and various types of cognitive loan than non-game based learning? Perspective from multimedia and media richness. In: Computers in Human Behavior (2017)
7. Blascovich, J., Loomis, J., Beall, A.C., Swinth, K.R., Hoyt, C.L., Bailenson, J.N.: Immersive virtual environment technology as a methodological tool for social psychology. In: Psychological Inquiry (2002)
8. de Melo, C.M., Gratch, J.: Beyond believability: quantifying the differences between real and virtual humans. In: International Conference on Intelligent Virtual Agents (2015)
9. Pena, J., Khan, S., Alexopoulos, C.: I am what I see: how avatar and opponent agent body size affects physical activity among men playing exergames. J. Comput.-Mediated Commun. (2016)
10. Heyselaar, E., Hagoort, P., Segaert, K.: In dialogue with an avatar, language behavior is identical to dialogue with a human partner. Behav. Res. Methods 49(1), 46–60 (2015). https://doi.org/10.3758/s13428-015-0688-7
11. Blascovich, J.: A theoretical model of social influence for increasing the utility of collaborative virtual environments. In: Proceedings of the 4th International Conference on Collaborative Virtual Environments (2002)
12. de Melo, C.M., Gratch, J.: Beyone believability: quantifying the differences between real and virtual humans. In: International Conference on Intelligent Virtual Agents, pp. 109–118 (2015)
13. Jacobs, N., Garnham, A.: The role of conversational hand gestures in a narrative task. J. Memory Lang. 291–303 (2006)
14. Alibali, M.W., Heath, D.C., Myers, H.J.: Effects of visibility between speaker and listener on gesture production; some gestures are meant to be seen. J. Memory Lang. 44, 169–188 (2001)
15. Ozyurek, A.: Do speakers design their cospeech gesture for their addressees? The effects of addressee location on representational gestures. J. Memory Lang. 46, 688–704 (2002)
16. Krauss, R.M., Dushay, R.A., Chen, Y., Rauscher, F.: The communicative value of conversational hand gestures. J. Experimental Soc. Psychol. 31, 533–552 (1995)
17. Rime, B.: The elimination of visible behavior from social interactions: effects on verbal, non-verbal and interpersonal variables. Euro. J. Soc. Psychol. 12, 113–129 (1982)

# Perceived Speed, Frustration and Enjoyment of Interactive and Passive Loading Scenarios in Virtual Reality

David Heidrich[1] , Annika Wohlan[2] , and Meike Schaller[2](✉)

[1] German Aerospace Center (DLR), Muenchener Straße, 82234 Weßling, Germany
david.heidrich@dlr.de
[2] German Aerospace Center (DLR), Linder Hoehe, 51147 Cologne, Germany
{annika.wohlan,meike.schaller}@dlr.de

**Abstract.** Long waits and disruptive loading breaks can evoke negative emotions, like frustration. While there is a lot of research on 2D-based loading scenarios, it is unclear how people react to loading screens in an immersive virtual reality (VR) environment. In this paper we conducted a user study to investigate the effects of interactive and passive loading screens on the users' loading screen experience (LSE) in VR. We measured perceived speed, enjoyment and frustration for long and short waiting times. Results show that interactive loading screens improved participants' LSE through increasing perceived speed and enjoyment, and decreased their frustration while waiting. Thus, previous findings of 2D-based research were confirmed. Therefore, our research provides a first approach for further investigations of different loading screens in VR.

**Keywords:** Virtual reality · Loading screens · User experience

## 1  Introduction

Virtual reality (VR) technology is increasingly used in the context of data visualization [14, 20]. Processing high amounts of data inside a VR-based real-time interactive system can introduce lag causing simulator sickness [19]. Hence, applications often reduce the complexity of the scene during processing by switching to a simple loading scenario, i.e. resulting in loading times. However, these disruptive loading breaks are undesirable for users as it can evoke negative emotions, like frustration [4, 6] and may therefore have a negative effect on the users' loading screen experience (LSE). We define LSE as the users' experience in a waiting situation while they receive feedback via a loading screen. The basic aspects of LSE are the perceived loading speed, the enjoyment, and the frustration caused by the perceived waiting time and the loading screen.

Since the intensity of emotions can be increased by a VR-based representation [17, 18], loading screens in VR could be perceived as even more negative or

© Springer Nature Switzerland AG 2020
C. Stephanidis et al. (Eds.): HCII 2020, LNCS 12428, pp. 343–355, 2020.
https://doi.org/10.1007/978-3-030-59990-4_26

positive compared to 2D-based loading screens. Additionaly, users cannot avoid the loading scenario in a virtual environment by simply turning away from the screen, thus avoiding the loading situation. This could intensify a negative LSE even more.

In the context of 2D-based loading screens, passive loading scenarios (e.g. animations or progress bars) have a bigger negative impact on the user experience than interactive loading screens [7,24]. However, this has not yet been reviewed in a VR-based context. So, it is unclear whether existing literature on 2D-based loading scenarios is applicable to a VR-based context. Interestingly popular VR-based applications, like *The Lab* [23], use passive loading screens. In order to clarify the influence of interactive and passive loading scenarios on LSE, we tested an interactive and a passive loading screen in a VR-based application.

**Contribution.** This article reports new findings on loading screens in VR. Thus, our contribution is twofold. 1) In a user study, differences in LSE between interactive and passive loading screens are measured by comparing an interactive and a passive loading scenario regarding perceived speed, frustration and enjoyment. 2) In the same user study, the influence of waiting duration (short and long) on perceived speed, frustration and enjoyment in interactive and passive loading scenarios is measured. The study results indicate a *significant* difference between interactive and passive loading screens in VR for both, short and long waiting times. Thus, the findings of 2D-based research are confirmed, i.e. interactive loading screens were better suited than passive loading scenarios. While this indicates that interactive loading screens might be generally better suited for VR-based applications, further research on the influence of immersion on the user experience during loading scenarios is needed.

## 2    Related Work

Forced breaks, i.e. loading screens, can have a negative impact on the user experience of an application. Users seem to dislike them that much, that e.g. in free-to-play video games, like Candy Crush, forced breaks are intentionally placed, resulting in users to voluntarily pay money to quit them [2]. However, in many applications loading screens cannot be avoided, e.g. when loading a big virtual environment. Visual Feedback on the loading progress while forced breaks has proven to be very helpful to avoid frustration [3] and to increase the users' tolerance regarding waiting times [15]. Additionally, latest research is proposing entertaining loading screen, e.g. interactive animations, as an even better solution than just visual feedback [7]. Interactive loading scenarios are perceived as faster and more enjoyable than a simple progress bar or non-interactive animation [7].

### 2.1    Perceived Waiting Time

Since the objective waiting time is mentally transformed into users' perceived waiting time, long waits can lead to a negative waiting experience [1]. In general, users tend to judge shorter waiting times to be more positive than longer ones [12]. Li and Chen [13] pointed out the importance of visual feedback design that has an influence on users' wait evaluations. For example, cartoon bars can improve users' wait experiences for short waiting times. Moreover, it has been found that the animation's speed is positively influencing users wait evaluations. Hence, with a faster animation, perceived waiting time is perceived shorter and users get more satisfied [22].

Kim et al. found that duration and the progress function affect the viewers' waiting time perception rather than the design of the loading symbol [11]. Hui and Tse [8] compared short, intermediate and long waiting times either with or without waiting-duration information. They found that waiting-duration information in longer waits results in a longer perceived waiting time, but increases the users' satisfaction while waiting. Contrary, Zhao et al. found that animation can also have a negative effect on user satisfaction during the wait for application loading. In their study they found that animated loading screens decrease users' satisfaction by affecting their duration estimation, thus creating an illusion of longer wait [24].

Nielsen [16] indicated that for waiting times longer than 10 s, users can not keep their attention on the dialog and want to perform other tasks. To that, Hurter et. al. [9] indicated that users desire to conduct extra activities than just waiting passively while waiting for a longer time. Hence, it is necessary to look more into long waits.

## 2.2  Immersive VR

The use of immersive VR technology, i.e. wearing a head-mounted display, is characterized by a high *immersion*. Immersion is achieved with objective system properties replacing sensory inputs from the real world with digital information [21]. That way, the users are disconnected from the real world. Since people tend to look for secondary tasks after a certain waiting duration [16], high immersion prevents the users from interacting with the real world. This is especially crucial in VR-based passive loading scenarios, where the only possible *interaction* is moving the hands around. Although developers could build a variety of interactions into their VR-based loading scenarios, like a balloon machine or a virtual dog [23], they are limited by the low computing power due to the loading process. Whether one should prioritize shorter loading times over less interactivity in a VR-based context is unclear.

Higher immersion can also intensify unpleasant experiences, like fear in horror games [18]. Especially negative emotions, like fear and anxiety [10], have been found to be stronger amplified by immersion than other emotions, like happiness [5]. In the context of VR-based loading screens, possible negative emotions evoked by the forced break could be intensified compared to a less immersive representation.

## 3    Hypotheses

As demonstrated above, perceived LSE is tested for various types of loading screens such as animations, progress bars and interactive screens. However, the question still remains about whether those effects are valid in a VR-based context. As VR technology offers a high immersive environment, loading scenarios might be perceived differently compared to mobile or desktop waiting time scenarios. Therefore, the main goal of our study is to address this question by investigating the effect of two types of loading screens (interactive and passive) in VR on users' LSE for two waiting times (short and long). We assume that the results of Hohenstein [7] on short waiting times also apply in a VR-based context.

> H1: During **short waiting times** in VR, peoples' LSE is better for interactive loading screens than passive loading screens.

Moreover, we assume that precisely because people look for alternative activities during long waiting times in the real world, interactive loading screens lead to higher LSE in a virtual environment, as they offer users possibilities to interact with their environment.

> H2: During **long waiting times** in VR, peoples' LSE is better for interactive loading screens than passive loading screens.

## 4    System Description

To test our hypotheses, we implemented two loading screens. Based on the work of Hohenstein et al. [7] on desktop-based interactive and passive loading screens, we chose a *passive animated Newton's cradle* and an *interactive Newton's cradle*. We implemented both loading screens into the existing visualization tool *IslandViz* [14]. This open-source tool visualizes OSGi-based software projects as islands on a virtual table in VR. On startup, IslandViz is importing the software system from a database and is then generating the island layout and the island shapes at runtime. This loading can take up to several minutes, depending on the size of the software project. During the loading time, the user enters the virtual environment, i.e. puts on the head-mounted display, and waits for the application to finish loading (Fig. 1). Then the islands are presented on the virtual table. The loading screens are developed with Unity 2019.1.6f1 using the SteamVR plugin version 2.3.2.

The Newton's cradles - both animated and interactive - consist of five balls each hanging from a thread, attached to the *Loading* lettering. In the passive version, the user cannot interact with the balls (Fig. 1). However, the first ball is swinging automatically after random time periods (2–5 s) causing a constant movement of the balls. In the interactive version, the user can grab a ball by touching it with the HTC Vive controller and holding the trigger button (Fig. 2).

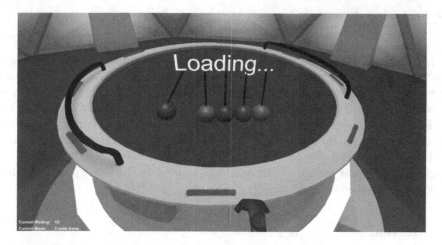

**Fig. 1.** Passive Loading Screen: The player is waiting in front of the virtual table watching the animated Newton's Cradle with automatically swinging balls.

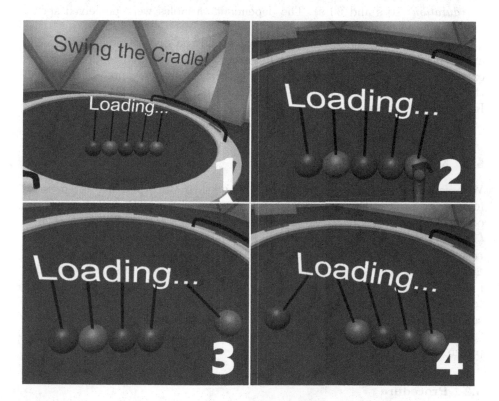

**Fig. 2.** Interactive Loading Screen: The participant is standing in front of the interactive Newton's cradle (1), touches the right ball (2), grabs and moves the ball (3), and releases the ball to make the whole cradle move (4).

The balls interact like a real cradle, i.e. the first ball bumping into the second ball causing the last ball to bounce up and vice versa.

We also manipulated the loading routine of the IslandViz to always take a predefined time. That way, although the loading of the software system did only take a few seconds, the loading screens are always displayed for the time period set by the experimenter. Additionally we added an experimenter interface, so we could change the current loading screen and start a condition, while the participant is immersed. When a loading screen condition is initiated, a *Start* button appears over the table. To start the loading screen, the user has to press the button, i.e. touch the button with the HTC Vive controller and press the trigger button. That way, the user is always located at the same position, when the loading condition is starting.

## 5  Method

The study employed a 2 × 2 within-subjects repeated-measures design. The two independent variables were *loading screen* (interactive and passive) and *waiting duration* (10 s and 60 s). The dependent variables were perceived speed, enjoyment and frustration.

The experiment consisted of two sessions: in the first session, participants experienced both loading screens for 10 s each. In the second session, participants experienced both loading screens for 60 s each. Thereby we wanted to find out if there are differences concerning users' LSE during long and short waiting times. In addition, we argue that 10 s waits in VR might be too fast to perceive the whole environment, especially the interactive loading scenario.

### 5.1  Measures

We assessed participants demographic data and their VR experience as a possible influencing variate. Perceived speed, enjoyment and frustration were measured on a 7-point Likert scale: '*How fast did you experience loading the application?*' (1 = very slow - 7 = very fast), '*How much fun did you have loading the application?*' (1 = very much - 7 = not at all), '*How frustrated were you while you were waiting?*' (1 = not at all - 7 = very much). Participants had to provide information about those three items every time after they had seen a loading screen. Since we have two waiting times, there are two measures of each perceived speed, enjoyment, and frustration for every loading screen. In the end of the study, we asked participants to rank both loading screens according to their preference.

### 5.2  Procedure

All questionnaires and experimental stimuli were administered to the participants in a laboratory setting. For the experiment, we used the HTC Vive VR system, consisting of the head-mounted display, two controllers and two base stations which were installed in our VR lab. Participants had enough space to move in the VR environment. The whole examination including instructions and debriefing took approximately 20 mins.

The part of the study that took place in VR was about 10 mins on average and consisted of two sessions, respectively 10 and 60 s. The order of the loading screens within the first session was randomized for each participant and remained the same for the second session as well.

After each loading screen, participants were asked to rank their experience about their perceived speed, enjoyment and frustration. Each question was displayed within the virtual environment while the participants had to answer verbally. After experiencing the screens in both sessions, participants were asked to rank the two loading screens according to their preference. The study was ended by open questions and the opportunity to share feedback, concerns or ideas regarding this evaluation.

### 5.3  Participants

In total, 25 persons (12 female, 13 male) participated in our study with a mean age of 28.6 ($SD = 5.58$). The participation was voluntary and participants were recruited through a local call for participation. All participants have a background in software engineering and have already worked with this tool, i.e. all participants know why the loading time exists (depending on the size of the visualized software project). 3 (12%) participants stated to have no VR experience. 14 (56%) participants stated to have little VR experience, and 6 (24%) stated to have some experience. 2 (8%) participants stated to use VR technology in their daily working life.

## 6  Results

We predicted that being confronted with an interactive loading screen compared to being confronted with a passive loading screen would improve participants' LSE, thus increase their perceived speed and enjoyment, and decrease their frustration. We performed statistical analysis using the Statistical Package for the Social Sciences (SPSS) Version 22.9 for Windows. For every method, the error probability was set to a significance level alpha = 5%.

### 6.1  Perceived Speed

A repeated measures analysis of variance (ANOVA) with a Greenhouse-Geisser correction determined that mean perceived speed levels showed a statistically significant difference between the interactive and passive loading scenario, $F(1, 24) = 55.40$, $p < .001$, $\eta^2 p = .70$ (Fig. 3). Moreover, there is a significant difference in the mean enjoyment levels between the short and long waiting time, $F(1, 24) = 142.73$, $p < .001$, $\eta^2 p = .86$. There is no interaction between loading screen and waiting duration, $p > .05$.

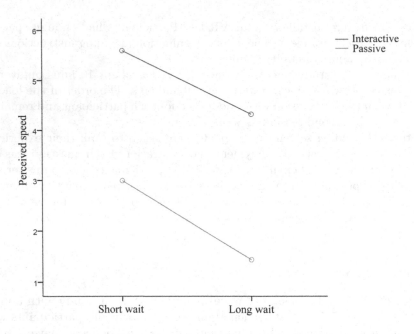

**Fig. 3.** Effects of interactive and passive loading screens on perceived speed. *Note*: The vertical axis reflects the degree of perceived speed, higher scores on this measure represent faster perceived loading times. The horizontal axis represents the waiting duration, whereas short wait represents the 10 s wait and long wait represents the 60 s wait.

Regardless the waiting time, participants were more likely to perceive waiting times faster when confronted with an interactive loading screen ($M = 4.3$, $SD = .22$) than with a passive loading screen ($M = 2.8$, $SD = .16$).

## 6.2   Enjoyment

Another repeated measures ANOVA revealed statistically significant differences in the mean enjoyment levels between the two loading scenarios ($F(1, 24) = 77.42$, $p < .001$, $\eta^2 p = .76$) and the two waiting times ($F(1, 24) = 13.54$, $p = .001$, $\eta^2 p = .36$) (Fig. 4). There is no interaction between loading screen and waiting duration.   Regardless the waiting time, participants were more likely to enjoy being confronted with an interactive loading screen ($M = 4.5$, $SD = .24$) than with a passive loading screen ($M = 2.2$, $SD = .20$).

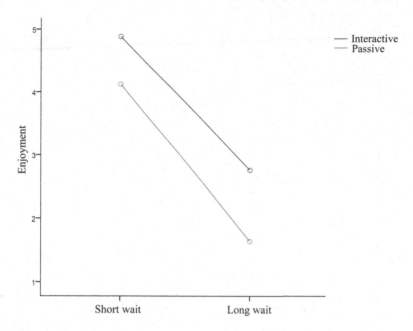

**Fig. 4.** Effects of interactive and passive loading screens on enjoyment. *Note*: The vertical axis reflects the degree of enjoyment, higher scores on this measure represent more enjoyment. The horizontal axis represents the waiting duration, whereas short wait represents the 10 s wait and long wait represents the 60 s wait.

## 6.3 Frustration

A final repeated measures ANOVA revealed statistically significant differences in the mean frustration levels between the two loading scenarios ($F(1, 24) = 42.38$, $p < .001$, $\eta^2 p = .64$) and the two waiting times ($F(1, 24) = 32.11$, $p = .001$, $\eta^2 p = .61$) (Fig. 5). There is no interaction between loading screen and waiting duration. Regardless the waiting time, participants were more likely frustrated being confronted with a passive loading screen ($M = 4.1$, $SD = .26$) than with an interactive loading screen ($M = 2.5$, $SD = .22$).

Overall, the results suggested that participants estimated LSE to be better for the interactive loading condition regardless the waiting duration. **Hence, our hypotheses H1 and H2 are confirmed.**

## 6.4 Preference

Only one participant (4%) preferred the animated Newton's Cradle over the interactive Newton's Cradle.

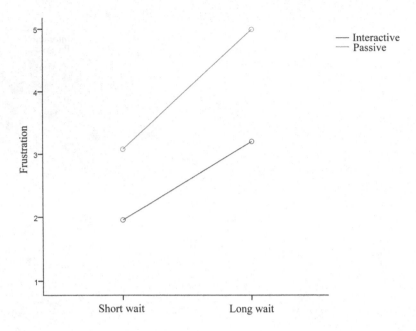

**Fig. 5.** Effects of interactive and passive loading screens on frustration. *Note:* The vertical axis reflects the degree of frustration, higher scores on this measure represent more frustration. The horizontal axis represents the waiting duration, whereas short wait represents the 10 s wait and long wait represents the 60 s wait.

## 7    Discussion

Previous research investigated various feedback types that might influence users' loading screen experience (LSE) in a 2D-based environment [8,13]. According to previous research users prefer interactive loading screens over passive loading screens [7]. In this study, we investigated interactive and passive loading screens in a VR-based environment to validate those results for VR-based applications. Additionally, we added a short and a long waiting time condition for each loading screen. Overall, the results suggested that participants estimated LSE to be better for the interactive loading condition regardless the waiting duration. In addition, the participants indicated that they favoured the interactive cradle over the animated cradle. Hence, previous research on 2D-based loading screens seem to be valid for VR-based environments and our hypotheses H1 and H2 are confirmed.

Interestingly, our findings are contradictory with popular VR-based video games, like The Lab [23]. However, it is uncertain whether these loading screens are actually conscious design decisions or whether these loading screens are pre-determined by the game engines and/or software architecture. Since our findings indicate significant differences between the loading screens, the interactivity of a loading screen could be an important factor to consider for good user experience design. This might be particularly critical for applications with many loading screens and is independent of the waiting duration. Even for short waiting times of a few seconds, interactive loading screens are perceived faster and more enjoyable. Hence, based on our findings, we recommend to include interactive loading screens in VR-based applications to reduce frustration and to increase perceived speed and enjoyment during loading.

## 7.1 Limitations

VR-based applications can be very variable and range from a black void with no embodiment at all to a photo-realistic environment with full body tracking. We would argue that, in a *default* VR application, one can at least see the movement of the controllers in the virtual environment. Hence, a loading scenario in VR is never a strictly passive but rather an interactive experience. Technically this means our passive loading scenario really was a less-interactive loading screen. However, this logic could also be applied to e.g. a desktop-based loading scenario, where the users hand is represented as a mouse courser.

Secondly, the repeated measure in our study design might have led to some sort of learning effect in the long waiting time condition. Since experiencing a loading screen multiple times during a play session is a realistic use case and all participants have experienced the same loading screens prior to the long conditions, we believe that comparing the two loading screens within the long wait time condition is still valid. However a conclusion about the difference between the short wait condition and the long wait condition is not possible.

Thirdly, our participants were not interacting with the software visualization tool after the loading. This might have weakened the strength of the measured LSE. However, all participants have used the IslandViz application before and knew what they were waiting for. This probably would have been a much bigger limitation with participants that are familiar with IslandViz.

Fourthly, we tested our loading screens in the context of software visualization, thus in a context of work. A different context where participants might be more motivated to see what comes after the loading screen, e.g. in video games, the measured effects might also be stronger.

Fifthly, our study was conducted with a relatively small sample of only computer scientists which does not cover a broad population. Hence, a future study should definitely have a bigger sample size and cover a broader population.

## 7.2 Future Research

First, we must compare VR-based loading scenarios with 2D-based loading scenarios. This will answer the question whether there is *any* influence of immersion, i.e. being locked away from the real world, on the LSE. Other possible VR-related factors influencing the LSE could be embodiment and presence.

Second, it is necessary to test other loading screens in a VR-based environment. Since we did not indicate the remaining waiting time in our study, another loading screen could be progress bars that indicate the remaining waiting time. This is a viable alternative to static loading screens [15]) and, in a desktop-based context, people should prefer a progress bar over a passive Newton's Cradle [7].

# 8 Conclusion

This study is a first step to understand the influence of different loading screens in VR on users' perceived speed, enjoyment and frustration. However, further analysis is necessary for understanding how loading screens can serve as effective "time shortening"- tools in a virtual environment. As for desktop or mobile applications, interactive loading screens in VR seem to be a generally better choice. Anyway, this study investigated the user's first contact with a loading screen. This raises the question of whether there is at all influence of immersion on the loading screen experience and whether those effects are accurate for repeated interactions.

# References

1. Antonides, G., Verhoef, P.C., van Aalst, M.: Consumer perception and evaluation of waiting time: a field experiment. J. Consumer Psychol. **12**(3), 193–202 (2002)
2. Blaszczynski, A., Cowley, E., Anthony, C., Hinsley, K.: Breaks in play: do they achieve intended aims? J. Gambl. Stud. **32**(2), 789–800 (2015). https://doi.org/10.1007/s10899-015-9565-7
3. Bouch, A., Kuchinsky, A., Bhatti, N.T.: Quality is in the eye of the beholder: meeting users' requirements for internet quality of service. In: Proceedings of the SIGCHI conference on Human Factors in Computing Systems (CHI 2000), pp. 297–304 (2000)
4. Ceaparu, I., Lazar, J., Bessiere, K., Robinson, J., Shneiderman, B.: Determining causes and severity of end-user frustration. Int. J. Hum.-Comput. Interact. **17**(3), 333–356 (2003). https://doi.org/10.1207/s15327590ijhc1703_3
5. Freeman, J., Lessiter, J., Pugh, K., Keogh, E.: When presence and emotion are related, and when they are not. In: Proceedings of the 8th Annual International Workshop on Presence (PRESENCE 2005), pp. 213–219. International Society for Presence Research (2005)
6. Gilleade, K.M., Dix, A.: Using frustration in the design of adaptive videogames. In: Proceedings of the 2004 ACM SIGCHI International Conference on Advances in Computer Entertainment Technology, pp. 228–232. ACM (2004)
7. Hohenstein, J., Khan, H., Canfield, K., Tung, S., Perez Cano, R.: Shorter wait times: the effects of various loading screens on perceived performance. In: Proceedings of the 2016 CHI Conference Extended Abstracts on Human Factors in Computing Systems, pp. 3084–3090. ACM (2016)
8. Hui, M.K., Tse, D.K.: What to tell consumers in waits of different lengths: an integrative model of service evaluation. J. Market. **60**(2), 81–90 (1996). https://doi.org/10.2307/1251932
9. Hurter, C., Girouard, A., Riche, N., Plaisant, C.: Active progress bars: facilitating the switch to temporary activities. In: CHI 2011 Extended Abstracts on Human Factors in Computing Systems, pp. 1963–1968. ACM (2011)

10. IJsselsteijn, W., Ridder, H.d., Freeman, J., Avons, S.E., Bouwhuis, D.: Effects of stereoscopic presentation, image motion, and screen size on subjective and objective corroborative measures of presence. Presence: Teleoper. Virtual Environ. **10**(3), 298–311 (2001). DOI: https://doi.org/10.1162/105474601300343621

11. Kim, W., Xiong, S., Liang, Z.: Effect of loading symbol of online video on perception of waiting time. Int. J. Hum. Comput. Interact. **33**(12), 1001–1009 (2017). https://doi.org/10.1080/10447318.2017.1305051

12. Lallemand, C., Gronier, G.: Enhancing user experience during waiting time in HCI: contributions of cognitive psychology. In: Proceedings of the Designing Interactive Systems Conference, pp. 751–760. ACM (2012)

13. Li, S., Chen, C.H.: The effects of visual feedback designs on long wait time of mobile application user interface. Interact. Comput. **31**(1), 1–12 (2019)

14. Misiak, M., Schreiber, A., Fuhrmann, A., Zur, S., Seider, D., Nafeie, L.: Islandviz: A tool for visualizing modular software systems in virtual reality. In: 2018 IEEE Working Conference on Software Visualization (VISSOFT), pp. 112–116, September 2018. https://doi.org/10.1109/VISSOFT.2018.00020

15. Myers, B.A.: The importance of percent-done progress indicators for computer-human interfaces. In: Proceedings of the SIGCHI Conference on Human Factors in Computing Systems (CHI 1985), pp. 11–17 (1985)

16. Nielsen, J.: Usability Engineering. Elsevier (1994)

17. Pallavincini, F., et al.: Is virtual reality always an effective stressors for exposure treatments? some insights from a controlled trial. BMC Psychiatry **13**(1), 52 (2013). https://doi.org/10.1186/1471-244X-13-52

18. Pallavicini, F., Ferrari, A., Pepe, A., Garcea, G., Zanacchi, A., Mantovani, F.: Effectiveness of virtual reality survival horror games for the emotional elicitation: preliminary insights using resident evil 7: Biohazard. In: International Conference on Universal Access in Human-Computer Interaction. pp. 87–101. Springer, Heidelberg (2018). https://doi.org/10.1007/978-3-319-92052-8_8

19. Pausch, R., Crea, T., Conway, M.: A literature survey for virtual environments: military flight simulator visual systems and simulator sickness. Presence: Teleoper. Virt. Environ. **1**(3), 344–363 (1992)

20. Romano, S., Capece, N., Erra, U., Scanniello, G., Lanza, M.: On the use of virtual reality in software visualization: the case of the city metaphor. Inf. Softw. Technol. (2019). https://doi.org/10.1016/j.infsof.2019.06.007

21. Slater, M., Wilbur, S.: A framework for immersive virtual environments (five): speculations on the role of presence in virtual environments. Presence: Teleoper. Virtual Environ. **6**(6), 603–616 (1997). https://doi.org/10.1162/pres.1997.6.6.603

22. Söderström, U., Bååth, M., Mejtoft, T.: The users' time perception: the effect of various animation speeds on loading screens. In: Proceedings of the 36th European Conference on Cognitive Ergonomics, p. 21. ACM (2018)

23. Valve: The lab. https://store.steampowered.com/app/450390/The_Lab/ (2016)

24. Zhao, W., Ge, Y., Qu, W., Zhang, K., Sun, X.: The duration perception of loading applications in smartphone: effects of different loading types. Appl. Ergon. **65**, 223–232 (2017). https://doi.org/10.1016/j.apergo.2017.06.015

# Augmented Riding: Multimodal Applications of AR, VR, and MR to Enhance Safety for Motorcyclists and Bicyclists

Caroline Kingsley[1]([✉]), Elizabeth Thiry[1], Adrian Flowers[1], and Michael Jenkins[2]

[1] Charles River Analytics Inc., Cambridge, MA 02138, USA
ckingsley@cra.com
[2] Pison Technology Inc., Boston, MA 02116, USA
mjenkins@pison.com

**Abstract.** Operating two-wheeled vehicles in four-wheel-dominant environments presents unique challenges and hazards to riders, requiring additional rider attention and resulting in increased inherent risk. Emerging display and simulation solutions offer the unique ability to help mitigate rider risk–augmented, mixed, and virtual reality (collectively extended reality; XR) can be used to rapidly prototype and test concepts, immersive virtual and mixed reality environments can be used to test systems in otherwise hard to replicate environments, and augmented and mixed reality can fuse the real world with digital information overlays and depth-based sensing capabilities to enhance rider situational awareness. This paper discusses the use of multimodal applications of XR and integration with commercial off the shelf components to create safe riding technology suites. Specifically, the paper describes informal and formal research conducted regarding the use of haptic, audio, and visual hazard alerting systems to support hands-on, heads-up, eyes-out motorcycle riding, as well as the use of an immersive mixed reality connected bicycle simulator for rapidly and representatively evaluating rider safety-augmenting technologies in a risk-free environment.

**Keywords:** Real world virtual reality applications · Applied augmented reality · Motorcycle heads up displays · Bicyclist alerting · Riding simulation · Multimodal alerting

## 1 Introduction

Operating two-wheeled vehicles in four-wheel-dominant environments presents unique challenges and hazards to riders, requiring additional rider attention and resulting in increased inherent risk. Accidents involving motorcycles or bicycles with traffic vehicles (e.g. passenger cars, SUVs, commercial trucks) are disproportionately fatal compared to other types of motorcycle or bicycle accidents. Additionally, a significant proportion of non-traffic motorcycle and bicycle accidents, compared to car accidents, are

© Springer Nature Switzerland AG 2020
C. Stephanidis et al. (Eds.): HCII 2020, LNCS 12428, pp. 356–367, 2020.
https://doi.org/10.1007/978-3-030-59990-4_27

attributed to unique hazards riders face such as potholes, inclement weather, uneven terrain, sand/gravel, construction zones, and even sharp turns and steep grades. As such, heightened risk requires heightened situational awareness and safety measures for riders on the road. While solutions such as smartphone-based GPS, and handlebar-mounted alerting systems, and on-bike and/or -vehicle warning systems (e.g. Waze, SmartHalo, V2X) offer ways to warn bicyclists and motorcyclists of upcoming hazards to prevent crashes, these technologies have limited market reach (i.e., expensive, platform-specific, emerging technology) and usability flaws. For example, smartphone touchscreens are less effective for cyclists wearing protective equipment (i.e., gloves) and also require riders to momentarily glance down from the road. As such, novel approaches to significantly enhance situational awareness while simultaneously supporting heads-up, eyes-out, hands-on riding while maintain a low barrier to adoption are required to effectively alert riders to the dynamic dangers associated with riding.

In recent years, the advances in the development of multimodal systems and visual display technologies have placed XR (extended reality; collectively referring to augmented reality, mixed reality, and virtual reality) and heads up displays (HUDs) at the forefront of information visualization and interaction. These novel technologies offer effective solutions that address situational awareness and safety across real-world and virtual environments, from navigation and real time hazard alerting to testing in simulated environments.

Unlike conventional solutions, augmented reality heads up displays (HUDs) offer the ability to significantly enhance situational awareness while simultaneously supporting heads-up, eyes-out, hands-on riding. While HUDs maintain the ability to visually augment users' field of view (FOV) with information at a relatively high level of specificity and detail, additional modalities, such as haptic and auditory displays, can also augment the riding experience while supporting attentive riding. Additionally, rather than assessing the detectability, usability, and viability of such alerting and communication modalities among the inherent dangers of live roadways, mixed reality (MR) and virtual reality (VR) offers the ability to replicate high fidelity, dynamic, and immersive riding environments that are also configurable and controllable, enabling us to test and validate augmented riding capabilities in realistic yet danger-free simulated environments.

This paper presents an overview of our work in the space of XR applied to transportation and enhancing the safety of two-wheeled riding (i.e. motorcycles and bicycles) with a specific focus on the practical application of AR for real time hazard alerting for motorcycle and bicycle riders and the use of VR for testing augmented alerting modalities in simulated riding environments.

## 2 Augmented Riding: On the Road

Although the automobile industry has invested in technology to improve driver safety (e.g., assisted braking, lane change hazard avoidance systems, obstacle detection capabilities, fully self-driving cars), motorcycle rider hazard avoidance technologies have remained largely unchanged over the past decades. Advances in object detection, terrain classification, and other computer vision technologies, as well as upcoming connected vehicle (CV) infrastructure present a new opportunity to enable unprecedented real-time hazard alerting for motorcyclists. However, no such technology has been adapted and made commercially available for motorcycles, despite the increasing prevalence of these types of systems on automobiles. One barrier to providing riders with routing information and real-time hazard alerts has been the inability to present this information en route in a format that does not distract riders. This barrier is quickly lowering, however, as portable display technology (e.g., smartphones, GPS devices, in-helmet audio headsets) and recent advances in consumer augmented reality (AR) displays (e.g., Microsoft HoloLens, Vuzix Blade) have paved the way for a number of motorcycle specific HUDs (e.g., NUVIZ, Crosshelmet, Everysight Raptor 2).

Compared to automobiles, warning systems present cyclists with unique challenges such as motorcycles or bicycles lacking large instrument clusters and center stacks to present warnings, and a heightened requirement to use both hands for operation. Furthermore, novel approaches to effectively provide warnings to riders must consider characteristics of the operational context (e.g., human factors, environmental factors, vehicle features) to ensure a successful solution presents hazard data to riders in a way that is not distracting but also facilitates rapid understanding to enable appropriate responses given the context of the environment.

### 2.1 Multimodal Applications for Rider Alerting

Outfitting riders with a HUD has the advantage of providing a display for the rider to receive information while not requiring space on the motorcycle itself. A HUD also has the benefit of allowing the technology to be ported to any motorcycle the rider may use. With a properly implemented hazard alert system, motorcyclists can benefit from many of the same technologies as cars with regard to routing and hazard alerts. Additionally, technologies that currently benefit drivers, including navigation on the center stack with applications such as Android Auto and Apple's Car Play, can be extended to riders through a HUD. Future technologies will include alerts that utilize communications between vehicles, their operators, and the infrastructure. Through properly designed HUDs and alert systems, riders on their motorcycles can see many of the same benefits as drivers in their cars.

Under a larger effort focused on addressing these gaps through the utilization of emerging technologies for enhanced motorcycle rider safety, we designed, developed, and tested a system to support safe riding by alerting riders while en route to upcoming hazards, sourced and validated through crowdsourcing techniques, through visual, audio, and haptic AR reporting and alerting modalities (Fig. 1). We have deployed this integrated, cloud-based system to various COTS AR devices, including the NUVIZ and Everysight Raptor HUDs, the SubPac M2X, Woojer, and bHaptics vibro-tactile wearables, and Bose Frames and in-helmet audio devices.

**Fig. 1.** Concept illustration of a modular multimodal hazard alerting suite for motorcyclists

To evaluate the effectiveness of these systems, we conducted both informal and formal evaluations, including a controlled live-riding human subjects research study to assess and validate the usability and acceptability of HUD- and audio-based hazard alerting on the NUVIZ HUD.

**Informal Evaluation.** Through informal internal evaluations, including usability evaluations of interfaces, hazard alerting testing while riding in the back of a vehicle, stationary on-motorcycle haptic and audio alerting testing, and user acceptance interviews, we were able to iteratively refine, develop, and deploy visual- and audio-based alerting capabilities on the NUVIZ AR HUD to support formal human use evaluation.

For example, to incorporate audio and haptic feedback into our system, we applied the same ecological interface design (EID) techniques that we used to design the visual symbology. EID is usually applied to visual displays by using simple graphical display elements that explicitly map relevant information to emergent geometrical properties of the interface (Vincente and Rasmussen 1990; Bennett and Flach, 2011). However, an expanding body of research sought to extend the EID concepts to non-visual display channels, specifically audio and haptic displays (Sanderson and Watson 2005; Watson and Sanderson 2007; Wagman and Carello 2003). Therefore, we applied proven EID techniques and initial user feedback to design audio and haptic display symbologies to further increase rider hazard awareness. These symbologies were designed to work in isolation or in combination with each other to provide more robust alerting capabilities. For example, a rider with only a Bluetooth speaker in their helmet could hear audio alerts about an upcoming hazard type, while a rider with an AR HUD and a speaker can benefit from visual and audio alerts by seeing upcoming hazards on the display as well as hearing audio alerts about the most prominent hazard within a particular distance.

Specifically, haptic alerts have been proven extremely effective at capturing and directing attention while minimizing user annoyance and distraction. Frequency, duration, pulse rate, and intensity can all modify the user's interpretation of and reaction to the alert. These metrics are especially important when designing a wide array of collision and hazard alerts. Relevant to motorcycle haptic alerting, we developed five alert categories that correspond to a distinct level of urgency and required response.

We informally tested the haptic outputs of these alerting categories in the intended use environment (i.e., on a motorcycle in an area with representative ambient noise) to assess test participants' reaction to each alert. During this test, participants sat on an idling Harley Davidson 2013 Sportster Xl1200 motorcycle while each alert was played through the SubPac haptic vest. We counterbalanced alert order over the participant group. While the alert was played, participants were asked to raise their hand to signify the moment they first perceived the alert. After the alert finished playing, participants verbally ranked each alert on three Likert scales regarding perceptibility, urgency, and annoyance. Using the results, we updated our alerting symbology guidelines from our prior empirically-informed hypotheses, as outlined in Table 1.

This informal idle motorcycle on-body haptics alerting evaluation is just one example of the many informal tests and evaluations we ran before our formal live riding evaluation.

**Formal Live Riding Evaluation Overview.** In our formal human subject's research study, conducted in partnership with Virginia Tech Transportation Institute (VTTI), participants rode their personal motorcycles with visual hazard alerts presented through the NUVIZ HUD and audio alerts through a Sena 2.0 Bluetooth helmet-mounted headset, both wirelessly synced via Bluetooth. Two types of hazards were utilized (obstacles

**Table 1.** Revised alerting symbology based on haptic testing results

| Hazard category | Hazard severity | TTC trigger | Modality | Alert type | Frequency | Duration | Intensity | Interburst interval | Other |
|---|---|---|---|---|---|---|---|---|---|
| Imminent | High | 2 | Abstract/ Haptic | Pulsed/ Continuous | 1700 Hz/75–150 Hz | 440 ms pulse/ Continuous | 30 dB above MT | 360 ms/ 100 ms | Quick onset, directional |
| Cautionary, action required | Moderate | 5 | Graded abstract | Pulsed | 1250 Hz | 400 ms–150 ms | 25 dB above MT | 300 ms–150 ms | |
| Cautionary, heightened attention | Low | 5 | Haptic | Continuous | 10 Hz | 440 ms | 15 dB above MT | 360 ms | Square wave |
| Hazard, cautionary | Variable | 5 | Auditory Icon | Single-Stage | N/A | Variable | 15 dB above MT | N/A | |
| Hazard, avoidance | Moderate | 5 | Speech/ Graded Haptic | Single-stage, Continuous | N/A 130–10 Hz | Variable/ 400 ms–150 ms | 20 dB above MT | 300 ms–150 ms | Directional sweep |

and surface hazards) and three visual hazard alert styles were paired with three audio alert styles. The visual alert styles included; full screen, single line, or map with audio components that accompanied the visual, shown in Fig. 2.

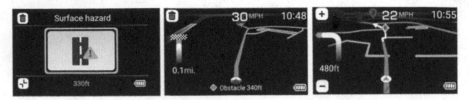

**Fig. 2.** Three different visual alert styles: full screen (left), single line (center), map (right) displayed on the NUVIZ HUD

In total, three audio styles were evaluated; voice-based, tone-based, or no audio. The evaluation focused on detection, timing, comprehension, distraction and preference of the three different visual and audio alerting types during on-road riding as well as overall rider perception and interest in the technology. Overall, full screen alerts were the most effective visual alert style (map-based alerts second) and voice-based alerts were the most effective audio style. Study results showed that full screen alerts had the fewest number of collision misses and were rated as the most preferred, easiest to detect and most understandable visual alert style by participants. We also found that to maximize understandability, the full screen alerts should be paired with voice-based audio rather than tone as voice alerts scored higher than tone in comprehension. Our findings indicated that the second best and most preferred alert type was using a map-based visual alerts with feedback indicating that participants appreciate having navigation information available as they approach a hazard. When asked specifically about whether or not they found the hazard alerts to be useful, participants said that overall, they were useful. Participants see more utility in hazard alerts that utilize voice alerts rather than tone. When participants that experienced voice audio alerts were compared to participants that experienced tone-based alerts, the voice-based audio increased usefulness by 23%. Participants indicated that a helmet mounted HUD device will make riding safer, and in general they would be likely to purchase one in the future (willing to pay on average $261.36). If participants purchased a device like this, the most likely reported use case would be for riding in areas they have never visited before or for long trips. In terms of improvements, participants indicated that they would like to see a larger, brighter, and more centrally located and helmet-agnostic display that is not affected by sunlight/glare with larger and less text. Additionally, some participants expressed interest in haptic alerting (not included in VTTI evaluation) and time-to-hazard display (rather than distance). Finally, participants were asked to provide the top three hazards they think would be most useful for riders to be alerted about. The top two hazards mentioned were gravel in roadway and potholes/rough roads, and third most were traffic related, specifically crash ahead and traffic delays.

While our study examined the usability and acceptance of the system with motorcyclists, a key component to future iterations and testing emergent technology for two-wheeled riders is continuous testing and evaluation of improvements and iterations.

Test courses are expensive to maintain and may not allow for full breadth of real-world conditions and limited engagements may fail to address potentially confounding human factors of technology use while en route. Novel technologies such as virtual, augmented, and mixed reality (XR) offer a low cost solution to conduct research and acquire feedback based on realistic operational scenarios. From a design standpoint, the use of AR when generating graphical display elements enabled us to acquire feedback based on operational scenarios (e.g. color display indoors vs. outdoors) during our iterative user-centered design process. This allowed us to then rapidly modify display components (e.g. display symbology) to be more intuitive and useful for end users as well as efficient in communicating hazard information to riders. Collectively, these technologies each present unique components and solutions to various aspects of research and development for emerging cyclist safety technologies. Mixed reality for example, can serve as a development-testing environment and can be used to inform future usability and fundamental research studies as it relates to effectiveness and performance. Graphical display elements as well as non-visual display elements (e.g. audio, haptic displays) can be evaluated through simulations to acquire informal user feedback based on realistic operational scenarios, while reducing the costs, risks, and time associated with human subjects live riding studies.

## 3  Mixed-Reality Riding Simulation

Mixed reality simulation environments offer unique capabilities to recreate immersive and dynamic real-world spatiotemporal situations that are otherwise expensive or unsafe. Mixed reality also supports multiple layers of information such as real-time geographical location, audio-visual alerts, and multimodal plugins, which allows for data and information driven end-to-end immersive experiences that meet the objectives of real-world operational environments. From a design and development standpoint, mixed reality allows for more dynamic rapid-prototyping, can serve as a development-testing environment, and can be used to inform future usability and fundamental research studies as they relate to on the road, multimodal heads-up, eyes-out riding system effectiveness and performance. However, for simulation environments to be effective, they require validation and the design of environments to be built with human and operational factors in mind. On the road data can be implemented and be used to inform the development of simulations with high levels of realism.

Under a related effort focused on connectivity for enhanced bicycle rider safety, we designed, developed, and implemented an immersive mixed reality connected bicycle simulator for rapidly and representatively evaluating rider safety-augmenting technologies in a risk-free environment. This MR bicycle simulator allows participants to enter into an immersive virtual reality (VR) urban environment and control a virtual bicycle by riding on a stationary VR-ready exercise bike (to enhance realism) and then interacting with various virtual and real-world tracked objects and variables.

### 3.1  Testing and Evaluation Safety

Adapting road safety technologies to cyclists, presents a need for safe testing and evaluation solutions. An ideal solution should provide an accurate model of hazards unique to motorcycles and bicycles in order to assess their potential to endanger riders to allow proper identification, characterization and assessment of hazards without risking rider safety.

There has been an increased focus on technology to increase the safety of cyclists (such as the Trek Cyclist Safety System), motivated by Federal, State, and Munici-pal level initiatives, such as Vision Zero (www.pedbikeinfo.org/topics/visionzero.cfm). Unfortunately, effective and ecologically valid human factors and performance testing of cycling technology is often prohibitively difficult and expensive to perform due to: (1) a lack of underlying infrastructure to support testing, (2) risks associated with conducting controlled riding studies that accurately portray the types of hazardous situations these novel technologies seek to mitigate, and (3) difficulty providing empirical evidence of the cost-benefit of this technology for the chosen environment to promote stakeholder engagement. This is particularly true for technology reliant on advanced infrastructure (e.g., DSRC V2X technology) that only have limited deployment in controlled settings in the United States. These challenges also apply to testing motorcyclist technology and if anything, have a heightened risk not only due to vehicles on the roadways but also the operation of the motorcycle itself.

## 3.2  NeXuS Bike Simulator

Mixed-reality simulations are a solution that allow for human subject evaluations and refinement of these technologies in ecologically valid environments, leveraging native control interfaces, and allowing developers to prototype a variety of alternate configu-rations to gather data on the factors most necessary for positive health out-comes. To overcome these challenges to the performance of ecologically valid cyclist safety testing, we developed the Native Cyclist Experience Using a Stationary (NeXuS) Bike Simu-lator system. A mixed reality, man-portable, and self-contained bicycle simulator that allows users to don a virtual reality (VR) head-mounted display (HMD) while riding a stationary exercise bicycle to immerse themselves in a simulated virtual environment (VE) designed to emulate real-world riding conditions (see Fig. 3).

**Fig. 3.** Cyclist performing an evaluation using the NeXuS bike simulator

In order to maximize the benefits of simulated evaluation testing, the NeXuS simulator consists of three core focus-areas: cycling realism, a scenario control module, and a dynamic agent-based traffic system. In order to ensure the actions taken by participants transfer between the real world and VE, participants control the simulator through use of a stationary bike that translates their force from pedaling to the simulator. Furthermore, to prevent participants from developing simulator sickness we use angular velocity and acceleration reported by the HMD to allow them to control their angular velocity through leaning and changes in weight distribution.

Live riding studies can be dangerous, and difficult to replicate and control. For example, live riding studies at multiple sites may have different conditions across sites that could affect the results of the study (e.g., friction, weather, lighting conditions, etc.). The scenario control module of the NeXuS simulator allows experimenters to modify and alter these environmental conditions within the simulation to provide a consistent repeatable experimental environment (see Fig. 4). The scenario control module also allows experimenters to generate a number of roadway hazards that would present hazards in a life-riding study, such as vehicles performing a right hook, and cutting off a cyclist, or vehicles running a red light as the cyclist approaches.

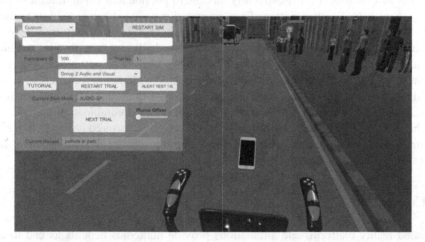

**Fig. 4.** Screenshot of scenario control module from NeXuS simulator to configure experiments

While dangerous cycling conditions can be evaluated under highly controlled environments, such as closed course tracks, they are difficult to evaluate in the dynamic traffic conditions a cyclist might encounter on their daily commute. The NeXuS bike simulator allows developers to simulate these conditions with its dynamic agent-based traffic system. For this traffic system, each vehicle in the simulator independently travels through the city going towards shifting randomized goals. Because the end state for each vehicle is fully randomized, traffic in the simulated environment is uniformly distributed ensuing participants encounter traffic wherever they are in the simulation. Each vehicle in the simulation independently incorporates agent-avoidance behavior, where they will change lanes to attempt to pass vehicles or cyclists if needed (determined by the scenario control module), and will make roadway decisions based on the state of city infrastructure systems (e.g., traffic lights).

### 3.3 Rapid Prototyping Using Simulation

The evaluation and refinement process for new technology can be prohibitively time consuming and expensive. It can require development of a prototype, development of a test track, and technological refinement. Simulator-based evaluations allow for this process to occur in a low-cost simulated environment that allows for incremental technological refinements of the technology.

Using the NeXuS bike simulator, we performed human subject evaluations of cyclist safety alerting technology powered by vehicle to vehicle (V2), and vehicle to infrastructure (V2I), connected vehicle (CV) technology. While it is expected that CV technology will be implemented in the United States in the near future, CV technology currently exists in limited-scope implementations that do not support the requirements of rigorous safety evaluations necessitated by new technology.

Using the NeXuS bike simulator we prototyped our CV-enabled cyclist alerting system as an extensible cyclist-capability module to provide new capabilities and technology to cyclists. We then designed and executed a within-subject study to evaluate and determine the optimal modality over which to alert cyclists to maximize safety and reduce distractibility. Riders rode around the simulated VR city on a pre-planned bicycle path and were presented with common cycling hazards (e.g., potholes, cars cutting cyclists off at intersections, cars overtaking from behind with minimal clearance) at controlled random intervals. Riders were then presented with various alert modalities (e.g., visual, audio, haptic handlebars or wrist wearable, or some combination thereof) and their hazard response and response time were objectively measured (compared against a control condition without the added alerting). Initial results from this evaluation were promising, indicating value in a larger follow-on study.

## 4   Conclusion

Extended reality platforms and simulations provide numerous benefits as end-to-end systems to design, develop, test, validate, and deploy practical XR systems as well as numerous emerging COTS devices—with particular relevance in the transportation domain. Transportation via vehicle, motorcycle, and bicycle places humans in largely the same environments with similar hazards and risk levels, yet the context of the equipment in use, system operation, user perceptibility, and viable communication displays varies greatly. For example, driving an automatic transmission car only requires one foot for gas and braking and at least one hand on the wheel for steering (albeit one hand on the wheel is more common for manual transmission driving), whereas riding a bicycle involves two feet pedaling and two hands on the handlebars for hand-braking, and riding a motorcycle typically involves one foot for shifting, one foot for braking, one hand for the throttle, and both hands for the hand brakes. Additionally, driving in a car is generally less affected by background noise than riding a bicycle, both of which are generally quieter contexts than riding a motorcycle due to the motorcycles' noise pollution itself. As such, these unique configurations, variables, and contexts of user transportation on public roadways provide significant opportunity both for the use of novel display modalities for communicating information as well as the use of immersive simulations to evaluate these display solutions in hard-, expensive, and risky-to-replicate environments. Motorcycle and bicycle riding in particular demands hands-on, head-up, eyes-on riding while maintaining awareness of dynamic hazards among the periphery and incoming information such as turn-by-turn navigation. Augmented reality HUDs and HMDs and dashboard displays, integrated vibrotactile haptic feedback (e.g., handlebars, on-body/jacket, on-seat, on-steering wheel), and spatial audio provide robust alternatives to standard smartphone-based directions and alerting to effectively reach two-wheeled transportation modes from a safety, viability, and usability standpoint. Immersive simulations provide opportunities to test in varied scenarios that may otherwise not be feasible due to cost or saturation of the underlying technology. Simulations are also a cost-efficient solution for refinement of algorithms without concern for protocol or hardware standards driving the system. In addition to high fidelity environment replication and en route situation awareness-enhancement, XR technologies can be used as rapid prototyping environments to allow users to use technology before it is commercially available to offer feedback regarding usability, acceptability, and utility of the system. There are numerous applications for the use of multimodal display technologies at every point of the journey, from concept testing, to development, to rider deployment. As connected vehicle infrastructure continues to develop, along with the technological sophistication of motorcycles and bicycles, we are entering a new realm where vehicles can truly communicate with each other. Critical to the success of V2X however, is the effective communication of information on the road and among the vehicles, to the rider.

**Acknowledgements.** Research discussed in this paper was funded by the US Department of Transportation (DOT) Federal Highway Administration (FHWA).

# References

Vincente, K.J., Rasmussen, J.: The ecology of human-machine systems II: Mediating direct perception in complex work domains. Ecol. Psychol. **2**, 207–249 (1990)

Bennett, K.B., Flach, J.M.: Display and Interface Design: Subtle Science, Exact Art. CRC Press, USA (2011)

Sanderson, P.M., Watson, M.O.: From information content to auditory display with ecological interface design: prospects and challenges. In: Proceedings of the Human Factors and Ergonomics Society Annual Meeting, vol. 49, no. 3, pp. 259–263. Sage CA: Los Angeles, CA: SAGE Publications (2005)

Watson, M.O., Sanderson, P.M.: Designing for attention with sound: challenges and extensions to ecological interface design. Hum. factors **49**(2), 331–346 (2007)

Wagman, J.B., Carello, C.: Haptically creating affordances: the user-tool interface. J. Exp. Psychol. Appl. **9**(3), 175 (2003)

# Virtual Environment Assessment for Tasks Based on Sense of Embodiment

Daiji Kobayashi[1] ⓘ, Yoshiki Ito[1](✉), Ryo Nikaido[1](✉), Hiroya Suzuki[2](✉),
and Tetsuya Harada[2](✉)

[1] Chitose Institute of Science and Technology, Hokkaido, Japan
{d-kobaya,b2160160,b2161360}@photon.chitose.ac.jp
[2] Tokyo University of Science, Tokyo, Japan
8119534@ed.tus.ac.jp, harada@te.noda.tus.ac.jp

**Abstract.** The quality of a virtual environment for a specified task based on the concept of sense of embodiment (SoE) was assessed in this study. The quality of virtual reality (VR) is evaluated based on the VR system or apparatus's performance; however, we focused on VR users executing tasks in virtual environments and tried to assess the virtual environment for the tasks. We focused on the user's sense of agency (SoA) and sense of self-location (SoSL), which were considered as components of the SoE. The SoA was measured based on the surface electroencephalogram of two body parts and our SoE questionnaire. We analysed the surface electroencephalogram waveforms using signal averaging and determined the observable latent time from the analysed waveforms for estimating the state of SoA. To assess the different virtual environments, we built two virtual environments composed of different versions of SPIDAR-HS as a haptic interface and a common head-mounted display. The experiment was executed in two virtual environments in addition to the reality environment. In the three environments, the participants executed the rod tracking task (RTT) in a similar way, and their EMG and subjective data were measured during the RTT. From the results, we considered the task performance based on the participants' SoA and SoSL, and the quality of the two virtual environments were compared. Furthermore, the relation between the quality of the virtual environment and the factors related to the characteristics of haptic and visual interfaces was revealed to some extent.

**Keywords:** Virtual reality · Sense of embodiment · Haptics · Electromyogram · Virtual environment assessment

## 1 Introduction

VR systems, including user interfaces indicating visual, kinaesthetic, or haptic stimuli, are generally used for human learning or refining trainee's skills; however, users are accustomed to experiencing pseudo stimuli in a virtual environment. Thus, a higher quality of virtual environment affects user's immersion in their tasks for training and takes advantage of the virtual environment.

© Springer Nature Switzerland AG 2020
C. Stephanidis et al. (Eds.): HCII 2020, LNCS 12428, pp. 368–382, 2020.
https://doi.org/10.1007/978-3-030-59990-4_28

To practically design such an effective virtual environment, elaboration of virtual reality (VR) systems could be required not only for VR tasks but also to validate and provide quality assessment of VR systems. Although the VR system engineer tries to enhance the overall VR system performance and assess the system performance to include its reliability from the view of hardware and software of the VR system, it is necessary to address the challenge of enhancing the user performance in the virtual environment. Therefore, this study aimed to develop an assessment method for measuring the quality of the environment in which a user performs a specified task. In this regard, we assumed that the quality of the virtual environment (QoVE) for a task could depend on the task specification. In view of this, we focused on not only the user's performance but also the user's sense of measuring the QoVE rather than the performance of the VR apparatus.

In recent years, the sense of agency (SoA) in virtual environments, has been mainly shaped by visual interfaces such as head-mounted displays (HMDs) for VR, which are characterized by electroencephalogram (EEG) analysis [1]. Measuring EEGs is a way to understand the relationship between performance and neurophysiological alteration; however, because of the various artefacts that can exist on EEGs when the participant executes tasks with physical activities such as moving their head and arms, they can be difficult to measure and understand. Instead of using an EEG, we tried to evaluate the virtual environment through user performance using surface electroencephalograms (EMGs) on their hands and arms [2, 3]. The virtual test environment was constructed using a SPIDAR-HS as a haptic device and a head-mounted display (HMD). As a result, we found that measuring an EMG was an effective method of observing the results of efferent signals to muscles for physical activities.

To compare the trends of EMG waveforms during tasks in different environments, we tried to apply signal-averaging and moving average methods to clarify the trends that may be affected by the task environment. As a result, we found that the reaction latency included in the trend of averaged EMG waveforms in the case of virtual environments was significantly longer than that in a reality environment. Therefore, we verified the usefulness of EMG waveform data in this study.

We tried to utilize Gonzalez-Franco's questionnaire to measure sense of embodiment (SoE), which consists of three senses: SoA, sense of self-location (SoSL), and sense of body ownership (SoBO) [4]. Although the questionnaire was modified for our experimental tasks, the score of respective senses in the virtual environment was significantly lower than the scores in the reality environment. For the purposes of our study, we named our test the Rod-Tracking Task (RTT) and it is described later in this paper.

Considering the results of surface EMG wave analysis and the SoE questionnaire, there was a theory that Synofzik's SoA model is a two-step account of agency and consists of judgement and feeling of agency [5]. Specifically, efferent signals to muscles could be generated quickly by the RTT in the real environment when afferent signals inducing SoA corresponded to and were possibly represented by feeling of agency; however, the conflicted afferent signals by the RTT in the virtual environment complicated the representation by the feeling of agency and required higher cognitive ability by the judgement of agency; it then took more time to generate the efferent signals to the motor organ. With regard to the SoSL and SoBO, we found that the scores of SoSL and SoBO in the virtual environment, which were assessed based on the answers to the different questions for evaluating SoA, were lower than in reality as well as the trend of SoA score. In this regard, Balslev pointed out that proprioceptive inputs affect the sense of agency [6]. Therefore, the SoSL as well as the SoBO affected by the user's proprioception are relevant to the SoA.

From above, we assumed that it is possible to measure the quality of the task environment, both virtual and reality, subjectively and objectively using the SoE questionnaire and the surface EMG. However, we have little experience in assessing the different virtual task environments using our method. Therefore, to clarify the validity of our assessment method for QoVE, we tried to apply the method to the two types of virtual environments for the RTT in this study.

## 2 Method

### 2.1 Rod Tracking Task for Different Environments

We applied the RTT, a well-understood and feasible test both of virtual environments and reality [2, 3], to assess the quality of the virtual environment,. Thus, the RTT was set up in virtual environments with different characteristics to assess the QoVE and we carried out experiments with participants undertaking the RTT. We then assessed the respective QoVE for the RTT based on the participants' performances and experiences in the RTT measuring EMG and SoE via questionnaires. The experimental procedures in both virtual environments were almost the same.

When executing the RTT, the user is required to grasp a rod with their right hand and attempt to move the rod between the ends of a curved slit in a panel, without contacting the sides of the slit. The panel is installed in front of the user and is rotated anticlockwise 45 ° with respect to the user, as shown in Fig. 1 [2]. In this regard, successful execution of the RTT requires a certain level of skill, and the task must interest participants to some extent.

To observe the participant's intentional and unintentional muscle tension while manipulating the rod, surface EMG on the abductor digiti minimi muscle (near the base of their fifth finger), and flexor carpi ulnaris muscle (on the lower arm) was sampled and recorded at 2 kHz using a multi-telemeter (Nihon-Koden WEB-9500). Figure 1 shows an experimental scene and two electrodes for the EMG as attached to the participants' lower arm and near the base of their fifth finger [2].

To test the RTT in the real environment, we used the experimental equipment for the RTT composed of several experimental apparatuses, as shown in Fig. 2. To record the movement of the right hand, a motion sensor (Leap Motion) connected to a personal computer (DELL XPS 8700) was used. The movement was sampled and recorded as three-dimensional coordinates at approximately 100 Hz. Additionally, the signal recording contact between the rod and the slit edges was sent to the PC via a USB I/O terminal (Contec AIO-160802AY-USB) and to a multi-telemeter as well as illuminating a red LED indicator for the benefit of the participant.

We applied different versions of the SPIDAR-HS apparatus as haptic interfaces for creating different virtual environments. In the two types of virtual environments, the visual scenes represented by the HMD (HTC VIVE PRO) were different depending

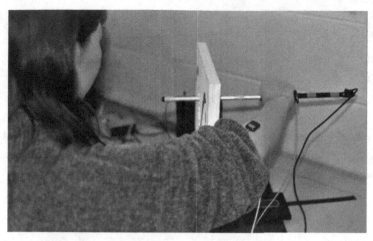

**Fig. 1.** Experimental scene of a participant executing the RTT in a real-world environment.

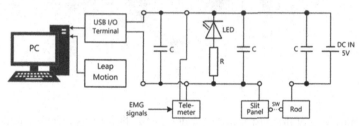

**Fig. 2.** Connection diagram of the devices recording the participant's performance in reality [1].

on the version of SPIDAR-HS, as shown in Fig. 3. Specifically, Fig. 3 shows type A (a), which was implemented in SPIDAR-HS version 2 (SPIDAR-HS2) and type B (b) implemented in SPIDAR-HS version 3 (SPIDAR-HS3), indicating augmented visual information representing the lean of the rod by a yellow line and in good condition by a green indicator. The movement of the rod was recorded using the SPIDAR-HS apparatus; however, the surface EMG on the same two body parts was recorded using a multi-telemeter, as shown in Fig. 2. The details of the physical characteristics of SPIDAR-HS are described later in this document.

The slit width, the size of the panel, and the rod diameter of 10 mm were determined based on the required difficulty of the task and were similar to the specifications applied in our previous study [2, 3]. The RTT needed to be difficult enough that an inexperienced participant was required to repeatedly practice approximately ten times before successfully executing the task in reality. The sine-curved slit and the direction of installation of the panel were also chosen from several patterns based on the required difficulty of the RTT. Furthermore, our previous study found that moving the rod in section B of the slit was the most difficult region for the participants and produced the greatest number of contacts because it was difficult to see the gap between the rod and the slit, as shown in Fig. 4 [2, 3]. Therefore, we also focused on the participant's performance in moving the rod in section B and analysed the performance data in this study.

 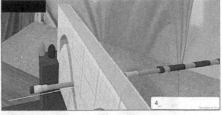

(a)  Type A implemented in SPIDAR-HS2          (b) Type B implemented in SPIDAR-HS3

**Fig. 3.** The real slit panel and the rod for the RTT in different types of virtual environments. (Color figure online)

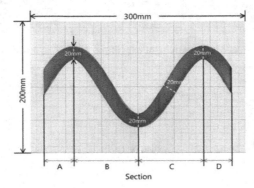

**Fig. 4.** Specifications of the slit panel. The width of the slit panel is 20 mm, and the track is divided into four sections (A–D) for convenience [2].

## 2.2  Virtual Environments for Assessment

To assess the two virtual environments as well as the reality environment, the RTT for participants was executed in different virtual environments using different versions of SPIDER-HS. The SPIDER-HS and the HMD were controlled by our multiple software running Unity on the Microsoft Windows 10 professional operating system.

SPIDER-HS consisted of controllers and eight motor modules including a motor, a threaded pulley, and an encoder for reading the yarn winding of each. The motor modules also presented a force sense as well as detecting the positions and angles of the two end effectors at the edge of the rod, as shown in Fig. 5. The workspace of the SPIDAR-HS was 2.5 m$^2$ [7].

Although SPIDAR-HS2 for the RTT was developed in a previous study [8] to determine the effect of parallel control for two end effectors, SPIDAR-HS3 was created for this study using a new control circuit for improving the precision of positioning and making the force sense more powerful. Thus, the minimum force for the moving the end effector at a constant speed of SPIDAR-HS3 was 2.54 N relative to 0.97 N in the case of SPIDAR-HS2.

Considering the precision of positioning by different versions of SPIDAR-HS, we measured the relative position error. Specifically, we moved an end effector in a straight

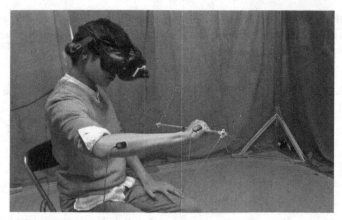

**Fig. 5.** The experimental scene represents the participant grasping a rod on eight strings to eight motor modules of SPIDER-HS [7].

line starting at spatial coordinates (0, 0, 0), where the point was the centre of the workspace and arriving at (−0.5 m, −0.5 m, +0.5 m) at a constant speed corresponding to over 10 s, and then the spatial coordinates detected by the SPIDAR-HS were measured, and the axial range error per moving distance was calculated. The measurement of the axial range error of the end effector was repeated ten times for the respective SPIDAR-HS system, and we calculated the averaged axial range error as shown in Table 1. Table 1 shows that the axial range error of the end effector per moving distance of SPIDAR-HS3 was approximately a third of the axial range error by SPIDAR-HS2.

**Table 1.** Averaged axial range error per moving distance (n = 10).

| Type of SPIDAR-HS | x-axis [%] | y-axis [%] | z-axis [%] |
|---|---|---|---|
| SPIDAR-HS2 | 5.36 | 11.8 | 12.0 |
| SPIDAR-HS3 | 1.33 | 3.38 | 3.12 |

Except for the control circuit, the two versions of the SPIDAR-HS apparatus are composed of the same components. In this regard, the task environment of SPIDAR-HS2 is the same as that of SPIDAR-HS3. Furthermore, SPIDAR-HS2 and SPIDAR-HS3 have characteristics that occasionally indicate unnatural behaviour such as vibration of the virtual end effector due to complicated factors [3]. Thus, for executing the RTT using the SPIDAR-HS2 or the SPIDAR-HS3, the participants were required to overcome the unnatural behaviours by trial practice before the experiment.

The physics of the rod and slit panel were similar to those of the real objects, especially visually, but were somewhat different in terms of haptic sense because of being in virtual environments. For instance, the rod as the end effector of both versions of SPIDAR-HS did not fall owing to gravity during the task. The other task conditions and specifications of SPIDAR-HS are shown in Table 2, and further details of the specifications of SPIDAR can be found in references [7, 8].

**Table 2.** Specifications of SPIDAR-HS2 and SPIDAR-HS3 were used for the RTT.

| Details | SPIDAR-HS2 | SPIDAR-HS3 |
|---|---|---|
| Stick length (length of the rod in VR) | 400 mm | |
| Stick diameter (width of the rod in VR) | 10 mm | |
| Path size (size of the slit panel in VR) | 200 × 300 × 20 mm | |
| Path thickness (width of the slit in VR) | 20 mm | |
| Required force for moving vertical upward | Over 0.97 N | Over 2.54 N |

In the experiment, both versions of the SPIDAR-HS apparatus recorded the spatial coordinates of two end effectors, and the visions seen by the participants through the HMD were recorded by a personal computer (PC). Thus, these data were utilized in this study as well as the other data obtained in the experiment.

### 2.3 Questionnaire for Investigating the SoE in RTT

In our previous study, to consider the performance in the virtual environment, we constructed a questionnaire for assessing the SoE for the RTT based on Gonzalez-Franco and Peck's embodiment questionnaire for investigating avatar embodiment in an immersive virtual environment [9]. The foci of the embodiment questionnaire were covered by six characteristics: body ownership, agency and motor control, tactile sensations, location of the avatar's body, the avatar's external appearance, and response to external stimuli. For the RTT trials in a virtual environment, however, the participants could not look at their hand either in the real environment or on their avatar in the virtual environment while executing the RTT; therefore, we selected nine statements from the embodiment questionnaire for the RTT as executed in both reality and the virtual environment. In addition, to construct a questionnaire for assessing the SoE for the RTT, we arranged the relevant statements based on the characteristics of the RTT. The nine questionnaire statements were composed of three categories, SoBO, SoA, and SoSL; these are shown in Table 3.

The questionnaire was administered to the participants at the end of the experiment. In this regard, Mar and Tabitha suggested that it should be clear that the questions should be related to the participants' experience during the experiment; therefore, we instructed the participants to answer the questionnaire while recalling the situations described in each item using a seven-point Likert scale. The Likert scale ranged from strongly disagree $(-3)$ to disagree $(-2)$, somewhat disagree $(1)$, neither agree nor disagree $(0)$, somewhat agree $(+1)$, agree $(+2)$, and strongly agree $(+3)$.

In this study, the SoE score for the RTT was composed of the SoBO, SoA, and SoSL scores, which were estimated as follows:

**Table 3.** Nine modified questionnaire items were used to investigate the participants SoE in RTT [3].

| Category | Question |
| --- | --- |
| SoBO | 1. It felt as if the real or the virtual rod I saw was moved by someone else |
| | 2. It seemed as if I might have more than one hand |
| | 3. I felt a sensation that I did not move my hand when I saw the rod moved, or I felt a sensation that I move my hand when I saw the rod stopped |
| SoA | 4. It felt like I could control the real or the virtual rod as if it was my own rod |
| | 5. The movements of the real or the virtual rod was caused by my behaviour |
| | 6. I felt as if the movements of real or the virtual rod was influencing my behaviour |
| | 7. I felt as if the real or the virtual rod was moving by itself |
| SoSL | 8. I felt as if the rod in my hand was located where I saw the real or the virtual rod |
| | 9. I felt as if the rod in my real hand were drifting toward the rod I see or as if the rod I see were drifting toward the rod in my real hand |

$$\text{SoBO score} = (- Q1 - Q2 - Q3)/3.$$
$$\text{SoA score} = (Q4 + Q5 - Q6 - Q7)/4$$
$$\text{SoSL score} = (Q8 - Q9)/2,$$
$$\text{SoE score} = (- Q1 - Q2 - Q3)/3 + (Q4 + Q5 - Q6 - Q7)/4 + (Q8 - Q9)/2.$$

In other words, we assume that the SoBO, SoA, and SoSL have equal impact on the SoE, ranging from $- 3$ to $+ 3$, and the SoE score reflects the participants' experience.

## 2.4 Experiment in the Real Environment

A criterion for evaluation is required to assess the QoVE for the RTT. In this regard, our previous study pointed out that the participants were affected by various factors arising from the design of the virtual environment, such as a sudden vibration of the rod on contact with the sides of the slit. On the other hand, the performance in a normal situation without the unnatural feedback to the participant refers to the performance in the real environment, and we defined that the performance in the real environment was a criterion for assessing the QoVE for the RTT. Thus, we observed the performance when the participant unconsciously contacted the rod with the slit edges or the sides of the slit in section B mainly because of a lack of skill. However, within-subjects design, such as the participants performing the RTT in both the reality and the virtual environment, could affect the performance in the real environment. In other words, the participants could obtain RTT skills in the virtual environment, and the particular skills could affect the RTT in the real environment. Therefore, in this experiment, we selected 13 participants who had no prior experience of RTT and were right-handed male students ranging from 21 to 22 years of age. The participants gave their informed consent for participation in advance.

A trial of the RTT in this experiment was to pull the rod from the far end of the slit to the end closest to the participant along the track. Subsequently, the participant had to push the rod from the close end to the far end and had to avoid contact of the rod with the sides of the slit except at both ends. The speed at which the participant moved the rod was chosen by the participant; the participant was asked to hold his head as steady as possible during the trial. This trial is shown in Fig. 1.

The participants repeated the trial five times, and we recorded the movement of the participant's hand using a video camera and a Leap Motion sensor. The contact signal, which was generated when the rod and slit edges came into contact, was recorded by both the PC and the multi-telemeter. The participants' opinions about the most difficult section in the slit panel and the reason why the rod came into contact with the slit edges were recorded in an interview after every trial, if the participants could recall.

The SoE questionnaire included some questions about phenomena which could arise in the virtual environment and the questions affected the participants' experience of the RTT only in the reality environment. At the end of the experiment, the participants reported their experiences; however, the SoE questionnaire was not applied to the participants for this experiment. Instead, the SoE questionnaire was applied to the participants who performed the RTT in both the virtual using SPIDAR-HS2 or SPIDAR-HS3 and the reality environment described later.

### 2.5 Experiment in Virtual Environment Using SPIDAR-HS2

To assess the QoVE for the RTT in two types of virtual environments, we selected nine students ranging from 21 to 24 years of age as participants. The participants were different from those participating in the reality environment experiment. The participants were right-handed and had no prior experience doing the RTT in a real environment but had executed other tasks in a virtual environment using the other SPIDAR-HS's apparatus and the HMD. In this regard, the participants were used to performing the RTT in virtual environments, and we assumed they had accrued sufficient skill to perceive the virtual environment to some extent. The participants gave their informed consent for participation in advance.

A trial of the RTT in this experiment was the same as in the reality environment experiment; that is, the participants pulled the end effector like the rod from the far end of the slit to the near end along the track and then pushed the rod from the near end to the far end. During both parts of the task, they had to avoid contact of the rod with the slit edges or the sides of the slit except at both ends. The speed at which the participant moved the rod was decided by the participant; however, the participant was required to hold their head as steady as possible during the trial in the virtual environment.

Before the experiment, the respective participants were measured by the EMG on the two body parts for ten seconds. After measuring the EMG in the resting state, the participants repeated the trial until their performance was improved for a total of up to five trials. The participants could take a rest and relax as required.

During the RTT, the changing position and the spatial coordinates of the two end effectors and the vision that the HMD indicated were recorded by the PC. The participant's surface EMG was recorded in a similar way to the experiment in a realistic environment. The contact signal, which was generated by the PC when the rod was in contact with the slit edges, was also recorded by the multi-telemeter via a USB I/O terminal. In addition, the participants' opinions, especially about the experience executing the

RTT, were recorded in interviews, and the participants answered the SoE questionnaire at the end of the experiment about the virtual environment.

After the experiment, the participants could take a rest as required and tried to repeat the RTT in the reality environment five times, in the same way as mentioned above. Therefore, we obtained the participants' EMG data while executing the RTT and the participants' opinions, especially regarding the experience executing the RTT in the virtual and the real environment. After the trials in the reality environment, participants answered the SoE questionnaire based on their experience in the RTT in reality.

### 2.6   Experiment in Virtual Environment for SPIDAR-HS3

Five months after the former experiment using SPIDAR-HS2, we executed the experiment using SPIDAR-HS3 in a similar way to the former experiment. The participants were seven students ranging from 21 to 24 years of age, and every participant had an experience in the RTT using the SPIDAR-HS2 and gave their informed consent for participation in advance. The procedure of the experiment was the same as the experiment using the SPIDAR-HS2; therefore, we obtained the participants' EMG data and answers to the SoE questionnaire in both the virtual and the real environment and their opinions about the execution of the RTT in the two environments.

## 3   Results

### 3.1   Contacts

To grasp the difficulty of the RTT in the respective environments, the number of contacts in Section B (see Fig. 4) by the participants were averaged for each of the five trials. Figure 6 shows the average number of contacts in reality and the two types of virtual environments from the first to the fifth trial, respectively, indicating that the number of contacts in the case of using SPIDAR-HS3 is less than that of using SPIDAR-HS2; however, it is not significant in all trials. Further, it is represented that more contacts are not avoided in both the virtual environment and the reality environment.

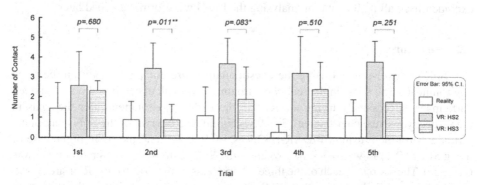

**Fig. 6.** Number of contacts in section B during a trial in reality (n = 13), the virtual environment using SPIDAR-HS2 (n = 9), and SPIDAR-HS3 (n = 7).

The total number of contacts was 326 in the virtual environment using SPIDAR-HS2 and 278 in the case of using SPIDAR-HS3, whereas 100 of 326 contacts in SPIDAR-HS2 and 83 of 278 contacts in SPIDAR-HS3 resulted from unnatural representation by the SPIDAR-HS systems. In this regard, we identified the direct factors of every questionable contact in the virtual environment based on their opinions, their vision in the HMD, and their performance recorded by the PC and the video camera.

As a result, the identified and specified direct factors were classified into three categories: unnatural visual and haptic representation by SPIDAR-HS systems, interruption of the visual scene in the HMD, and scrambling out of the rod from the slit. Specifically, unnatural visual representation (UVR) included the case in which the participant could not recognize the contact and the correct rod's lean visually (see Table 4). Unnatural haptic representation (UHR) included the case in which the participant could not recognize that the rod contacted the slit edges haptically and the rod touched the slit edges. The latter two cases were from unidentified software bugs.

**Table 4.** Unnatural contacts in the two virtual environments.

| Haptic interface | UVR | UHR | Others | Total |
|---|---|---|---|---|
| SPIDAR-HS2 | 63 (63%) | 31 (31%) | 7 (7%) | 100 (100%) |
| SPIDAR-HS3 | 29 (35%) | 43 (52%) | 11 (13%) | 83 (100%) |

We assumed that the unnatural contacts in the virtual environment could be a factor in decreasing the SoE. In particular, the participants could be involved in more complicated situations by the unnatural contacts and it was difficult to understand the reason for the contacts; therefore, the unnatural contacts could make the SoA score decrease. Furthermore, it could show as a decreased SoE score since the unnatural contacts deteriorate the QoVE for the RTT.

On the other hand, we considered the unnatural contacts to be an abnormal situation for the RTT; therefore, intended surface EMG data at the unnatural contacts were excluded from all EMG data for analysing the EMG waveform described later.

### 3.2 SoE Score

The SoE scores estimated using the questionnaire were compared between the three environments. The score in the reality environment was calculated based on the sixteen participants successively trying the RTT in the reality environment after the trials in the respective virtual environment. The 95% confidence interval of the SoE score in the virtual environment using SPIDAR-HS2 was $-0.9 \pm 3.4$, and the score in the case of using SPIDAR-HS3 was $0.3 \pm 3.4$, whereas the SoE score in the real environment was $6.8 \pm 2.0$. The scores of each of the three components of the SoE for the RTT are shown in Fig. 7. Figure 7 also shows that the SoSL score has a strong effect on the SoE score in virtual environments.

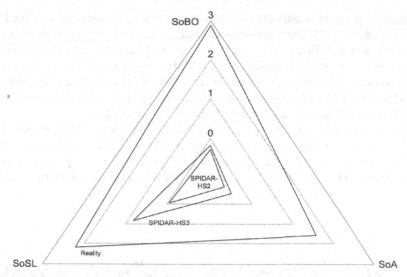

**Fig. 7.** Comparison of the average scores of the three components of SoE for the RTT trials in the three different environments.

The SoA scores in the two virtual environments indicate that the participants results were complicated by the conflicted afferent signals in those virtual environments. However, the SoSL score in the RTT using SPIDAR-HS3 was higher than in the case of using SPIDAR-HS2. In this regard, the axial range error by SPIDAR-HS3 was improved from SPIDAR-HS2, as shown in Table 1; thus, the two SoSL scores were consistent with the improved axial range error. In other words, the SoE questionnaire could reflect QoVE for the RTT by the physics of the SPIDAR-HS system in a virtual environment.

### 3.3 Surface EMG Waveforms

In our previous study, we found a difference in the trend of root-mean-square (RMS) EMG waveforms in the reality and the two virtual environments. Although the consideration of the difference of those waveforms was insufficient for measuring the QoVE for the RTT based on the surface EMG, we found that the latent time after the rod contact with the slit edges or the sides of the slit was different between the reality and the virtual environment [3]. More concretely, we assume that the latent time relates to the time required to grasp the situation when the rod contacts the slit edges and gets off the rod from the slit edges. Thus, shorter latent time refers to generating efferent signals to the motor organ by SoA more quickly; furthermore, the shorter latent time could enhance the QoVE.

Therefore, we investigated the details of the EMG waveforms for two seconds after the participants contacted the rod with the slit edges in section B; however, the EMG waveforms generated by unnatural contacts were eliminated. With regard to the performance after the contact, the participants tried to manipulate the rod using their fingers rather than arms; thus, we focused on the surface EMG near the base of their fifth finger to find the latent time until the muscle around the fifth finger activates. Specifically, we eliminated the personal equation of the EMG by calculating the proportion of the EMG waveform to the averaged amplitude of the EMG measured under the participant's resting state before each trial. In addition, multiple samples of standardized EMG waveforms were rectified to observe the amplitude of the EMG, which represents the intensity of muscle activities. In accordance with the above method, we obtained the samples of standardized rectified EMG (SR-EMG) extracted from every intended two-second duration from which the participants contacted the rod with the slit edges.

To investigate the characteristics of the SR-EMG waveforms, the samples of the waveforms were averaged out per participant in both the virtual environment using respective versions of SPIDAR-HS and the real environment, and we applied the moving-average method to the averaged SR-EMG to clarify the trend of the waveform, as shown in Fig. 8.

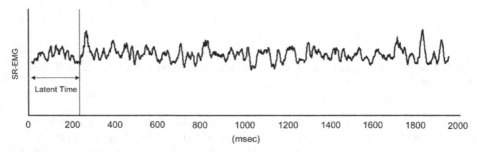

**Fig. 8.** An example of the averaged SR-EMG waveform in the virtual environment using SPIDAR-HS2 made from seven intended EMG waveforms by participant-A, and applied using the moving-average method. This represents a latent time until the muscle activity was intensified.

To compare the latent time between the task environments, the latent time of the averaged SR-EMG waveform for each participant was determined using the moving-average method when necessary. Thus, we analysed the EMG data of only seven participants who performed the RTT in both virtual environments using SPIDAR-HS and compared the latent time between every experiment. As a result, the averaged latent times for the respective seven participants (P1 to P7) in the virtual environment using the respective versions of SPIDAR-HS are shown in Table 5.

**Table 5.** Averaged latent times (ms) in the RTT for seven participants in two different environments.

| Haptic interface | P1 | P2 | P3 | P4 | P5 | P6 | P7 | Mean |
|---|---|---|---|---|---|---|---|---|
| SPIDAR-HS2 | 300 | 284 | 245 | 129 | 293 | 674 | 240 | 309.3 |
| SPIDAR-HS3 | 203 | 58 | 110.5 | 114 | 99.5 | 113 | 73.5 | 110.2 |

As shown in Table 5, the average latent times for every participant in the virtual environment using SPIDAR-HS2 was longer than the latent times when using SPIDAR-HS3 and was statistically significant (p = .023). According to our previous study, therefore, the latent time until the muscles are activated significantly refers to the necessary time for recognizing the situations when the rod contacts the slit edges unexpectedly. Therefore, in accordance with Synofzik's SoA model, the intended situations in the virtual reality using SPIDAR-HS3 were perceived by feelings of agency rather than judgement of agency shortened the latent time. In other words, SPIDAR-HS3 could enhance the QoVE in the RTT rather than the SPIDAR-HS2.

## 4 Discussion

We tried to assess the QoVE for the RTT based on the user's viewpoints and compared the performance using surface EMG and the result of a subjective SoE questionnaire concerning two virtual environments examining different versions of SPIDAR-HS; however, the effect of visual information implemented in SPIDAR-HS on the participants' performance was also a concern. From these points, our knowledge of this study is explained as follows.

From the result of the SoE questionnaire, the SoSL, which is a component of SoE in SPIDAR-HS3, was better than SPIDAR-HS2, as shown in Fig. 6. In this regard, SPIDAR-HS3 had a lower axial range error than SPIDAR-HS2 (see Table 1). Furthermore, the augmented visual information represented in the HMD could be useful for the participants in perceiving the lean of the rod. Regarding the visual representation in the RTT using the SPIDAR-HS3, about half of the participants said that they referred to the yellow line indicating the lean of the rod. In other words, they relied on the augmented visual information because in the virtual environment using both versions of SPIDAR-HS it was difficult to recognize the positional relation between the rod and the edges of the slit panel. Therefore, the SoA score of the two virtual environments was considerably lower than the SoA score in the real environment, as shown in Fig. 7.

With regard to the SoA, we focused on the averaged latent time of the SR-EMG waveform in this study. As the result, the latent times of SR-EMG waveforms in the case of using SPIDAR-HS3 was shorter than the case of using the SPIDAR-HS2. This also indicates that the SoA in the virtual environment using SPIDAR-HS3 was superior to the quality of the virtual task environment in the RTT. Further, the SoSL and the SoA by SPIDAR-HS3 decreased the number of contacts (see Fig. 6); however, we could not consider the SoBO score in this study.

Consequently, we concluded that the QoVE of the virtual environment using SPIDAR-HS3 was a quality task environment for the RTT from multiple measures such as their performance represented by SR-EMG and the result of the subjective SoE questionnaire. However, we applied SPIDAR-HS3, which is a developing haptic system for the experimental quality assessment of a virtual environment. Thus, the overall evaluation results of the virtual environment indicate the QoVE using the SPIDAR-HS3 should be improved for applying the use of the RTT.

## 5 Conclusion

In this study, we tried to assess the quality of the virtual environment for a specified task, such as RTT, based on the concept of SoE. In particular, we focused on the SoA and the SoSL, which were the components of SoE in the RTT, measured the participants' surface EMG, and applied our SoE questionnaire. With regard to the surface EMG, we proposed a method for analysing the EMG waveforms using signal averaging and determined the observable latent time from the analysed waveforms for estimating the SoA. As a result, we considered task performance based on the participant's SoA and SoSL, and the quality of the two virtual environments were compared. Furthermore, the relation between the quality of the virtual environment and the factors related to the characteristics of haptic and visual interfaces was revealed to some extent.

## References

1. Jeunet, C., Albert, L., Argelaguet, G., L'ecuyer, A.: Do you feel in control?: towards novel approaches to characterise, manipulate and measure the sense of agency in virtual environments. IEEE Trans. Visual. Comput. Graph. **24**(4), 1486–1495 (2018)
2. Kobayashi, D., et al.: Effect of artificial haptic characteristics on virtual reality performance. In: Yamamoto, S., Mori, H. (eds.) HCII 2019. LNCS, vol. 11570, pp. 24–35. Springer, Cham (2019). https://doi.org/10.1007/978-3-030-22649-7_3
3. Kobayashi, D., Shinya, Y.: Study of virtual reality performance based on sense of agency. In: Yamamoto, S., Mori, H. (eds.) HIMI 2018. LNCS, vol. 10904, pp. 381–394. Springer, Cham (2018). https://doi.org/10.1007/978-3-319-92043-6_32
4. Kilteni, K., Groten, R., Slater, M.: The sense of embodiment in virtual reality. Presence **24**(4), 373–387 (2012). Fall 2012
5. Synofzik, M., Vosgerau, G., Newen, A.: Beyond the comparator model: a multifactorial two step account of agency. Conscious. Cogn. **17**(1), 219–239 (2008)
6. Balslev, D., Cole, J., Miall, R.C.: Proprioception contributes to the sense of agency during visual observation of hand movements: Evidence from temporal judgments of action. J. Cognit. Neurosci. **19**(9), 1535–1541 (2007)
7. Tsukikawa, R., et al.: Construction of experimental system SPIDAR-HS for designing VR guidelines based on physiological behavior measurement. In: Chen, J.Y.C., Fragomeni, G. (eds.) VAMR 2018. LNCS, vol. 10909, pp. 245–256. Springer, Cham (2018). https://doi.org/10.1007/978-3-319-91581-4_18
8. Suzuki, H., et al.: Implementation of two-point control system in SPIDAR-HS for the rod tracking task in virtual reality environment. In: Yamamoto, S., Mori, H. (eds.) HCII 2019. LNCS, vol. 11570, pp. 47–57. Springer, Cham (2019). https://doi.org/10.1007/978-3-030-226 49-7_5
9. Gonzalez-Franco, M., Peck, T.C.: Avatar embodiment. towards a standardized questionnaire. Front. Robot. AI, 22 June 2018. https://doi.org/10.3389/frobt.2018.00074. Accessed 10 Mar 2020

# Camera-Based Selection with Cardboard Head-Mounted Displays

Siqi Luo[1], Robert J. Teather[2(✉)], and Victoria McArthur[3]

[1] School of Computer Science, Carleton University, Ottawa, ON, Canada
[2] School of Information Technology, Carleton University, Ottawa, ON, Canada
rob.teather@carleton.ca
[3] School of Journalism and Communication, Carleton University, Ottawa, ON, Canada

**Abstract.** We present two experiments comparing selection techniques for low-cost mobile VR devices, such as Google Cardboard. Our objective was to assess the feasibility of computer vision tracking on mobile devices as an alternative to common head-ray selection methods. In the first experiment, we compared three selection techniques: air touch, head ray, and finger ray. Overall, hand-based selection (air touch) performed much worse than ray-based selection. In the second experiment, we compared different combinations of selection techniques and selection indication methods. The built-in Cardboard button worked well with the head ray technique. Using a hand gesture (air tap) with ray-based techniques resulted in slower selection times, but comparable accuracy. Our results suggest that camera-based mobile tracking is best used with ray-based techniques, but selection indication mechanisms remain problematic.

**Keywords:** Mobile VR · Selection · Google Cardboard

## 1 Introduction

Combining cheap and lightweight cardboard-style HMDs with mobile devices makes VR more accessible to people than ever before. Devices such as Google Daydream (which includes a plastic HMD shell for a mobile phone and a touchpad controller) or Google Cardboard allow users to employ their mobile phone as a VR head-mounted display (HMD). Considerably more affordable than a dedicated VR device, Daydream is priced at around $140, and Cardboard is around $20. Both devices make it possible for more people to experience VR using ordinary mobile phones. However, one drawback is that Cardboard and similar devices do not offer complex interaction techniques. Google Cardboard is, simply put, a cardboard box to contain the mobile, with two focal length lenses. See Fig. 1. The mobile's built-in inertial measurement unit (IMU) tracks head orientation. The button located on the side of Google Cardboard can be pressed to provide input. Their low cost and simplicity may attract a larger user base than conventional HMDs.

However a major limitation of these devices is that mobile sensors offer low-fidelity tracking, and cannot provide absolute 6DOF position and orientation tracking. As a

© Springer Nature Switzerland AG 2020
C. Stephanidis et al. (Eds.): HCII 2020, LNCS 12428, pp. 383–402, 2020.
https://doi.org/10.1007/978-3-030-59990-4_29

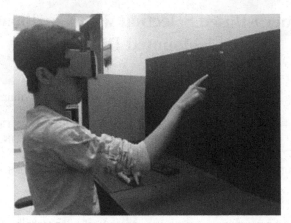

**Fig. 1.** Participant performing a selection task wearing a Google Cardboard HMD. The button can be seen at the top-right of the device. No other external controller is provided.

result, interaction on mobile VR is more limited than with trackers offered by high-end HMDs. Many past VR interaction techniques require absolute position tracking; without it, only a few of these techniques are compatible with mobile VR [17]. Browsing the Google Play store, one can see that the variety of applications is limited. Most are "look-and-see" type applications, which involve a fairly passive user experience of watching videos in 3D, sometimes using IMU-based head tracking to allow the user to look around the scene [17]. Additionally, there is little research on whether using the type of button provided on cardboard devices is the best design alternative for selecting targets in mobile VR applications. Yet, most mobile VR applications rely on the Cardboard button, to confirm selections despite measurably worse user feedback than alternative approaches [23].

Our research evaluates selection methods using "Cardboard-like" HMDs. Specifically, we investigate the performance of the built-in cameras available on virtually all modern cellphones. With appropriate software, such cameras can track the hands to provide absolute pose information, and also support gesture-recognition [5, 8]. Hence, they may provide a good alternative to head-based selection using a button. Although the tracking quality is low, such research can guide future mobile device development, for example, if higher resolution or depth cameras are required to provide better mobile VR interaction.

## 2  Related Work

### 2.1  Mid-air Interaction

Interaction in 3D is more complex than 2D depending on the task [12]. Interaction in 2D only requires up to 3 degrees of freedom (DOF) including translation in the $x$ and $y$-axes and rotation around the $y$-axis. In contrast, full 3D interaction requires three additional DOFs: $z$-axis translation and two more rotational DOFs around the $z$- and $x$-axes. Consequently, 2D interaction initially appears to be a poor match for 3D scenarios

[1]. However, while additional DOFs can give more freedom and support a wider range of interaction techniques, they can also be a source of frustration [14]. Previous work suggests minimizing the required DOF for manipulating virtual objects. Generally, the more simultaneous DOFs required, the greater the difficulty to control the interaction technique [2, 7, 20]. Additionally, the absence of tactile feedback and latency and noise common to motion tracking systems also impacts user performance [5, 15]. Current alternatives involve using DOF-limiting techniques in 3D environments [18, 20], for example, modelled after mouse control.

## 2.2 Selection in 3D

Selection is a fundamental task in VR [1, 12], typically preceding object manipulation. Improving selection time can improve overall system performance [1]. Many factors impact selection accuracy, including the target's size and distance [6], display and input device properties [19], object density [22], etc. For example, input device degrees of freedom (DOF) influence selection, with lower DOFs generally yielding better performance [4]. Display size and resolution also affect performance [15].

Two major classes of VR selection techniques include ray-casting and virtual hands [12]. Past research has shown that virtual hands tend to perform better in high accuracy (and nearby) tasks [13]. This is likely due to a combination of proprioception [14] and good visual feedback. However, in mobile VR scenarios, good 3D pose data is unavailable; such devices only provide head orientation, but not position. As a result, users lose the advantage of head motion parallax depth cues which have long been known to be beneficial in 3D selection [2] and other 3D tasks [11]. Since they rely on good depth perception, virtual hand techniques may offer demonstrably worse performance than ray-casting when implemented on mobile VR platforms.

## 2.3 ISO 9241 and Fitts' Law

Our experiments employ the ISO 9241-9 standard methodology for pointing device evaluation [9]. The standard is based on Fitts' law [6] and recommends the use of throughput as a primary performance metric.

Fitts' law models the relationship between movement time ($MT$) and selection difficulty, given as index of difficulty ($ID$). $ID$, in turn, is based on target size ($W$) and distance to the target ($D$):

$$ID = \log_2 \frac{D}{w} + 1 \tag{1}$$

Throughput is calculated as:

$$TP = \frac{\log_2\left(\frac{De}{We} + 1\right)}{MT} \tag{2}$$

where

$$W_e = 4.133 \times SD_x \tag{3}$$

$W_e$ is *effective* width and $D_e$ is the *effective* distance of movements. $D_e$, is calculated as the average movement distance from the previous selection coordinate to the current one. $W_e$ is calculated as the standard deviation $(SD_x)$ of the distance between the selection coordinate and the target, multiplied by 4.133 [9]. This adjusts target size post-experimentally to correct the experiment error rate to 4% (i.e., 4.133, or $\pm2.066$ standard deviations from the mean, corresponding to 96% of selections hitting the target). We adopted a previously validated methodology for extending the standard into 3D scenarios [19, 20]. This approach projects cursor/ray selection coordinates into the target plane and uses the projected coordinates in calculating both $D_e$ and $W_e$, to address the angular nature of ray-based selections. An alternative approach proposed by Kopper et al. [10] instead employs the angular size of motions and targets to address this problem. We employ the projection method, since we also use a direct touch (i.e., virtual hand) technique, where angular measures do not apply.

## 3 Common Methodology

We first present the elements common to our two experiments, with experiment-specific details appearing in subsequent sections. The first experiment compared three different selection techniques using a mobile VR HMD, while the second focused on evaluating different selection indication mechanisms.

### 3.1 Participants

The same 12 participants took part in both experiments. There were 2 female and 10 male participants, aged between 18 and 30 years old (mean $\approx$ 22.67 years old). Two were left-handed. All had normal or corrected-to-normal stereo vision.

### 3.2 Apparatus

We used a Samsung Galaxy S8 smartphone as the display device. The device has a 5.8 in. screen at 1440 x 2960 pixel resolution and 12-million-pixel main camera. We used a Google Cardboard v2 as the HMD (Fig. 1). The Cardboard has a button on the right side; pressing the button taps the touchscreen inside the HMD.

The virtual environment (Fig. 2, right) was developed using Unity3D 5.5 and C# and presented a simple selection task based on ISO 9241-9 [9]. The overall target sequence is seen in Fig. 2 (left). A single target sphere appeared per selection trial at a specified position. Target distance was always fixed at 0.8 m.

Targets were initially displayed in blue and became pink when intersected by the selection ray/cursor. The first target appeared at the top of the ring cycle. Upon clicking, the target disappeared, and the next target appeared, whether the target was hit or not. An example of two subsequent targets being selected, and the pink highlight are shown in Fig. 3

**Fig. 2.** (Left) ISO 9241-9 selection task target ordering pattern. (Right) The participant's view of a target in the virtual environment.

**Fig. 3.** (Left) Bottom target prior to selection, partially cut off by the edge of the HMD field of view. (Right) Subsequent (top-left) target depicting cursor intersection highlight. (Color figure online)

**Selection Techniques.** Our study included three selection techniques: head ray, finger ray, and air touch. All three techniques were evaluated in Experiment 1, while only head ray and finger ray were included in Experiment 2.

Head ray is a typical interaction technique used in mobile VR. Selection is performed using a ray originating at the user's head [24] and the ray direction is controlled by the orientation of their head via the mobile IMU (see Fig. 4). A black dot in the center of the viewport provided a cursor to use for selection. When the participant turned their head, the cursor always remained in the center of the viewport.

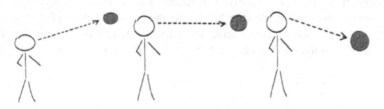

**Fig. 4.** Target selection with head ray.

The finger ray technique is similar to image-plane interaction [16], and also uses a ray originating at the head. The direction of the ray is controlled by tracking the user's index fingertip with the mobile camera (Fig. 5). A black dot at the end of the ray acts as a cursor.

**Fig. 5.** Target selection with finger ray.

Finally, air touch is a representative virtual hand selection technique and mimics real-world selection. With air touch, the user must physically tap the targets in space, and thus requires depth precision (Fig. 6).

**Fig. 6.** Target selection with air touch.

We used the Manomotion SDK[1] to acquire the hand position for both the finger ray and air touch conditions. Manomotion uses the built-in RGB camera on the back of a smartphone to track the user's hand, providing coordinates for their fingertips and the center of their palm. Since the built-in mobile device RGB camera has no depth sensor, hand depth was determined by the SDK's proprietary algorithms. The further the hand moves from the camera the larger the reported z value. To ensure the farthest targets were still reachable with the air touch condition (which required directly touching targets and hence precision in depth) we iteratively adjusted a scale factor between the VE and the Manomotion-provided depth coordinates. In the end, a distance of 2 m in the VEs mapped to approximately 70 cm of actual hand motion in reality. This ensured that the farthest targets (2 m into the screen) were still reachable from a seated position with air touch.

### 3.3  Procedure

After describing the experiment and obtaining informed consent, participants completed a demographic questionnaire. Next, they sat down and put on the HMD. We then gave them instructions about how to control each selection technique and gave them about a

---

[1] www.manomotion.com.

minute to practice using the system. These practice trials were not recorded. The first target sphere would appear at the center of the viewport. After selecting the first target, the formal test began. Participants confirmed each selection by using the current selection indication method. Targets appeared in the VE following the ring pattern common to ISO 9241-9 evaluations, as described above [9]. Upon completing one condition, which consisted of 72 trials (12 × 2 × 3), participants were given approximately 1–2 min to rest before beginning the next condition. After all conditions were completed, they filled out a preference questionnaire.

### 3.4 Design

The dependent variables in both experiments included movement time (s), error rate (%) and throughput (bit/s). Movement time was calculated from the beginning of a selection trial when the target appears, to the time when the participant confirms the selection by pressing the button on the secondary touchpad. Error rate was calculated as the percentage of trials where the participant missed the target in a given block. Throughput was calculated according to Eq. 2 presented in Sect. 2.3. Finally, we also collected subjective data using questionnaires and interviews after each participant completed the experiment.

## 4 Experiment 1: Selection Techniques

This experiment compared performance of the selection techniques described above: air touch, finger ray, and head ray. Experiment-specific details that differ from the general methodology sections are now described.

### 4.1 Experiment 1 Apparatus

For experiment 1, we also used a secondary touchpad device: a Xiaomi cellphone to provide an external selection indication method. This device was connected by Bluetooth to the Galaxy S8. To indicate selection, participants tapped the thumb-sized "A" button displayed on the device's touchscreen (see Fig. 7).

**Fig. 7.** Xiaomi cellphone used as secondary selection device in experiment 1.

While we note that this is not a realistic selection indication mechanism, we decided to include it to ensure consistency between the selection techniques. This avoids conflating selection technique and selection indication method, for example, using gestures to indicate selection with air touch and finger ray, and the Cardboard button with head ray. We considered using the Cardboard button instead, but this would prevent using the right hand with both finger ray and air touch. Since most people are right-handed, using the left hand to perform hand postures while using the right hand to press the button would certainly provide unrealistic performance results.

### 4.2  Experiment 1 Design

The experiment employed a $3 \times 2 \times 3$ within-subjects design with the following independent variables and levels:

*Technique:*    Head ray (HR), finger ray (FR), air touch (AT);
*Object depth:*  Close (1.3 m), medium (1.7 m) and far (2 m);
*Object size:*   Large (0.7 m) and small (0.4 m);

For each selection technique, participants completed 6 blocks (3 object depths $\times$ 2 object sizes) of 12 selection trials, for a total of 72 selections with each selection technique per participant. Across all 12 participants this yielded 2592 trials in total. Each selection trial required selecting one target sphere. Within a block, both target depth and target size were constant. Target size and selection technique order was counterbalanced according to a Latin square, while depth increased with block.

### 4.3  Result and Discussion

**Movement Time.** Movement time $(MT)$ is the average time to select targets. Mean movement time for each interaction method is shown in Fig. 8.

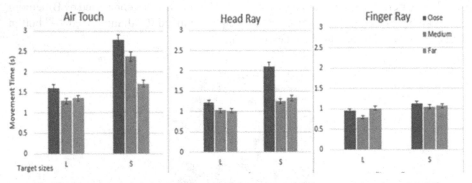

**Fig. 8.** Movement time by target size, depth, and technique. Error bars show $\pm 1$ *SD*.

Movement time was analyzed with repeated-measures ANOVA. The result ($F_{2,18}$ $= 33.98$, $p < .001$) revealed that technique had a significant main effect on MT. Post hoc testing with the Bonferroni test (at the $p < .05$ level) revealed that the difference between all techniques was significant. Overall, air touch was worst in terms of speed. Finger ray was faster than both air touch and head ray. Air touch offered significantly higher movement times than the other selection techniques at 1.86 s, about 50% slower than head ray and almost twice as long as finger ray. Head ray was, on average, slower than finger ray.

There were also significant interaction effects between techniques and target size ($F_{2,18} = 3.91, p < .05$), as well as technique and target depth ($F_{4,36} = 2.98, p < .05$). Predictably, smaller size targets generally took longer for all techniques. However, the interaction effect indicates that this was most pronounced with air touch, where small targets were about 60% worse than large targets. Target size only affected the ray-based techniques when the distance was close or medium ($p < 0.05$).

We had expected that camera noise would affect movement time for both camera-based techniques. It is encouraging that despite camera noise, finger ray still offered faster selection than head ray. As noted earlier, the combination of large target size and closer target distances (close and medium) resulted in a significant difference in movement time between head ray and finger ray. This is likely because head ray required more head motion than finger ray, which allowed subtle finger or arm movements. This is consistent with previous research that also found that finger-based selection was faster than the head when reaching a target [18].

**Error rate.** A selection error was defined as missing the target, i.e., performing the selection while the cursor is outside the target. Error rate is thus calculated as the percentage of targets missed for each experiment block (i.e., 12 selections). Average error rates for each condition are seen in Fig. 9.

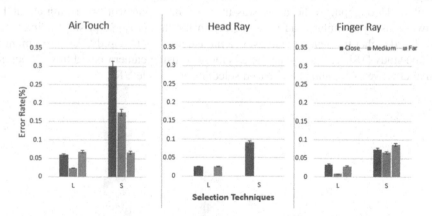

**Fig. 9.** Error rate by target size, depth, and technique. Error bars show ±1 *SD*.

A repeated measures ANOVA revealed that there was a significant main effect for technique ($F_{2,18} = 385.52, p < .001$). Post hoc testing with the Bonferroni test (at the $p < .05$ level) revealed significant differences between each technique. Air touch had a significantly higher error rate than other two selection techniques, five times that of head ray, and around double that of finger ray. Moreover, there were significant two-way interactions between selection technique and both target size ($F_{2,18} = 4.28$, $p < .001$) and target depth ($F_{4,36} = 5.58$, $p < .001$).

Specifically, post hoc testing with the Bonferroni test (at the $p < .05$ level) revealed that target depth did not yield a significant difference with finger ray or head ray. However, there were significant differences between target depth with air touch. Overall, medium distance (1.7 m) had the lowest error rate across all selection techniques. Close targets had a worse error rate, likely due to ergonomic reasons; close targets falling partially outside the field of view required more movement to select them and often required an unnatural arm pose. Air touch and head ray performed significantly worse at close distance than other two depths. Notably, as seen in Fig. 9 the single worst condition was the combination of close small targets with air touch.

Finally, post hoc testing revealed a significant difference ($p < .001$) between target sizes with air touch, but not the other two selection techniques. The error rate increased dramatically with the smaller target size when using the air-touch interaction method. Generally, smaller target sizes had higher error rate for each selection technique. Air touch was much less precise and had a higher error rate than the other two selection techniques, especially with different target depths. This indicates that participants had difficulty identifying the target depths when using direct touch. As expected, this was not a problem with the ray techniques.

**Throughput.** The average throughput for each technique is seen in Fig. 10. ANOVA revealed that selection techniques had a significant main effect on throughput ($F_{2,18} = 90.63$, $p < .001$). Post hoc testing with the Bonferroni test (at the $p < .05$ level) showed that all three techniques were significantly different from one another. Finger ray had the highest throughput at 2.6 bit/s, followed by head ray (1.8 bit/s) and then air touch (1.0 bit/s). Throughput for head ray was in line with recent work reporting about 1.9 bit/s when using a similar head-based selection method [18]. Surprisingly, finger ray also offered higher throughput than using a mouse in a HMD-based VR environment in the same study [18]. This suggests there is merit to using camera-based finger tracking as an alternative to common head-based selection in mobile VR.

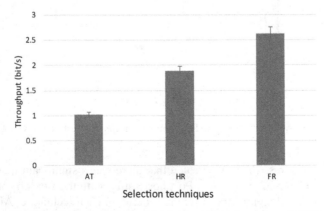

**Fig. 10.** Throughput for each technique. Error bars show ±1 SD.

**Subjective Data and Interview.** Participants completed a questionnaire ranking the techniques. Overall, air touch ranked worst of the three (see Fig. 11).

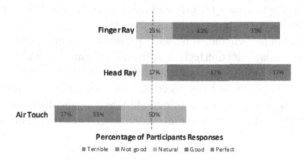

**Fig. 11.** Participants preference for each technique.

We interviewed each participant after Experiment 1 to solicit their feedback about each selection technique. Most participants mentioned physical fatigue with air touch. They found it difficult to hit targets because they needed to adjust their hand position forward and backward constantly to find the target position in depth. This result is similar to previous research on visual feedback in VR [21], which reported the same "homing" behavior, and found that highlighting targets on touch increased movement time but decreased the error rate. As reported in previous work [20, 21], stereo viewing appeared to be insufficient for participants to reliably detect the target depth with air touch, necessitating the use of additional visual feedback. Although we added colour change upon touching a target and used a room environment to help facilitate depth perception, it seemed participants still had difficulty determining target depth.

Fatigue was high at the largest target depth; participants had to stretch their arms further to reach the targets, which made their upper arms and shoulder even more tired. Head ray also yielded some neck fatigue, especially for close targets, as these increased the amount of required head motion compared to farther targets; targets could be potentially partially outside the field of view.

Despite poor performance, two participants reported that they preferred air touch because the found "it is interesting to really use my finger to touch the targets rather than just turning around my head and touching the button." They found the latter "very boring." All participants found that head ray was most efficient. "It is very easy to control and fast." However, 10 participants said they would choose finger ray as their favorite because "it is convenient to move my fingers slightly to hit the target. I did not even need to move my head and arms."

## 5   Experiment 2: Selection Indication

This experiment compared different methods of indicating selection, or phrased differently, methods of "clicking" a target. The first experiment showed that finger ray and head ray can offer higher selection performance than air touch in mobile VR scenarios.

However, for reasons described earlier, we used an external touchpad as the selection indication mechanism. To address this limitation, Experiment 2 compared two alternative methods of selection indication, including hand gestures and buttons. We also included the Experiment 1 data for the head and finger ray conditions to compare the touchpad selection indication mechanism as a baseline.

In consideration of participants' time, and to maintain a manageable experiment size, we decided to remove one selection technique from Experiment 1. We ultimately decided to exclude air touch from this experiment for two reasons: 1) our informal observations prior to conducting Experiment 1 suggested that air touch would be less effective than the other techniques (as confirmed by our results), and 2) air touch was found to work less reliably with the gesture selection indication method detailed below than finger ray and head ray.

### 5.1 Experiment 2 Apparatus

This experiment used only the head ray and finger ray techniques from Experiment 1. The separate Xiaomi smartphone, used to indicate selection in Experiment 1, was not used in Experiment 2. Instead, Experiment 2 used two new selection indication methods: the Google Cardboard button (CB) and the *tap* hand gesture (HG).

The tap gesture is similar to that performed when selecting an icon on a touchscreen device, see Fig. 12 (left) [17]. It requires bending the finger at the knuckle to indicate a selection. We originally considered a pinch gesture instead, which involved bending index finger and thumb like a "C" shape, then closing the fingertips. However, the tracking SDK was unable to reliably detect the pinch gesture, yielding longer selections times than tap.

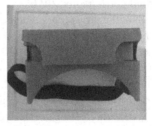

**Fig. 12.** Left: Tap gesture. Right: Modified Google Cardboard with both left and right-sided button

Participants used their dominant hand to perform the tap gesture. When using the finger ray selection method, they had to keep the pointing finger stable until the tap gesture was performed. In the Cardboard button condition, participants pressed the capacitive button built into the cardboard frame. Since we had expected most participants would be right-handed, we added a second button on the left side of Google Cardboard (see Fig. 12 right). Adding the left-side Cardboard button ensured that participants could always select with their dominant hand, and indicate selection using their non-dominant hand. This ensured consistency with the hand gesture condition, which also always used the dominant hand.

## 5.2 Experiment 2 Design

To investigate interactions between selection technique and selection indication, this experimented included both finger ray and head ray, and the two selection indication methods described above. The experiment employed a $2 \times 2 \times 3 \times 2$ within-subject design with the following independent variables:

*Technique:*       Head Ray (HR), Finger Ray (FR);
*Indication:*      Cardboard button (CB), hand gesture (HG);
*Object depth:*    close (1.3 m), medium (1.7 m) and far (2 m);
*Object size:*     large (0.7 m) and small (0.4 m);

Participants completed Experiment 2 immediately following Experiment 1. As noted previously, our analysis also includes the data for the touchpad selection indication method from Experiment 1 as a comparison point for the two new selection indication mechanisms. Each block consisted of 12 selection trials for each combination of target size and target depth. Selection techniques and selection indication mechanisms were counterbalanced according to a Latin square. Overall, there were 12 participants $\times$ 2 pointing methods $\times$ 2 selection indication mechanisms $\times$ 2 target sizes $\times$ 3 target depths $\times$ 12 selections = 3456 trials in total.

## 5.3 Results and Discussion

Results were analyzed by using a repeated-measures ANOVA. Since Experiment 1 exclusively used the touchpad as a selection indication mechanism, we included this data as a basis of comparison with the two new selection indication mechanisms. Specifically, our analysis includes data from Experiment 1 for the finger ray and head ray, for all dependent variables. These are depicted as "FR + TP" and "HR + TP" in the results graphs below (i.e., TP indicates "touchpad" selection indication). On all results graphs, two-way arrows ($\leftarrow \rightarrow$) indicate a pairwise significant difference with post hoc test at 5% significance level. The best performing conditions is highlighted in red.

**Completion Time.** Mean completion time for each condition is seen in Fig. 13 Completion time was analyzed using a repeated-measures ANOVA, which revealed a significant interaction effect ($F_{5,36} = 22.44$, $p < .001$) between selection technique and selection indication method. Post hoc testing with the Bonferroni test (at the $p < 0.05$ level) revealed a significant difference between head ray and finger ray when using the Cardboard button.

There was no significant difference between head ray and finger ray when using hand gestures, nor when using the touchpad. There were significant differences between touchpad and both the Cardboard button and hand gesture when using finger ray as the selection technique. Finger ray with either the Carboard button or hand gesture yielded higher times compared to the Experiment 1 touchpad. Finger ray also took longer with Carboard button than head ray with Carboard button. This surprised us, given how fast finger ray was in Experiment 1. This highlights the importance of investigating selection indication mechanisms in conjunction with pointing techniques. Results separated by target depth and size are seen in Fig. 14.

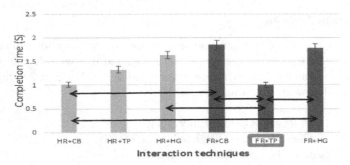

**Fig. 13.** Average completion time for combinations of techniques. Error bars show ±1 SD (Color figure online)

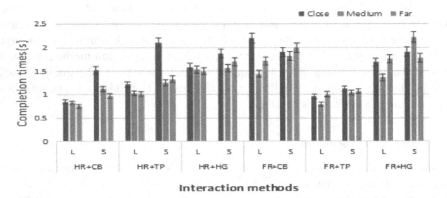

**Fig. 14.** Completion time in depths, sizes, and combination techniques. Error bars show ±1 SD

Generally, smaller target size yielded slower selections. Like Experiment 1, the medium target depth yielded faster completion times compared to the far and close target distances. The finger ray + touchpad still offered the fastest selection times overall, for every target size and depth combination. This suggests that finger ray itself is a promising technique, if a suitable selection indication method is used with it. However, as discussed earlier, the touchpad is an impractical solution.

**Error rate.** Average error rates for each technique combination is shown in Fig. 15. Repeated-measures ANOVA revealed a significant interaction effect between selection technique and selection indication method ($F_{5,36} = 16.57, p < .001$). Post hoc testing with the Bonferroni test (at the p < . 05 level) revealed significant differences between finger ray + Carboard button and using head ray with all three selection indication mechanisms. The highest error rate was with the finger ray + Cardboard button condition. Notably, hand gestures worked better with both finger ray and head ray, than either selection method worked with Cardboard button, which had highest error rate for both selection methods.

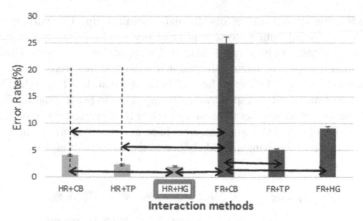

**Fig. 15.** Average error rate for combinations of techniques. Error bars show ±1 SD

According to a Bonferroni post hoc test (at the $p = .05$ level), target depth did not have significant effects on error rate with the exception of the finger ray + hand gesture combination. Finger ray + Cardboard combination had a higher error rate in each combination of target depths and sizes than other interaction methods.

**Throughput.** Average throughput for each technique combination is shown in Fig. 16. Repeated measures ANOVA ($F_{5,36} = 70.08, p < .001$) indicated a significant interaction effect between selection technique and selection indication method. Post hoc testing with the Bonferroni test (at the $p < .05$ level) revealed a significant difference in throughput between each selection indication method when using head ray. With head ray, the hand gesture performed worst, and Cardboard button performed best. With finger ray, throughput was much lower with the hand gesture and Cardboard button than with the touchpad.

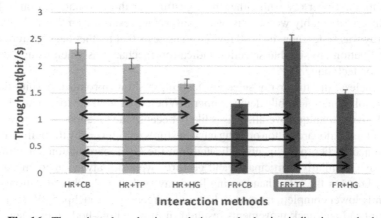

**Fig. 16.** Throughput by selection technique and selection indication method.

**Subjective Results.** Based on our questionnaire results (Fig. 17), participants rated finger ray + Cardboard button worst, and head ray + Cardboard button best. Both finger ray + touchpad and head ray + touchpad were ranked positively. During post-experiment interviews, participants indicated that hand gestures were more convenient than pressing the button or touchpad. Several indicated that they found the Cardboard button was sometimes a bit unresponsive, requiring them to press it harder. Some also mentioned that the HMD was not tight enough when they pressed down on the button, requiring them to hold the HMD with their other hand at times. No participants mentioned any physical fatigue in Experiment 2, likely due to the absence of air touch.

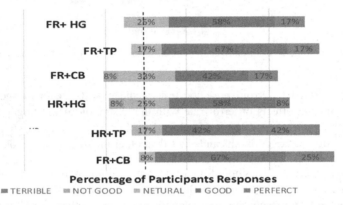

**Fig. 17.** Questionnaire score from each participant for each combination of techniques

## 5.4  Discussion

When using the finger ray with either hand gestures or the Carboard button, selection performance was notably worse. This was a surprising result, given that in Experiment 1, finger ray was faster than the head ray. This suggests that neither hand gesture nor the Cardboard button are suitable selection indication mechanisms when using the finger ray selection technique.

Nevertheless, the finger ray selection technique shows promise, despite the poor tracking resolution of a mobile device camera. When used with the touchpad, the finger ray was the best technique overall. This is likely because the touchpad was more reliable than either the Cardboard button or using hand gestures recognized by the built-in camera. As mentioned earlier, participants indicated that the Cardboard button sometimes felt unresponsive. Similarly, participant hand gestures were not always recognized by the tracking SDK. On the other hand, using head ray with the Cardboard button yielded significantly lower completion times, in line with finger ray + touchpad. We suspect this is because head and neck movements resulted in more whole-body movement, unlike finger ray which only required finger movement. For example, when using the head ray, participants had to turn their bodies slightly to face the target. During such movement, it is faster to press the Cardboard button (since it is positioned on the HMD) rather than tapping the touchpad (which is fixed on the table).

Finger ray + hand gesture took longer than head ray + hand gesture. This result surprised us, as we had expected the hand gesture would be a "natural fit" with the finger ray technique. After all, the hand gesture was performed with the same hand being used to point at targets with finger ray. Our expectation was that participants could perform the hand gesture as soon as the ray intersected the target, which may thus be faster than pressing the Cardboard button. This result can likely be explained by the comparative lack of camera sensitivity. The participants' tap gestures were frequently not recognized on the first try; multiple hand gestures thus increased the time required to select targets. It is possible that a better camera or a different gesture may improve this result.

We were also surprised by the significantly higher error rate for finger ray with the Cardboard button. This may be because of the so-called "Heisenberg" effect [3] in 3D selection, where the selection indication mechanism sometimes moves the pointing device at the instant of selection, which results in missing the target. In this case, when pressing the Cardboard button, the HMD often moved slightly, which moved the selection ray. However, when using head ray, participants frequently used their other hand to hold the Google Cardboard, so the error rate was notably better. In contrast, when using the finger ray, participants used one hand to direct the ray, and the other to press the button. As a result, the error rate increased in that condition. Due to the overall better movement time and accuracy with the Cardboard button, throughput was also higher with the finger ray condition.

In terms of subjective preference, the head ray + Cardboard button was rated best. This indicates that smooth operation during pointing is an important factor for users. Further, there was no physical fatigue was reported during Experiment 2; it seems the air touch technique used in Experiment 1 was the primary cause of the physical fatigue reported by participants. This suggests that finger ray yields much lower fatigue than air touch. In general, head ray + Cardboard button still had the advantage on both throughput, error rate and completion time. However, both head ray and finger ray with hand gestures were not far off, and could be a potential alternative in the future, especially with advances in camera-based tracking.

## 5.5 Limitations

Our experiments were conducted in "idealized" lab settings with specific lighting levels and background colour chosen to increase contrast. This provided optimal conditions for the camera tracking SDK, despite which, we still observed constant cursor jitter during the experiment. In real-world usage scenarios, there would clearly be worse tracking interference. Hence, our results should be viewed as a "best-case" with current technologies.

Due to the jitter observed in the camera-based selection techniques, target size was also limited. It would be impossible to hit very small targets; previous work has shown that once jitter approaches half the target size, selection accuracy falls dramatically [24]. Through pilot testing, we modified the experiment conditions to account for this problem, but clearly these results would not generalize to smaller targets that would be possible in standard VR systems.

Finally, we note that the depth of the virtual hand in the air touch condition was calculated using a scale factor between the Manomotion SDK and the VE coordinate system. However, the scale factor in the z-axis was fixed in our study, which was not ideal for all participants. For example, one participant with shorter arms had difficulty in reaching the farthest targets. Customizing this ratio for each participant would provide a better user experience.

## 6 Conclusions

In this paper, we compared potential selection techniques for low-cost mobile VR. Our objective was to assess if alternatives to common head-based selection methods were feasible with current technology, employing computer vision tracking approaches on mobile devices. To this end, our study employed only a smartphone and a cardboard HMD. In the first experiment, we compared air touch, head ray and finger ray in selection tasks. Overall, air touch performed worst, and finger ray performed best. However, since this experiment used an unrealistic selection indication mechanism (to improve experimental internal validity), we conducted a second experiment to compare selection indication methods. Results of Experiment 2 indicated that the secondary touchpad worked very well with finger ray, despite its impracticality. The built-in Cardboard button worked well with head ray.

Our results suggest that finger ray is promising for mobile VR, even when tracked by a single camera. Despite tracking imprecision, the technique performed well when used with an external touchpad. Future research could focus on further investigating potential selection indication methods to use with finger ray. For example, different gestures that are more reliably detectable may yield better performance than the tap gesture used in our experiment. Such gestures would also work in practical contexts, unlike the touchpad used in our first experiment.

In contrast, direct touch techniques like air touch performed very poorly; single-camera hand tracking seems to be out of reach for current mobile device cameras. Our results indicate that higher DOF techniques yield lower performance, consistent with previous results [2, 20]. Direct touch might be possible with more powerful future mobile devices supporting more robust vision-based tracking software, or if depth cameras become common on mobiles. Overall, from Experiment 1, there seems to be greater promise for ray-based selection techniques employing mobile camera tracking than virtual hand techniques.

## References

1. Argelaguet, F., Andujar, C.: A survey of 3D object selection techniques for virtual environments. Comput. Graph. **37**(3), 121–136 (2013)
2. Arsenault, R., Ware, C.: The importance of stereo and eye-coupled perspective for eye-hand coordination in fish tank VR. Presence Teleoperators Virtual Environ. **13**(5), 549–559 (2004)
3. Bowman, D., Wingrave, C., Campbell, J., Ly, V.: Using pinch gloves (tm) for both natural and abstract interaction techniques in virtual environments. In: Proceedings of the HCI International. Springer, New York (2001)

4. Bowman, D.A., Hodges, L.F.: An evaluation of techniques for grabbing and manipulating remote objects in immersive virtual environments. In: Proceedings of the ACM Symposium on Interactive 3D graphics - SI3D 1997, pp. 35–38. ACM, New York (1997)
5. Erol, A., Bebis, G., Nicolescu, M., Boyle, R.D., Twombly, X.: Vision-based hand pose estimation: a review. Comput. Vis. Image Underst. **108**(1–2), 52–73 (2007)
6. Fitts, P.M.: The information capacity of the human motor system in controlling the amplitude of movement. J. Exp. Psychol. **47**(6), 381–391 (1954)
7. Hand, C.: A survey of 3D interaction techniques. Comput. Graph. Forum **16**(5), 269–281 (1997)
8. Hernández, B., Flores, A.: A bare-hand gesture interaction system for virtual environments. In: Proceedings of the International Conference on Computer Graphics Theory and Applications (GRAPP), pp. 1–8. IEEE, New York (2014)
9. ISO, ISO: 9241–9 Ergonomic requirements for office work with visual display terminals (VDTs) - part 9: requirements for non-keyboard input devices international standard. Int. Organ. Stand. (2000)
10. Kopper, R., Bowman, D.A., Silva, M.G., McMahan, R.P.: A human motor behavior model for distal pointing tasks. Int. J. Hum. Comput. Stud. **68**(10), 603–615 (2010)
11. Kulshreshth, A., LaViola Jr, J.J.: Evaluating performance benefits of head tracking in modern video games. In: Proceedings of the ACM Symposium on Spatial User Interaction - Sui 2013, pp. 53–60, ACM, New York (2013)
12. LaViola Jr., J.J., Kruijff, E., McMahan, R.P., Bowman, D., Poupyrev, I.P.: 3D User Interfaces: Theory and Practice. Addison-Wesley Professional, USA (2017)
13. Lin, C.J., Ho, S.-H., Chen, Y.-J.: An investigation of pointing postures in a 3D stereoscopic environment. Appl. Ergon. **48**, 154–163 (2015)
14. Mine, M.R., Frederick P., Brooks, J., Sequin, C.H.: Moving objects in space: exploiting proprioception in virtual-environment interaction. In: SIGGRAPH 1997: Proceedings of the ACM Conference on Computer Graphics and Interactive Techniques, pp. 19–26. ACM, New York (1997)
15. Ni, T., Bowman, D.A. Chen, J.: Increased display size and resolution improve task performance in information-rich virtual environments. In: Proceedings of Graphics Interface 2006, pp. 139–146. CIPS, Toronto (2006)
16. Pierce, J.S., Forsberg, A.S., Conway, M.J., Hong, S., Zeleznik, R.C., Mine, M. R.: Image plane interaction techniques in 3D immersive environments. In: Proceedings of the Symposium on Interactive 3D Graphics - SI3D 1997, pp. 39–43. ACM, New York (1997)
17. Powell, W., Powell, V., Brown, P., Cook, M., Uddin, J.: Getting around in google cardboard–exploring navigation preferences with low-cost mobile VR. In: Proceedings of the IEEE VR Workshop on Everyday Virtual Reality (WEVR) 2016, pp. 5–8. IEEE, New York (2016)
18. Ramcharitar, A., Teather, R.J.: EZCursorVR: 2D selection with virtual reality head-mounted displays. In: Proceedings of Graphics Interface 2018, pp. 114–121. CIPS, Toronto (2018)
19. Teather, R.J., Stuerzlinger, W.: Pointing at 3D targets in a stereo head-tracked virtual environment. In: Proceedings of the IEEE Symposium on 3D User Interfaces, pp. 87–94. IEEE, New York (2011)
20. Teather, R.J., Stuerzlinger, W.: Pointing at 3D target projections using one-eyed and stereo cursors. In: Proceedings of the ACM Conference on Human Factors in Computing Systems - CHI 2013, pp. 159 – 168. ACM, New York (2013)
21. Teather, R.J., Stuerzlinger, W.: Visual aids in 3D point selection experiments. In: Proceedings of the ACM Symposium on Spatial User Interaction - SUI 2014, pp. 127–136. ACM, New York (2014)
22. Vanacken, L., Grossman, T., Coninx, K.: Exploring the effects of environment density and target visibility on object selection in 3D virtual environments. In: Proceedings of the IEEE Symposium on 3D User Interfaces - 3DUI 2007, pp. 117–124. IEEE, New York (2007)

23. Yoo, S., Parker, C.: Controller-less interaction methods for Google cardboard. In: Proceedings of the ACM Symposium on Spatial User Interaction - SUI 2015, p. 127. ACM, New York (2015)
24. Zeleznik, R.C., Forsberg, A.S., Schulze, J.P.: Look-that-there: exploiting gaze in virtual reality interactions. Technical report, Technical Report CS-05 (2005)

# Improving the Visual Perception and Spatial Awareness of Downhill Winter Athletes with Augmented Reality

Darren O'Neill, Mahmut Erdemli[✉], Ali Arya, and Stephen Field

Faculty of Engineering and Design, Carleton University, Ottawa K1S 5B6, Canada
{darren.oneill,mahmut.erdemli,ali.arya,stephen.field}@carleton.ca

**Abstract.** This research study addresses the design and development of an augmented reality headset display for downhill winter athletes, which may improve visual perception and spatial awareness, and reduce injury. We have used a variety of methods to collect the participant data, including surveys, experience-simulation-testing, user-response-analysis, and statistical analysis. The results revealed that various levels of downhill winter athletes may benefit differently from access to athletic data during physical activity, and indicated that some expert level athletes can train to strengthen their spatial-awareness abilities. The results also generated visual design recommendations, including icon colours, locations within the field-of-view, and alert methods, which could be utilized to optimize the usability of a headset display.

**Keywords:** Augmented reality · Interface · Winter sports · Spatial awareness · Visual perception

## 1 Introduction

Spatial awareness is essential to the performance and success of any athlete but is especially important to winter sports athletes participating in high-speed sports, where a split-second decision could be the difference between success and failure, which could result in injury [17]. Having faster reaction times to terrain changes and visual stimuli may give them a competitive advantage in competition and similarly having good peripheral awareness may also improve their chances of success both at training runs or competition. Downhill winter athletes have, "underlined the importance of visual perception for optimal performance, even though they seem to rather unconsciously perceived visual information during races" [22] (Fig. 1).

Wearable solutions can offer a technological advantage in sports and games. Particularly, users can benefit from Augmented Reality (AR) headsets for live monitoring of useful information such as biometric and spatial data without losing the sight of "the real world" [5]. For example, an AR headset is used for pocket billiards that gives related visual information to the player and increases

© Springer Nature Switzerland AG 2020
C. Stephanidis et al. (Eds.): HCII 2020, LNCS 12428, pp. 403–420, 2020.
https://doi.org/10.1007/978-3-030-59990-4_30

**Fig. 1.** Participants standing on the platform during experience simulation testing

the activity's success rate by calculating the distance to the target with dashed lines between six pockets and the balls for an effective shot [25]. Physical status and biometrics information can also help as visual information to improve the visual and spatial perception of an athlete during an activity in order to increase athletes' performance [15]. Yet there are no studies on how to effectively present such data on a Head-Mounted Display (HMD) screen for extremely fast-paced sports such as downhill skiing. In this paper, we investigate how downhill winter athletes can benefit from AR HMD by determining the usability and evaluating different visual design choices to deliver the data to the athletes. Our investigation is based on designing a prototype and running an empirical study of athletes with different skill levels. We propose recommendations and guidelines for AR displays, including icon colours and location within the field-of-view, and the alert methods that could be utilized to optimize the usability of a headset display.

## 2    Related Work

### 2.1    Use of Biometric Data for Athletes

In recent years, several wearable products have been released for tracking individual performance. Athletes who participate in sports like running and cycling are already utilizing wearable devices, such as wristbands and headset displays, to read biometric data during and after performing a physical activity [5]. Athletes can improve visual and spatial perception and receive constant feedback with monitoring physical conditioning with biometrics such as pedometers, accelerometers, and heart rate monitors. The most common type of wearable device for athletes is a set of systems classified as personal information systems that "help people collect and reflect on personal information" [16]. Personal information systems can facilitate opportunities to share information between coaches and athletes [24]. Biometric sensors embedded in wearables devices can provide coaches and athletes with performance data [3].

## 2.2   Mixed-Reality Technologies in Exercise and Skiing

Schlappi et al. [22] suggest that HMDs and AR can improve the user's daily lifestyle, and will "help fill in the blind spots and take the guesswork out of everyday living by supplementing the myopic vantage of real-time experience with a continuous, informatics mode of perception." Delabrida et al. [8] created an AR application that can be used with mobile phones and printed 3D HMDs to measure the distance between objects. Recon-jet goggles are specifically designed for cyclists or long-distance runners. Recon Jet includes a dual-core processor, dedicated graphics, Wi-Fi, Bluetooth, GPS, webcam and a sensor designed to turn on the device when it is equipped [7]. Oakley Airwave snow googles deliver informatics on the HUD speed, such as air time, vertical feet, navigation, and a buddy tracking for snowboarders moving down the slope [1]. There has been other a qualitative analysis based on interviews with top-level downhill winter athletes [22], however there is no study yet that investigates their effectiveness and various usability aspects.

Head-mounted displays can have negative effects on the safety level of an athlete when the virtual action is not necessary for athlete's posture. One rule is "not disturbing the user's behavior with a virtual object and allowing freehand interaction" [10]. Icons and numerical data should be displayed in the periphery and not the center of the user's field-of-view, as it could cause major visual distractions and occlude environmental objects like signs, lights, trees, and rocks. While the main benefit of using HMD is that the user can see navigation information regardless of where they are looking, the display may block or at least cover a certain portion of the real-world view [12].

## 2.3   AR Display Design

When designing dynamic icons for visual screens, Gestalt principles related to the visual perception of 2-dimensional images can be adapted for AR headset displays. In a study on Gestalt theory in visual screen design, Chang et al. [6] list the key Gestalt principles. From the eleven Gestalt principles Chang listed, our icon most benefited from having a single focal point, and using visual perception of closure and simplification using uncluttered graphics. HUD used a single icon, to prevent complex graphics and ambiguous conclusions. Too many focal points are likely to confuse learners and diffuse their interest. Int he core of our graphics by using x and y graph, we allowed the individual perceive the form as a complete together and apply closure with the icon and the screen.

The location, colour, and opacity of elements, or icons within visual displays, especially head-mounted displays, can affect the visual perception of the person viewing the display. Schömig et al. [23] suggested a landmark-based map when using head-mounted displays. Landmarks can suggest any place, object, or feature that persists in the Heads-Up Display (HUD). Navigational scenarios, route-finding, or distance to turn would aid users to help with navigation.

Albery [2] developed and tested SORD, a complex HMD system used by military pilots. It was composed of a tactile vibrating vest, a helmet-mounted visual display, and 3D audio headphones. This multi-sensory HMD system allows pilots to bring their attention to external visual tasks outside the cockpit, without having to continuously return their attention to the aircraft attitude instruments on the dashboard. The SORD allows the pilot to monitor airspeed, altitude, heading, bank and pitch of the aircraft in real-time, reducing the pilot's workload by eliminating the requirement to frequently monitor the cockpit displays and controls [2]. It was found through extensive research that HMD symbology used by pilots, "should be designed to support an efficient instrument scan, support other mission-related symbology, and utilize as little of the display as possible" [9].

Ito et al. [11] compared the effects of icons and text using Heads Up Display, and the results indicated that presenting more letters increases the speed of information. On the other hand, spending long durations looking at the displayed information has a potential risk of accidents. Bartram et al. [4] conducted visual perception test studies, which indicate that blinking icons will capture the end-user's attention more effectively translational or rotational motion. Nozawa et al. [20] created an initial study as a VR experience for indoor ski training with a slope simulator. The results show that athletes overcome cybersickness by following a pre-recorded expert motion animation and learn by copying their body gestures.

As far as we know, there has not been a user study evaluating the visual design requirements of augmented headset displays for winter downhill athletes. Our goal is to investigate the visual design requirements for an augmented headset display for downhill winter athletes. Based on our research gaps, we identify the following research questions.

- Can downhill winter athletes benefit from access to spatial orientation data and other important athletic information during physical activity?
- How can sensory modalities and technologies be integrated to transmit unobtrusive data to the athletes?
- What visual format is the most effective to deliver the data to the athletes?

## 3   Research Design

### 3.1   Prototype Design and Development

Visual perception, dynamic visual acuity, and quick reaction times to visual stimuli are proven to be essential to the success of high-level athletes [18, 21, 22]. A wearable headset could likely deliver essential information most effectively to the athlete while not being overly obtrusive and distracting during physical activity. The increased awareness could potentially improve success rates and prevent injury.

Military aircraft pilots and extreme downhill winter athletes have many similarities, which may indicate that AR headsets and principals utilized by military pilots could also be utilized to benefit downhill winter athletes. Similar to downhill winter athletes, pilots as well have the capacity to use their visual senses to receive necessary data without being distracted [19]. Recent HUD and HMD design principles that are used by the military AR displays, and their icons could potentially be utilized to support efficient variable scanning for downhill winter athletes because pilots similar to athletes require spatial orientation data and other biometric information. A comparison with the results of military pilots [19] can contribute to new recommendations or guidelines for downhill winter athletes (See Fig. 2).

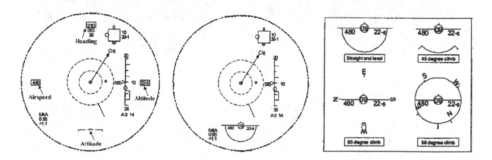

**Fig. 2.** Icons used by military aircraft pilots (left) during flight [9]

To investigate the use of AR HMD, we developed a prototype (See Fig. 3) consisting of three different parts; an Augmented Reality Head Mount Display, a projection of a video that displays a first-person view snowboarding experience, and a physically elevating platform to simulate a downhill skiing experience.

During the experiment, the slope changing platform moved in sync with the changing angle of the hill in the downhill simulation video, played on the screen. The platform began each simulation at a position of 5 (a beginner level slope, green circle) and for the slope angle test changed to 14 (blue square) for the 2nd interval, 20 (black diamond) for the 3rd interval and back down to 9 (green circle) for the final 4th interval. The angle intervals were selected based on the information by Kipp [13,14] which revealed that in North America; green circle slope gradient: 6–25% (0–11), blue square slope gradient: 25–40% (12–18), black diamond slope gradient: 40% (19 and up).

The main structure of the slope changing platform was constructed using rectangular steel pipes welded together to support the weight of participants who weighed within the range of 100–250 lbs. The structure had to be sturdy to withstand the force of participants leaning side to side while the platform changed its slope. A remote control, electric car jack powered by a heavy-duty, 15-amp, DC motor with a lift capacity of 4000 lbs was sourced and used to power the changing angle of the platform.

A wireless headphone was worn by participants to create background sound effects such as mountain wind sounds on the loop, which also helped cancel the sound of the electric motor and platform moving.

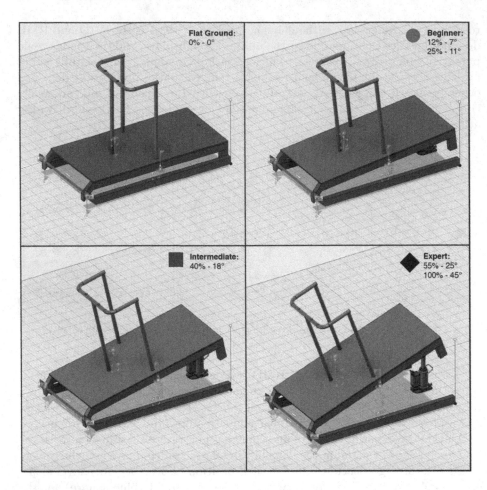

**Fig. 3.** Isometric plans of the testing platform design and associated equipment (Color figure online)

During the experience, participants wore both wireless headphones and a Moverio BT-200TM smart glasses headset 10). Participants stood on a specially designed, slope changing platform, in front of a large, high-resolution television screen to test an icon developed, within a simulated downhill winter environment. The Moverio BT-200TM was powered by a smartphone controller, which was stored in a pocket on the safety harness during test simulations and linked by a charging cord, which was somewhat obtrusive. Icons and symbols currently have been used for AR displays by military aircraft were used as guidelines for developing the icon for this research 5). The icon was designed to be simple and easy to read, displaying only the athlete's downhill speed, vertical altitude on the mountain, and, most importantly, the current slope angle of the hill (Fig. 4).

**Fig. 4.** The Moverio BT-200 smart glasses worn by participants (Google images)

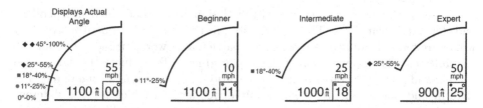

**Fig. 5.** Icon designed for Moverio headset display experience testing (Color figure online)

## 3.2  Research Approach

**Experiment Design.** The visual icon (See Fig. 5) displayed in the HMD was used to assess different variables of visual perception during four test scenarios:

1. Icon color scenario,
2. Icon location scenario,
3. Icon alert method,
4. Icon slope angle indication method.

During the first scenario, participants stood on the platform and viewed the icon displayed in the headset, which changed to a different colour every 7 s, based on the visual spectrum of colours or ROYGBIV (Red, Orange, Yellow, Green, Blue, Indigo, Violet). The hypothesis for this first scenario was that participants would rank the green icon as having the best clarity and visibility, because green iconography has been previously tested, used and proved the most effective by military pilots [19]. A secondary hypothesis was that the red icon would also be ranked high due to its stark contrast to the outdoor winter environment with colors of green, white and blue.

The other two scenarios to evaluate icon alert mode and icon location followed a similar structure, allowing participants to input their feedback immediately after viewing the different icons.

During the fourth scenario (icon slope angle indication method), the participants determined the angle of the slope changing platform, stood on during the simulation, by shouting out the estimated angle value at three specific time intervals during the 90-s simulation video.

During the first part of the fourth scenario, participants were instructed to make an educated guess and attempt to determine the exact slope angle value of the platform at that specific interval, using only their spatial awareness abilities and without any icon displayed in the headset for assistance.

During the second part of the fourth scenario, the participants watched the same 90-s simulation video, but this time the headset displayed an icon that indicated the angle value of the slope changing platform in real-time. Instead of using spatial awareness abilities and making an educated guess, the participants looked at the icon displayed in the headset to read and state aloud the slope angle value at the same time intervals.

The experiment was a within-subject design using predefined independent and dependent variables. The test results are the subjective quantitative data which are formed from the answer of the participants were questioned about the usability and visual perception of the displayed icon. To evaluate the data of the success and preference rate of the four scenarios above, we used visual perception criteria and operationalize them as our dependent variables for the different scenarios. The independent variables of this test were the different scenarios manipulating the color, location, method of alert the icon. (See Figs. 6, 7 and 8) To assess the accuracy of spatial awareness of athletes, we ran the scenario one final time without using the headset and the visual information.

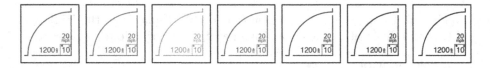

**Fig. 6.** Icon colours (from left) red, orange, yellow, green, blue, indigo, violet (Color figure online)

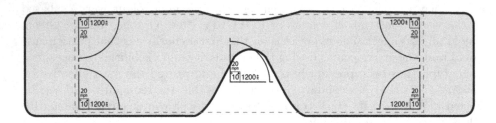

**Fig. 7.** Icon locations in the headset display: top-left, bottom-left, middle, top-right, bottom-right

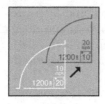 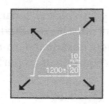 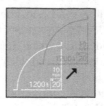 

**Fig. 8.** Icon alert methods: colour changing, expanding, blinking, rotating (Color figure online)

**Participants.** During the study, 34 participants with previous experience skiing, or snowboarding at different skill-levels were recruited. Participants were recruited through university and online winter sport athlete communities. Participants were initially screened for their expertise in snowboarding/skiing and using Head-Mounted Displays. The requirements were to be between 18–55 years of age with good vision and not having any injuries or disabilities. The user testing was conducted at the researchers' university in a private room with no external noise or visual distractions present.

**Apparatus.** The key features of the prototype are as follows:

1. Moverio BT-200: an augmented reality headset used to display the icon during the testing.
2. Wireless Headphones: played ambient mountain wind sounds during the entire simulation,
3. Remote Controlled Electric Car Jack: used by the primary researcher to change the slope, angle of the platform during the simulations, had an angle range of 5 to 20,
4. High-Resolution Screen: used to play first-person, simulation videos recorded on real ski hills by the primary researcher, positioned close to participants to feel more immersive,
5. Safety Harness: a belt worn by participants, clipped to the safety railing to prevent falling,
6. Safety Enclosure: padded metal railing that participants held onto during the simulation,
7. Rubber Floor Pad: gave the participants extra foot grip when the platform angle was steep,
8. GoProTM Cameras: two cameras were used to record front and side views during testing.

**Procedure.** The experience simulation testing sessions were conducted over a 31-day period, with 34 total participants, and were approximately 30 min in length from start to finish.

Each participant first read and filled out a consent form, followed by a pre-test survey and skill level assessment. After all the required surveys and documents were completed, the participants were instructed to step onto the slope changing platform and were fitted with the equipment to perform the experience simulation testing.

After each scenario was completed, the participants were given the post-test survey sheet to rank the different coloured icons from the best, most visible and easy to read icon displayed by the headset, to the worst and least visible icon (1 = best, 7 = worst). Participants finished the study with a 1-on-1 interview with the researcher, which allowed each participant to express any final comments.

# 4   Results

## 4.1   Data Collection

The participants were combined of 9 male skiers, 8 female skiers, 9 male snowboarders and 8 female snowboarders, for a final ratio of 52% male and 48% female for both skiers and snowboarders. A specific number of male and female, ski and snowboarding participants were selected to ensure that the sample size and results of participant testing sessions were reflective of the Canadian Ski Council ratios. Statistics gathered and presented by the Canadian Ski Council showed that in 2015 a total of 58.3% of Canadian downhill skiers were male and 41.7% female, and of all Canadian snowboarders 61% were male and 39% female.

Pre-test survey indicating participants' skill level in ski or snowboarding resulted as 10 participants (29%) at the beginner level, 7 participants (21%) at the intermediate level, 16 participants (47%) at the expert level and 1 participant (3%) at the competitive level.

When participants were asked if they knew what "augmented reality" was 23 participants (68%) indicated yes and then wrote their own definition, while 11 participants (32%) indicated that they did not know what AR was. Next, when participants were asked if they have used AR devices during an activity, 29 participants (85%) indicated that they had never used an AR device before, 3 participants (9%) indicated that they had rarely use AR devices, 1 participant (3%) indicated that they sometimes use AR devices, and 1 participant (3%) indicated frequent use of AR devices (Figs. 9 and 10).

The first part of the post-test survey determined the effect of each of the four variables (icon colour, location, alert method, and slope indicator). The post-test survey asked questions that related to the icon display features and system usability, which the participants responded to using a Likert-scale such as:

– The level of difficulty to determine the slope angle value with/without icon assistance,
– Determine if the interface helped you to determine the slope more quickly or accurately,
– Indicate if it would be useful to have access to this data while riding,

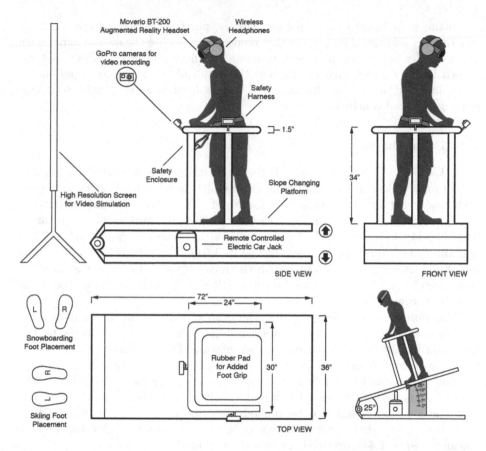

**Fig. 9.** An orthographic drawings of the setup for snowboarding simulation testing

**Fig. 10.** The Moverio BT-200 smart glasses worn by participants (Google images)

- What group of athletes would benefit from this display? (beginner, expert, etc.),
- What senses are best suited for transmitting this data to the athletes?
- How natural and comfortable viewing the headset felt during testing?
- How much you learned about augmented reality during the study?

In the second part of the post-test survey, participants were given a chance to write any additional comments that could help improve the overall experience and usability of the AR headset and icon. The user response analysis method helped identify issues with the display that multiple participants experienced, while also confirming some findings and results from other data collected, supporting the final conclusions.

## 4.2  Data Anaysis

**Icon Colour.** The seven icon colours used during this research study were; red, orange, yellow, green, blue, indigo and violet (See Fig. 6). Participants gave each colour a rank, 1 for best to 7 for worst. The results indicated that the orange coloured icon was ranked with the best average rank or mean (mean = 2.91 ± 1.44), followed closely by the red icon in second (mean = 2.94 ± 1.82) and the green icon in third (mean = 2.97 ± 1.51). The yellow icon was ranked fourth (mean = 3.5 ± 2.08), blue in fifth (mean = 4.0 ± 1.60), indigo in sixth (mean = 5.56 ± 1.26) and violet (mean = 6.12 ± 1.37) was ranked as the worst overall in seventh place.

Therefore, this data suggests that the orange, red and green coloured icons were preferred by participants and appeared most clear and visible to the participants during the downhill winter sport simulation. Therefore using these colours to display information could help to optimize the optics of a headset display.

The hypothesis for this first scenario was that participants would rank the green icon as having the best clarity and visibility. A secondary proposition was that the red icon would also be ranked high due to its stark contrast. However, the results show that orange was the most preferred color for an icon (mean = 2.91 ± 1.44), over red (mean = 2.94 ± 1.82).

The least visible to the participants was the yellow icon in fourth (mean = 3.5), blue in fifth (mean = 4.0), indigo in sixth (mean = 5.56) and violet (mean = 6.12) which was ranked as the worst overall in terms of visibility and clarity of viewing. The cool blue coloured icons (blue, indigo, and violet) were all ranked as the worst overall in fifth, sixth and last place, which may be a result of participants being visually exposed to sky and snow during the downhill winter simulation, however additional studies would have to be conducted to confirm this proposition.

**Icon Location.** Participants gave each of the five locations a rank, 1 for best to 5 for worst. The subjective results showed that participants selected the top-right icon as the best overall (mean = 2.15 ± 1.05), followed by bottom-right location (mean = 2.53 ± 1.26) and top- left location (mean = 2.74 ± 1.31) in third. Both the middle location (mean = 3.41 ± 1.56) and the bottom-left location (mean = 3.65 ± 1.04) of the test icon were the least desired by participants and were ranked fourth and fifth respectively, with a mean greater than 3.40.

These results suggest that on average, participants preferred viewing icons on the right-hand side of their vision since the top-right location was ranked first and the bottom-right location was ranked second, which could indicate that many of them are right-eye dominant. However, more tests would have to be conducted to confirm this. The top-level locations were ranked first and third, while the bottom-level locations were ranked second and fifth, suggesting that the participants preferred the icons to be located within the top segment of the field-of-view compared to the bottom segment of the field-of-view.

**Icon Alert Mode.** Participants gave each of the four alert modes a rank, 1 for best to 4 for worst. The subjective responses from this study showed that the colour changing icon alert method was ranked best overall (mean $= 1.71 \pm 1.06$), followed by the blinking icon alert method in second (mean $= 2.35 \pm 0.88$), the rotating icon alert method in third (mean $= 2.53 \pm 1.11$) and finally the pulsing icon alert method (mean $= 3.41 \pm 0.70$) was ranked worst overall. The two icon alert methods with a blinking style were ranked first and second in terms of most attention grabbing while also not being overly distracting to the user.

**Icon Slope Angle Indicator - No Icon Indicator.** The data from this study showed that for the first, resting angle interval (5) 6 participants (18%) responded with the correct angle, and 22 participants (66%) responded within the range of 0–10). For the second angle interval (14) no participants responded correctly and only 6 (20%) participants answered within the range of 11–20. For the third angle interval (20) 2 participant) answered correctly and 4 participants (12%) answered within the range of 16–25. Finally, for the fourth angle interval (9) no participants answered correctly but a total of 12 participants (36%) answered within the range of 6–15 When the platform was at its resting position (5) 18% of participants responded correctly and 66% participants answered within a 10 range, compared to the remaining three angle intervals which has significantly less participants respond correctly or accurately within a 10 range.

The test without using headsets on the slope design showed that participants responded incorrectly without an assistive icon displayed in the AR headsets. The results suggest that it is difficult for an individual to use their visual and spatial awareness abilities to accurately determine the angle of the slope.

## 5    Discussion

### 5.1    Visual Interface Design Recommendations

Average participants preferred viewing icons on the right-hand side of their field-of-view, since top-right is ranked first and bottom-right second, which could indicate that many of the participants were right-eye dominant. However, the icons located within top-level icons were ranked first and third, while the icons located within the bottom-level were ranked second and fifth, suggesting that participants on average preferred the icon to be located at the top of the field-of-view compare to the bottom of the field-of-view. This could be a result of easier and more comfortable viewing of an icon located in the top-level, as one participant stated in their post-test comment, "[It was] difficult to change the focal point quickly in the bottom- right, [it was] easier to change focus when looking or glancing to the top-right away from the snow".

Another way to optimize the colour of a display icon would be to ensure that every athlete can read the information clearly by implementing a customization of the icon and allow the end-user to pick the icon colour that works best for them.

Icon alert method used to determine which of the four icon alert methods was most visible and most effective at capturing the participant's attention. The results showed that the icon alert method that was preferred and ranked as the best and most effective overall was the colour changing icon alert (mean = 1.71), followed by the blinking icon alert in second (mean = 2.35), the rotating icon in third (mean = 2.53) and finally the pulsing icon alert method (mean = 3.41) ranked as worst overall.

An icon in the field-of-view of a headset display that combines both colour and a highly detectable but non-distracting blinking effect will likely have a higher rate of detection, which was indicated by the data of the top-ranked, coloured, blinking icon alert method test conducted during the study (See Fig. 11). In conclusion, the results of the visual variable tests are as follows:

- Icon Colour: An icon using orange, red, or green colours that contrast the outdoor winter environment will allow for the best visibility and clarity for the athlete reading the data,
- Icon Location: An icon located in the top level of the field-of-view, and more specifically, the top-right, will allow for the best unobtrusive and natural viewing of the interface,
- Icon Alert Method: An icon that changes colour and blinks in the display will attract the visual attention of the athlete best during a downhill winter sport activity.

**Recommendations for Physical Design.** During the post-test survey when participants were asked to indicate if wearing the AR headset and viewing the icon felt natural and comfortable, 8 participants (23%) strongly-agreed, 19 participants (56%) agreed, 7 participants (21%) disagreed and thought that the display was not comfortable and natural and no participants strongly-disagreed. The comfort of the Moverio BT-200TM AR headset that was worn by participants during this thesis study was evaluated, and 79% of participants agreed that the Moverio BT-200TM felt natural and comfortable to wear.

There are a number of ways to increase the comfort of a headset such as reducing the weight, distributing the weight of the headset more evenly throughout the main body of the device, making it more adjustable, and altering the size of the headset display lens giving the field-of-view a larger surface area than standard ski goggles and more space to display an unobtrusive icon with information.

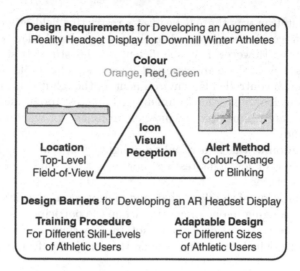

**Fig. 11.** Design requirements and barriers for developing an AR display for Downhill Winter Athletes

## 5.2   Suggestions for Study Improvement

The limitations related to the thesis study can be divided into two different categories; environmental limitations and technological limitations.

The first environmental limitation related to this study was the simulation room temperature of 21 °C compared to winter temperatures experienced on an actual ski hill. Another environmental limitation was the lack of physical precipitation (snow, rain, etc.) and wind or air pressure during the test. To overcome this wind limitation, participants wore wireless headphones playing realistic wind sounds. However, there was no physical wind sensation.

The lighting within the simulation room was also an environmental error, although some natural light came from windows, mostly overhead artificial florescent bulbs provided white light to participants.

The wireless headphones used to simulate wind sounds were large and bulky and would not be worn in real-life by athletes on the slopes.

The slope changing platform also was problematic because of technological limitations. Firstly, the remote control, electric car jack had a single rate of speed to lower and raise the platform to change its angle during the simulations. The platform required the primary researcher to raise and lower the platform using a remotely using a handheld controller.

As with any simulation training, there were some technical limitations due to the equipment that was available, compatibility of devices within the simulation system, and the current level of technology. The first limitation is the size and weight of the Moverio BT-200TM headset used during this study, which resembles reading glasses, not ski goggles and is weighted on the front with no back strap that ski goggles would have to add extra support.

Another limitation was the AR headset's restrictive field-of-view, which did not allow for an icon to be displayed in the far peripheral edges of the participant's field-of-view. However, it was sufficient for this study. It should be noted that limitations such as these are expected when designing, building, and testing prototypes within a controlled lab environment. Although it is nearly impossible to replicate an outdoor winter environment within a test room the measures that were taken adequately simulated a winter environment for what was required of this research study.

For the duration of this study, only day-light testing was conducted. There was no night- time or low-light testing; however, this could be explored in an entirely separate study.

One final limitation of the slope changing platform was that it simulated only the slope angle changing in one dimension while the participant stood with bent knees. Thus, there was no side to side motion of carving experienced on both skis or snowboard, which would have made the simulation feel more realistic.

## 6    Conclusion

The primary goal of this thesis study was to investigate the barriers and design recommendations for developing an AR display and icon system for downhill winter athletes. Insights from athlete participants and quantitative results from the simulation testing helped answer the first research question. It was determined that downhill winter athletes could benefit from access to spatial orientation and other important data during an activity.

Participant insights and responses to questions within the post-test survey also helped answer the second research question. Participant comments and responses to questions indicated that the visual modality and augmented reality headset displays were best to transmit unobtrusive data to downhill winter athletes during activity. When asked which of the five senses would be best suited to transmit data to downhill winter athletes during an activity, 27 participants (81%) ranked "sight" as the best-suited sense.

Finally, the inquired data helped to identify the icon colour, location in the field-of-view and visual alert that participants preferred for viewing, and answered the third research question. During testing, participants ranked orange (mean = 2.91), red (mean = 2.94) and green (mean = 2.97) all very closely, which indicated that the participants preferred icons using orange, red or green colours which appeared to contrast the outdoor winter environment and allow for best visibility and clarity while reading the data.

Similarly, participants ranked icon locations from best (1) to worst (5) and ranked top-right first (mean = 2.15), top-left second (mean = 2.74) and bottom-right third (mean = 2.53) which indicated that participants preferred icons in the top level of the field-of-view (more specifically the top right) which seemed to allow for the best unobtrusive and natural viewing of the icon. Participants also ranked alert methods from best (1) to worst (4) and ranked the blinking-color icon (mean = 1.71) and the blinking icon (mean = 2.35) as top two alerts, which indicated that icons which change colour and blink on/off in the display were preferred by participants and appeared to attract their visual attention most effectively during the testing.

Future research directions include (1) validating the results obtained from the simulations used in this study, in an outdoor ski hill environment, (2) exploring if the HMD would provide navigational instructions relative to the orientation of the head, (3) evaluating the downhill winter athlete's ability to interpret the icon while in a competitive scenario, (4) utilizing the augmented reality headset system and slope changing platform designed specifically for this study, to evaluate the visual perception and spatial awareness of different types of athletes.

# References

1. Oakley Airwave™ Snow Goggles. https://www.oakley.com/en-us/product/W0OO7049?variant=700285844411&fit=GLOBAL&lensShape=STANDARD
2. Albery, W.B.: Spatial Orientation Retention Device - Current Status, vol. 12, October 2017. http://www.dtic.mil/docs/citations/ADA441896
3. Baca, A., Dabnichki, P., Heller, M., Kornfeind, P.: Ubiquitous computing in sports: a review and analysis. J. Sports Sci. 27(12), 1335–1346 (2009)
4. Bartram, L., Ware, C., Calvert, T.: Moving icons: detection and distraction. In: Human-Computer Interaction: IFIP TC. Published by IOS Press on Behalf of the International Federation for Information Processing (IFIP, 13 International Conference on Human-Computer Interaction, pp. 157–165. INTERACT 2001, 9th-13th July 2001, Tokyo, Japan, Amsterdam (2001)
5. Bozyer, Z.: Augmented reality in sports: today and tomorrow. Int. J. Sport Cult. Sci. 3(4), 314–325 (2015). https://doi.org/10.14486/IJSCS392
6. Chang, D., Dooley, L., Tuovinen, J.E.: Gestalt theory in visual screen design: a new look at an old subject. In: Tuovinen, J.E., et al. (eds.) 02 Proceedings of the Seventh World Conference on Computers in Education Conference on Computers in Education: Australian Topics, vol. 8, pp. 5–12 (2002)
7. Cooper, D.: Recon jet review: expensive fitness glasses with potential to be better 9 July–17 October 2018. https://www.engadget.com/2015/07/17/recon-je
8. Delabrida, S., D'Angelo, T., Oliveira, R.A., Loureiro, A.A.: Wearable HUD for ecological field research applications. Mob. Netw. Appl. 21(4), (2016)
9. Geiselman, E.E.: Development of a non-distributed flight reference symbology for helmet-mounted display use during off-boresight viewing (1999)
10. Ishiguro, Y., Rekimoto, J.: Peripheral vision annotation. In: Proceedings of the 2nd Augmented Human International Conference on - AH 11 (2011)
11. Ito, K., Nishimura, H., Ogi, T.: Head-up display for motorcycle navigation. In: SIGGRAPH Asia 2015 Head-Up Displays and Their Applications, pp. 10:1–10:3. SA 2015, ACM, New York (2015). https://doi.org/10.1145/2818406.2818415
12. Jose, R., Lee, G.A., Billinghurst, M.: A comparative study of simulated augmented reality displays for vehicle navigation. In: Proceedings of the 28th Australian Conference on Computer-Human Interaction, OzCHI 2016, pp. 40–48. ACM, New York (2016). https://doi.org/10.1145/3010915.3010918
13. Kipp, R.W.: Alpine skiing. In: Human Kinetics Kotsios, Information Technology, vol. 23, no. 2, p. 157. Kumar, V.: 101 Design Methods: A Structured Approach for Driving Innovation in Your Organization, Wiley, Hoboken (2013)

14. Kuru, A., et al.: Snowboarding injuries. In: Mei-Dan, O., Carmont, M., (eds.) Champaign. Adventure and Extreme Sports Injuries. Springer, London (2012)
15. Kuru, A.: A thesis submitted to the graduate school of natural and applied sciences of middle east technical university. Ph.D. thesis (2006)
16. Li, I., Dey, A., Forlizzi, J.: A stage-based model of personal informatics systems. In: Proceedings of the 28th International Conference on Human Factors in Computing Systems - CHI 10 (2010)
17. Louis, M., Collet, C., Champely, S., Guillot, A.: Differences in motor imagery time when predicting task duration in alpine skiers and equestrian riders. Res. Q. Exerc. Sport **83**(1), 86–93 (2012). https://doi.org/10.1080/02701367.2012.10599828
18. Memmert, D., Simons, D.J., Grimme, T.: The relationship between visual attention and expertise in sports. Psychol. Sport Exerc. **10**(1), 146–151 (2009)
19. Nicholl, R.: Airline head-up display systems: human factors considerations. SSRN Electron. J. **04**, (2014)
20. Nozawa, T., Wu, E., Perteneder, F., Koike, H.: Visualizing expert motion for guidance in a VR ski simulator. In: ACM SIGGRAPH 2019 Posters, pp. 64:1–64:2. SIGGRAPH 2019, ACM, New York (2019). https://doi.org/10.1145/3306214.3338561
21. Pascale, M.T., Sanderson, P., Liu, D., Mohamed, I., Stigter, N., Loeb, R.G.: Detection of visual stimuli on monocular peripheral head-worn displays. Appl. Ergon. **73**, 167–173 (2018)
22. Schläppi, O., Urfer, J., Kredel, R.: Sportwissenschaft **46**(3), 201–212 (2016). https://doi.org/10.1007/s12662-016-0400-9
23. Schömig, N., Wiedemann, K., Naujoks, F., Neukum, A., Leuchtenberg, B., Vöhringer-Kuhnt, T.: An augmented reality display for conditionally automated driving. In: Adjunct Proceedings of the 10th International Conference on Automotive User Interfaces and Interactive Vehicular Applications, AutomotiveUI 2018, pp. 137–141. ACM, New York (2018). https://doi.org/10.1145/3239092.3265956
24. Wakefield, B., Neustaedter, C., Hillman, S.: The informatics needs of amateur endurance athletic coaches. In: CHI 2014 Extended Abstracts on Human Factors in Computing Systems, pp. 2287–2292. CHI EA 2014. ACM, New York (2014). https://doi.org/10.1145/2559206.2581174
25. Yeo, H.S., Koike, H., Quigley, A.: Augmented learning for sports using wearable head-worn and wrist-worn devices. In: 2019 IEEE Conference on Virtual Reality and 3D User Interfaces (VR), pp. 1578–1580, March 2019. doi: 10.1109/VR.2019.8798054

# Desktop and Virtual-Reality Training Under Varying Degrees of Task Difficulty in a Complex Search-and-Shoot Scenario

Akash K. Rao[1(✉)], Sushil Chandra[2], and Varun Dutt[1]

[1] Applied Cognitive Science Laboratory, Indian Institute of Technology Mandi, Mandi, Himachal Pradesh, India
akashrao.iitmandi@gmail.com

[2] Department of Biomedical Engineering, Institute of Nuclear Medicine and Allied Sciences, Defence Research and Development Organisation, Delhi, India

**Abstract.** Two-dimensional (2D) desktop and three-dimensional (3D) Virtual-Reality (VR) play a significant role in providing military personnel with training environments to hone their decision-making skills. The nature of the environment (2D versus 3D) and the order of task difficulty (novice to expert or expert to novice) may influence human performance in these environments. However, an empirical evaluation of these environments and their interaction with the order of task difficulty has been less explored. The primary objective of this research was to address this gap and explore the influence of different environments (2D desktop or 3D VR) and order of task difficulty (novice to expert or expert to novice) on human performance. In a lab-based experiment, a total of 60 healthy subjects executed scenarios with novice or expert difficulty levels across both 2D desktop environments (N = 30) and 3D VR environments (N = 30). Within each environment, 15 participants executed the novice scenario first and expert scenario second, and 15 participants executed the expert scenario first and novice scenario second. Results revealed that the participants performed better in the 3D VR environment compared to the 2D desktop environment. Participants performed better due to both expert training (performance in novice second better compared to novice first) and novice training (performance in expert second better compared to expert first). The combination of a 3D VR environment with expert training first and novice training second maximized performance. We expect to use these conclusions for creating effective training environments using VR technology.

**Keywords:** 3D virtual reality · 2D desktop environment · Order of training · Novice · Expert · Decision-making · Instance-based learning theory

## 1 Introduction

According to [1], accurate and coherent decision-making is very crucial to ensure safety and efficiency in complex sociotechnical systems. An individual's decision-making abilities are usually linked to her experience in the domain and her repository of stored rule-based heuristics [1]. Some of these rule-based heuristics may be exploited to make

© Springer Nature Switzerland AG 2020
C. Stephanidis et al. (Eds.): HCII 2020, LNCS 12428, pp. 421–439, 2020.
https://doi.org/10.1007/978-3-030-59990-4_31

decisions in dynamic tasks [1]. However, decision-making, as defined by [2], may involve interdependent decisions that rapidly change as a function of the feedback received by an individual [2]. To make decisions in stressful situations, more experiential and natural-istic training may be required [3]. However, creating naturalistic conditions for training in the real world might turn out be either expensive or dangerous or both. This fact may be particularly true about decision-making in the military domain, where the conditions are usually uncertain, complex, stressful, and non-deterministic [4].

In the absence of real-world environments, research has proposed synthetic environ-ments to provide customizable, flexible, benign, and cost-effective training platforms to hone people's decision-making skills (including those of military personnel) [4–6]. Synthetic environments are mostly created either on a two-dimensional, non-immersive desktop screen (2D desktop) or a 3-dimensional (3D), immersive virtual reality (VR) through a head mounted display (HMD) [5]. The 2D desktop environment has been the traditional method of training people in decision tasks [5]. In contrast, the 3D VR allows the possibility for individuals to immerse themselves in a virtual environment, move freely and seamlessly in the environment, and examine the environment descrip-tors governing the virtual environment from different perspectives [7, 8]. Thus, unlike the 2D desktop environment, the 3D VR environment may allow individuals to build a better mental model of the environment that causes efficient skill acquisition during training [8].

An aspect of the training in different 2D desktop and 3D VR virtual environments may be the order of task difficulty. Traditionally, environments have been designed to start with novice task difficulty and move towards expert task difficulty [9]. It is believed that this order of task difficulty from novice to expert will allow individuals to improve their decision-making in synthetic environments. However, very few studies have varied the order of task difficulty to investigate whether novice to expert is indeed the right order for effective training. For example, one study [9] has investigated the effects of 3D VR training under varying order of task difficulty in a dynamic decision-making context. Out of 30 healthy subjects in the experiment, half of them executed an easier (novice) scenario first and the other half executed a difficult (expert) scenario first [9]. Results showed the participants who faced the expert scenario first fared better than the participants who executed the novice scenario first. Thus, this study challenged the belief that the novice to expert progression may be best for training in synthetic environments. Although [9] evaluated the effects of task difficulty in a dynamic decision-making context in a VR environment, these authors did not compare the order of task difficulty between the 3D VR and 2D desktop environments. This research overcomes this literature gap by investigating the order of task difficulty in different 2D desktop and 3D VR environments.

Some theories of decision-making may provide researchers the ability to hypoth-esize the change in performance across both 2D desktop and 3D VR environments as well as due to different orders of task difficulty [10, 11]. For example, Instance-based Learning Theory (IBLT) [10], a theory of how individuals make decisions from expe-rience, has elucidated decision making in dynamic contexts very efficiently. According to IBLT, decision-making is a 5-step process: recognition of the situation, judgment based on experience, choices among options based upon judgments, execution of the chosen actions, and feedback to those experiences that led to the chosen actions [10].

As per IBLT, when the training in a 3D VR environment, the decision-maker would be able to store salient instances (or experiences) with depth cues from the task compared to those stored in a 2D desktop environment. The 3D instances may help decision-makers to enhance their performance in 3D VR environments compared to 2D desktop environments.

Furthermore, according to another decision-making theory (the desirable difficulty framework [11]), in any learning task or domain, difficult and complex stimuli imparted during training may facilitate better skill transfer at transfer [11]. Thus, the expert to novice order of task difficulty is expected to provide a better learning environment compared to the novice to expert order of training.

The primary objective of this research is to test these expectations in experiments with human participants where the environment (2D desktop or 3D VR) were varied as well as the order of task difficulty (novice to expert or expert to novice) were varied. The contribution of this work is novel because to the best of authors' knowledge, this study would be the first of its kind to study the combined influence of the environment as well as the order of task difficulty on human performance.

In what follows, first, we provide a brief overview of the research involving 2D desktop and 3D VR environments and their applications for training individuals. Next, we elucidate an experiment to investigate the performance and cognitive implications of 2D desktop and 3D VR training under two different orders of task difficulty (novice to expert or expert to novice). Finally, we detail the results and discuss the applications of our results on dynamic decision-making involving different environments and order of difficulty in the real world.

## 2  Background

Researchers have pointed out that training in synthetic environments have benefits beyond merely supplementing orthodox training frameworks [1]. According to [1], even low-fidelity synthetic environments may contain aspects of the real world that cannot possibly be created in traditional/naturalistic real-world settings [1]. In addition, they have also reasoned that given the extreme customizable/flexible nature of synthetic environments, these environments could potentially be used to assist individuals to prepare for situations which they have never encountered before or are not safe for humans to experience in the real-world [1].

There have been a few studies who have evaluated the effects of 2D, non-immersive desktop environments and 3D, immersive virtual reality (VR) environments on basic cognitive processes [9, 12–15]. For instance, researchers in [12] conducted an experiment to examine how 2D desktop environments and 3D VR environments would affect spatial learning, especially when the ambulatory locomotion was restricted in the 3D VR environment. Results showcased that the participants felt significantly more workload and motion sickness in the 3D VR environment compared to the 2D desktop environment; however, the performance across both the desktop and VR environments was not significantly different [12].

On the contrary, researchers in [13] reported better performance in the 3D VR environment compared to the 2D desktop environment in a spatial processing task [13].

Participants exhibited better learning rates and higher reaction times in the 3D VR environment compared to the 2D desktop environment [13]. Participants also reported higher mental demand and effort requirements in the 3D VR environment compared to the 2D desktop environment [13].

Similarly, researchers in [14] investigated the effects of 3D VR and 2D desktop environments on visuospatial processing and subsequently found out that 3D VR environment facilitated better encoding of visuospatial memory compared to 2D desktop environment [14]. Results indicated that participants in 3D VR environment were more accurate in placing the physical replicas of the objects in a real room (which had the same physical configuration as the virtual room) compared to the 2D desktop environment [14].

Some researchers have investigated the effectiveness of manned and unmanned interfaces in 2D desktop and 3D VR environments in an underwater dynamic decision-making task [15]. Results again revealed that subjects performed better in the VR task compared to the desktop task with participants reporting higher mental demand, effort, and frustration levels in the VR task compared to the 2D desktop task [15].

Another factor that may play a role in influencing individual performance in 3D VR or 2D desktop environments is the order of task difficulty. Although, to the best of authors' knowledge, there is no research concerning the influence of order of task difficulty on performance across both 3D VR or 2D desktop environments, there have been studies in language [16], cognitive [17], and VR literatures [9]. For example, [16] explored the effects of contextual interference in foreign vocabulary acquisition, retention, and transfer. Results indicated that a difficult French-English translation in training led to less forgetting and better learning across time compared to an easier French-English translation [16].

Similarly, a cognitive research study [17] found that difficult but successful retrievals were better for encoding of memory instances than easier successful retrievals. To substantiate this 'retrieval effort hypothesis' [17], these researchers setup conditions under which retrieval during practice was successful but difficult differentially. Results indicated that as the difficulty of retrieval during practice increased, final test performance increased [17].

A recent study has investigated the effects order of task difficulty; however, in only a dynamic decision-making VR context [9]. Out of 30 healthy subjects in the experiment, half of them executed an easier scenario first and the other half executed a difficult scenario first [9]. Results showed the participants who faced the difficult scenario first fared better than the participants who executed the easy scenario first. Though this study evaluated the effects of task difficulty in a VR environment, it did not shed light on comparison of performance in the VR environment with a 2D desktop environment.

As per IBLT [10] and the desirable difficulty framework [11], when the difficulty level of the task is significantly high, individuals would be able to amass superior experiences during training. This accumulation of superior experiences would subsequently allow the individual to obtain a finer mental representation of the task at hand, which would eventually lead to enhanced decision-making. In addition, the experiences gained in the 3D VR environment are expected to be more salient compared to those in the 2D desktop environment. Thus, the 3D VR environment is likely to yield superior performance compared to the 2D desktop environment.

# 3 Materials and Methods

## 3.1 Participants

A total of 60 participants (32 males and 28 females; mean age = 23.56 years, SD = 3.12 years) at the Indian Institute of Technology Mandi, India, and at the Department of Biomedical Engineering, Institute of Nuclear Medicine and Allied Sciences, Defence Research and Development Organization, India, participated in this experiment. The study was initially approved by an ethics committee at the Indian Institute of Technology Mandi and the Institute of Nuclear Medicine and Allied Sciences. Participation in the study was entirely voluntary. All the participants gave a written consent before they executed the tasks in the experiment. All the participants hailed from a Science/Technology/Engineering/Mathematics (STEM) background. All the participants received a flat payment of INR 100 irrespective of their performance in the task.

## 3.2 The Decision-Making Simulations

A terrain-based virtual search-and-shoot environment was designed for 2D desktop and 3D VR using Unity3D version 5.5 [18]. The 3D avatars of the enemies in the environment were designed using Blender Animation version 2.79a [19]. As shown in Fig. 1(a), the environment consisted of three headquarters located at different sites. The environment narrative was that the enemies had sieged and acquired these three headquarters and the central objective of the participant was to kill all the enemies and reacquire all the headquarters within the prescribed 10-min time limit. The participant's health in the environment was initialized to 100 and this health would decrease based on the difficulty level of the scenario. The total number of enemies in the environment were 15. Two different levels of task difficulty (novice and expert) were introduced in both the 2D desktop and 3D VR environments. The difficulty levels were introduced by making several physical changes and movement/behavior-based changes on the enemy's artificial intelligence. A detailed account of the different levels of difficulty introduced in the environment is as given in Sect. 3.4. As shown in Fig. 1(a), the 2D desktop environment was executed in a desktop computer at a resolution of 1920 × 1080 pixels and a field-of-view of 110°. As shown in Fig. 1(b), the 3D VR environment was executed using an android mobile phone and a Samsung GearVR [15] head mounted display (HMD) at a field-of-view of 110°. Both the fields-of-view in 2D desktop and 3D VR environment were consistent with the horizontal field-of-view of an healthy, adult human being [9]. The participants used a mouse and a keyboard to navigate and shoot in the 2D desktop environment, and they used a DOMO Magickey Bluetooth controller [20] to navigate and shoot in the 3D VR environment. The experiment was conducted in an isolated location, devoid of any external noise.

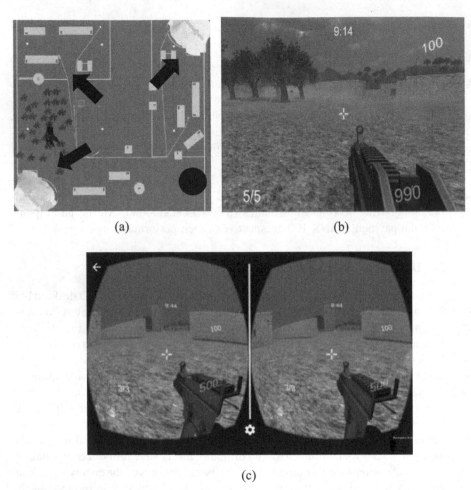

(a)                                    (b)

(c)

**Fig. 1.** (a) Overhead map of the terrain-based search-and-shoot environment designed in Unity 3D. The arrows indicate the locations of the headquarters and the circle indicates the initiation point of each participant in the environment. (b) The 2D desktop environment. (c) The android-based VR environment.

### 3.3  Experiment Design

In a lab-based experiment, a total of 60 healthy subjects executed both the novice and expert scenarios (within-subjects) in two between-subject environments, 2D desktop (N = 30) and 3D VR (N = 30), as shown in Fig. 2. Each environment provided two training scenarios of varying difficulty levels in the enemy's physical and behavioral/movement-based changes through the use of state machines and probabilistic networks. Within each environment, 15 participants executed the novice scenario first and the expert training second, and 15 participants executed the expert training first and the novice training second (see Fig. 2). Before the experiment began, each participant was thoroughly debriefed about the objectives to be achieved in the experiment. In addition, all the participants undertook a 5-min acclimatization session to get accustomed to the level of immersion and get familiar with the controls to be used in the task. After the execution of each task, several behavioral and cognitive variables were recorded. The behavioral variables recorded included the percentage of enemies killed (out of 15), the total amount of time taken to execute the simulation (in seconds), and the accuracy index (calculated by dividing the number of bullets taken by the participant to kill the same enemy in that specific scenario by the number of bullets needed to kill an enemy in a specific scenario). This index was multiplied by 100 to scale it between 0 and 100. In addition, we also acquired a computerized version of the NASA-TLX [21], which had six subscales (namely mental demand, physical demand, temporal demand, performance satisfaction, frustration level, and effort) to be indicated on a 10-point Likert scale by the participant. Overall, on account of IBLT [10], we expected participants to perform better in the 3D VR environment compared to the 2D desktop environment. Also, on account of [11], we also expected the participant's performance to improve due to training in varying order of difficulty. We carried out one-way ANOVAs to evaluate the main effect of the types of virtual environments (2D desktop and 3D VR) and the order of training (Novice First and Novice Second or Expert First and Expert Second) on the performance and cognitive measures. In addition, we also carried out mixed ANOVAs to evaluate the interaction effects of the variation in virtual environments and the order of training on the performance and cognitive measures.

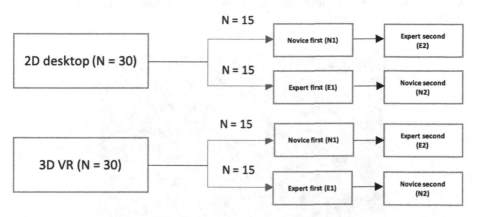

**Fig. 2.** The experimental design

## 3.4 The Variation in Task Difficulty

The variation in the physical attributes of the environment and the enemy with respect to the task difficulty is shown in Table 1. As shown in Table 1, the ammunition available in the novice scenario was 1000; in the expert scenario, it was kept to 500. On a similar note, the delay between successive shots by the enemy avatar was kept to 30 frames in the novice scenario compared to 15 frames in the expert scenario. The rate of decrease in the health of the enemy per shot was kept to 10 in the novice scenario and 8 in the expert scenario. Similarly, the rate of decrease of health of the player per shot was kept to 1 in the novice scenario compared to 2 in the expert scenario.

**Table 1.** The variations introduced in the physical attributes of the environment with respect to the task difficulties

| Attribute | Novice environment | Expert environment |
|---|---|---|
| Ammunition available to the participant | 1000 | 500 |
| Delay between successive bullets by the enemy avatar | 30 frames | 15 frames |
| Rate of decrease in health of the enemy per shot | 10 | 8 |
| Rate of decrease in health of the participant per shot | 1 | 2 |

As shown in Fig. 3, the whole physical mesh was divided into three different sections, namely, covered areas (in light blue), partially open areas (in purple), and open areas (in green). As shown in Fig. 3, the areas covered in light blue represent areas with good cover and hiding spots for the enemies/player. The purple areas represented the areas with medium cover/hiding spots and the green areas represented the fully open areas with no hiding spots/covers.

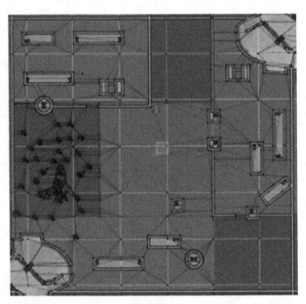

**Fig. 3.** Image illustrating different areas in the physical mesh. Different colors represent different levels of cover/hiding spots. (Color figure online)

We also used finite state machines to determine the actions associated with the behavior and movement of the enemy avatars in the environment. As mentioned, the total number of enemies in the environment were 15. The enemies were divided into two factions: Three assault group of enemies and 3 stealth enemies. Each assault group consisted of 4 enemies where one was the leader and the other three were followers. The leader would decide the behavior/movement to be executed and the followers would follow suit. The assault group of enemies were indifferent to the cost of the areas and were programmed to move to four possible destinations – randomly, towards the participant, towards a headquarter or stay at the same location. The probabilities of the assault group moving to all these destinations were varied according to the task difficulty (see Table 2). The stealth group, consisting of 3 enemies, acted alone. The stealth group were programmed in such a way to avoid confrontation with the participant and always walked in areas with minimum cost. These enemies also had probabilities attached to their movement, but the probability of moving towards the participant was excluded.

**Table 2.** The variation in the probabilities of movement with respect to the task difficulties

| Movement | Novice environment | Expert environment |
|---|---|---|
| Towards participant | 25 | 75 |
| Towards headquarters | 25 | 75 |
| Random | 75 | 25 |
| Static | 75 | 25 |
| Towards headquarters (stealth) | 30 | 70 |
| Random | 70 | 30 |

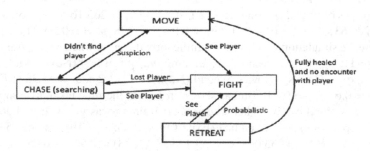

**Fig. 4.** The general state of the enemy avatar

As shown in Fig. 4, the behavioral layer of the environment design (i.e., the layer responsible for determining the enemy AI's current state and the subsequent goals) consisted of different states for the environment, namely move, chase, retreat and fight. The enemy avatar would initially begin moving from its initial location, and it would efficiently look for the participant's last known location (in the chase/search state). Once the enemy avatar would spot the participant (i.e., when the enemy avatar would locate the participant's avatar in its field-of-view), it would begin to engage in battle with the participant. Once the enemy avatar's health decreased to $\leq 50$ units, it would calculate the distance to the nearest area with the lowest cost and start running in its direction. The switching between all these states were probabilistic.

## 4 Results

### 4.1 Performance Measures

**Percentage of Enemies Killed.** Figure 5(a) shows the percentage of enemies killed across different kinds of virtual environments. As shown in the Figure, the percentage of enemies killed was significantly higher in the 3D VR condition compared to the 2D desktop condition (2D desktop: $\mu = 49.26\%$, $\sigma = 11.16\% < $ 3D VR: $\mu = 54.96\%$, $\sigma = 14.57\%$; $F(1, 59) = 4.98$, $p < 0.05$, $\eta_p^2 = 0.012$). Also, as shown in Fig. 5(b), the percentage of enemies killed was significantly higher when novice training was given second compared to when novice training was given first (Novice First (N1): $\mu = 42.26\%$, $\sigma = 11.02\% < $ Novice Second (N2): $\mu = 65.26\%$, $\sigma = 16.9\%$; $F(1, 58) = 38.4$, $p < 0.05$, $\eta_p^2 = 0.61$). As shown in Fig. 5(c), the percentage of enemies killed was significantly higher when expert training was given second compared to when expert training was given first (Expert First (E1): $\mu = 33.86\%$, $\sigma = 11.8\% < $ Expert Second (E2): $\mu = 45.3\%$, $\sigma = 12.65\%$; $F(1, 58) = 13.1$, $p < 0.05$, $\eta_p^2 = 0.56$). Furthermore, the interaction[1] between the type of virtual environments and the order of training significantly influenced the percentage of enemies killed ($F(1, 56) = 6.23$, $p < 0.05$, $\eta_p^2 = 0.1$). Overall, participants killed more enemies in the 3D VR environment compared to the 2D desktop environment when the order of training imparted was expert first and novice second.

**Time Taken.** The time taken to complete the simulation was not significantly different across the types of virtual environments (2-D desktop: $\mu = 265.16$ s, $\sigma = 65.24$ s ~ 3-D VR: $\mu = 283.63$ s, $\sigma = 64.19$ s; $F(1, 59) = 1.19$, $p = 0.28$, $\eta_p^2 = 0.023$). The time taken to complete the simulation was also not significantly different when novice training was given second compared to when novice training was given first (Novice First (N1): $\mu = 248.26$ s, $\sigma = 71.67$ s ~ Novice Second (N2): $\mu = 241.2$ s, $\sigma = 34.38$ s; $F(1, 58) = 0.24$, $p = 0.63$, $\eta_p^2 = 0.03$). But, as shown in Fig. 6, the time taken to complete the simulation was significantly higher when expert training was given second compared to when expert training was given first (Expert First (E1): $\mu = 191.4$ s, $\sigma = 53.06$ s < Expert Second (E2): $\mu = 233.06$ s, $\sigma = 55.13$ s; $F(1, 58) = 8.9$, $p < 0.05$, $\eta_p^2 = 0.11$). The interaction between the type of virtual environments and the order of training did not significantly influence the time taken to complete the simulation ($F(1, 56) = 0.28$, $p = 0.6$, $\eta_p^2 = 0.05$).

---

[1] For the analysis of the interaction effect, the performance/cognitive measures were averaged across both variations of order in the task difficulty, i.e., [Novice First (N1) + Expert Second (E2)]/2 and [Expert First (E1) → Novice Second (N2)]/2.

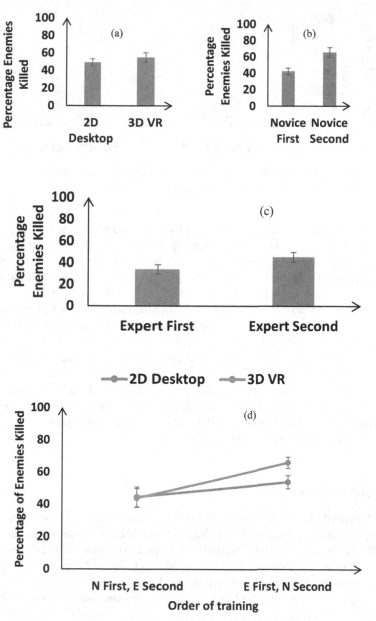

**Fig. 5.** Means and 95% confidence intervals of the percentage of enemies killed (a) across different types of virtual environments (b) across different novice training conditions (c) across different expert training conditions (d) across different types of virtual environments and task difficulties with training order. The error bars show 95% CI across point estimates.

**Percentage Accuracy Index.** Figure 7(a) shows the percentage accuracy index across different virtual environments. As shown in Figure, the percentage accuracy index was significantly higher in the 3D VR environment compared to the 2D desktop environment

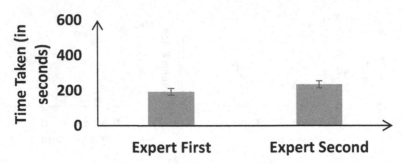

**Fig. 6.** Means and 95% confidence intervals of the time taken across different expert training conditions. The error bars show 95% CI across point estimates.

(2D desktop: $\mu = 55.03\%, \sigma = 12.94\% < $ 3D VR: $\mu = 49.66\%, \sigma = 8.67\%; F(1, 59) = 5.75, p < 0.05, \eta_p^2 = 0.4$). Also, as shown in Fig. 7(b), the percentage accuracy index was significantly higher when novice training was given second compared to when novice training was given first (Novice First (N1): $\mu = 46\%, \sigma = 13.59\% < $ Novice Second (N2): $\mu = 65.46\%, \sigma = 16.38\%; F(1, 58) = 25.06, p < 0.05, \eta_p^2 = 0.32$). As shown in Fig. 7(c), the percentage accuracy index was significantly higher when expert training was given second compared to when expert training was given first (Expert First (E1): $\mu = 34.83\%, \sigma = 7.28\% < $ Expert Second (E2): $\mu = 50.26\%, \sigma = 10.43\%; F(1, 58) = 44.1, p < 0.05, \eta_p^2 = 0.59$). Furthermore, the interaction between the type of virtual environments and the types of task difficulty significantly influenced the percentage accuracy index ($F(1, 56) = 7.6, p < 0.05, \eta_p^2 = 0.12$). Overall, as per our expectations, participants recorded a higher accuracy index in the 3D VR environment compared to the 2D desktop environment and when the order of training imparted was expert training first, novice training second (see Fig. 7(d)).

### 4.2 Cognitive Measures

**Mental Demand.** Figure 8(a) shows the self-reported mental demand across different virtual environments. As shown in Fig. 7(a), the mental demand was significantly higher in the 3D VR environment compared to the 2D desktop environment (2-D desktop: $\mu = 5.06, \sigma = 2.33 < $ 3-D VR: $\mu = 6.43, \sigma = 1.56; F(1, 59) = 7.1, p < 0.05, \eta_p^2 = 0.12$). Also, as shown in Fig. 8(b), the mental demand was significantly higher when novice training was given first compared to when novice training was given second (Novice First (N1): $\mu = 5.46, \sigma = 2.5 > $ Novice Second (N2): $\mu = 4.06, \sigma = 1.25; F(1, 58) = 7.5, p < 0.05, \eta_p^2 = 0.51$). Furthermore, as shown in Fig. 8(c), the mental demand was significantly higher when expert training was given first compared to when expert training was given second (Expert First (E1): $\mu = 6.73, \sigma = 2.03 > $ Expert Second (E2): $\mu = 5.13, \sigma = 1.38; F(1, 58) = 12.7, p < 0.05, \eta_p^2 = 0.23$). The interaction between the type of virtual environments and the order of training did not significantly influence the self-reported mental demand ($F(1, 56) = 1.22, p = 0.27, \eta_p^2 = 0.02$).

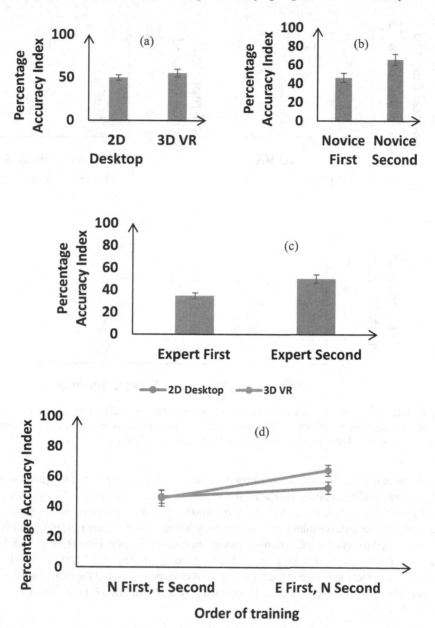

**Fig. 7.** Means and 95% confidence intervals of the percentage accuracy index (a) across different virtual environments (b) across different novice training conditions (c) across different expert training conditions (d) across different virtual environments and task difficulties with training order. The error bars show 95% CI across point estimates.

**Physical Demand.** The physical demand was not significantly different across the types of virtual environments (2D desktop: $\mu = 5.4$, $\sigma = 2.23 \sim$ 3D VR: $\mu = 5.73$, $\sigma = 1.65$; $F(1, 59) = 0.42$, $p = 0.52$, $\eta_p^2 = 0.04$). As shown in Fig. 9(a), the physical demand

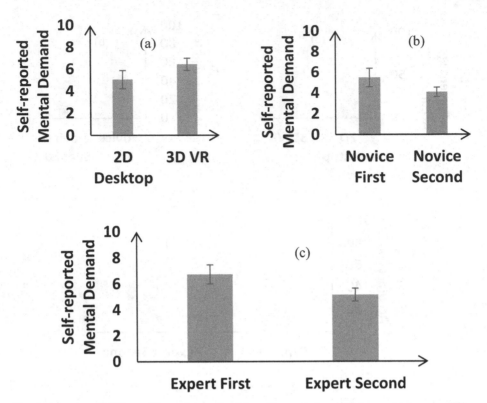

**Fig. 8.** Means and 95% confidence intervals of the self-reported mental demand (a) across different virtual environments (b) across different novice training conditions (c) across different expert training conditions. The error bars show 95% CI across point estimates

was significantly higher when novice training was given first compared to when novice training was given second (Novice First (N1): $\mu = 5.96$, $\sigma = 2.09 >$ Novice Second (N2): $\mu = 4.8$, $\sigma = 1.9$; $F(1, 58) = 5.11$, $p < 0.05$, $\eta_p^2 = 0.3$). Furthermore, as shown in Fig. 9(b), the physical demand was significantly higher when expert training was given first compared to when expert training was given second (Expert First (E1): $\mu = 5.3$, $\sigma = 2.08 >$ Expert Second (E2): $\mu = 6.83$, $\sigma = 1.96$; $F(1, 58) = 8.19$, $p < 0.05$, $\eta_p^2 = 0.18$). The interaction between the type of virtual environments and the order of training did not significantly influence the self-reported physical demand ($F(1, 56) = 0.15$, $p = 0.7$, $\eta_p^2 = 0.052$).

**Temporal Demand.** The self-reported temporal demand was significantly higher in the 2D desktop condition compared to the 3D VR condition (2D desktop: $\mu = 6.1$, $\sigma = 2.33 >$ 3D VR: $\mu = 4.9$, $\sigma = 1.82$; $F(1, 59) = 4.64$, $p < 0.05$, $\eta_p^2 = 0.59$) as shown in Fig. 10. The self-reported temporal demand was not significantly different when novice training was given first compared to when novice training was given second (Novice First (N1): $\mu = 5.87$, $\sigma = 2.15 \sim$ Novice Second (N2): $\mu = 6$, $\sigma = 1.28$; $F(1, 58) = 0.74$, $p = 0.85$, $\eta_p^2 = 0.008$). The temporal demand was also not significantly different when expert training was given first compared to when expert training was given second (Expert First

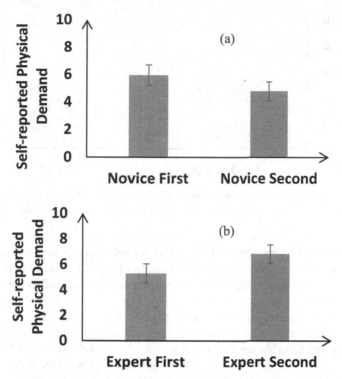

**Fig. 9.** Means and 95% confidence intervals of the self-reported physical demand (a) across different novice training conditions (b) across different expert training conditions. The error bars show 95% CI across point estimates

(E1): $\mu = 5.9$, $\sigma = 2.26 \sim$ Expert Second (E2): $\mu = 6.23$, $\sigma = 1.86$; $F\,(1, 58) = 0.39$, $p = 0.54$, $\eta_p^2 = 0.001$). The interaction between the type of virtual environments and the order of training did not significantly influence the self-reported temporal demand ($F\,(1, 56) = 1.03$, $p = 0.31$, $\eta_p^2 = 0.048$).

**Frustration Level.** The self-reported frustration level was not significantly different in the 2D desktop condition compared to the 3D VR condition (2D desktop: $\mu = 5.93$, $\sigma = 2.34 \sim$ 3D VR: $\mu = 5.66$, $\sigma = 1.66$; $F\,(1, 59) = 0.253$, $p = 0.61$, $\eta_p^2 = 0.004$). The self-reported frustration level was not significantly different when novice training was given first compared to when novice training was given second (Novice First (N1): $\mu = 5.73$, $\sigma = 2.24 \sim$ Novice Second (N2): $\mu = 5.9$, $\sigma = 1.44$; $F\,(1, 58) = 0.18$, $p = 0.73$, $\eta_p^2 = 0.04$). The frustration level was also not significantly different when expert training was given first compared to when expert training was given second (Expert First (E1): $\mu = 5.1$, $\sigma = 2.36 \sim$ Expert Second (E2): $\mu = 5.66$, $\sigma = 2.32$; $F\,(1, 58) = 0.88$, $p = 0.35$, $\eta_p^2 = 0.003$). The interaction between the type of virtual environment and the order of training did not significantly influence the self-reported frustration level ($F\,(1, 56) = 0.063$, $p = 0.81$, $\eta_p^2 = 0.001$).

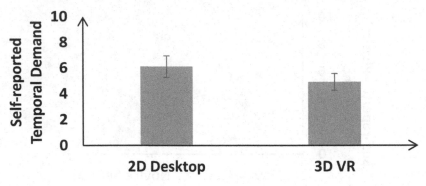

**Fig. 10.** Means and 95% confidence intervals of the self-reported temporal demand across different types of virtual environments. The error bars show 95% CI across point estimates.

**Performance Satisfaction.** Figure 11(a) shows the self-reported performance satisfaction across different virtual environments. As seen in Fig. 11(a), the performance satisfaction was significantly higher in the 3D VR condition compared to the 2D desktop condition (2D desktop: $\mu = 4.9, \sigma = 1.88 < 3$D VR: $\mu = 6.1, \sigma = 1.49; F(1, 59) = 7.58$, $p < 0.05, \eta_p^2 = 0.42$). Also, as shown in Fig. 11(b), the performance satisfaction was significantly higher when novice training was given second compared to when novice training was given first (Novice First (N1): $\mu = 5.33, \sigma = 1.98 <$ Novice Second (N2): $\mu = 6.93, \sigma = 1.5; F(1, 58) = 12.37, p < 0.05, \eta_p^2 = 0.18$). The performance satisfaction was not significantly different when expert training was given first compared to when expert training was given second (Expert First (E1): $\mu = 5.06, \sigma = 2.24 \sim$ Expert Second (E2): $\mu = 5.96, \sigma = 1.95; F(1, 58) = 2.74, p = 0.103, \eta_p^2 = 0.001$). The interaction between the type of virtual environment and the order of training did not significantly influence the self-reported performance satisfaction ($F(1, 56) = 1.89, p = 0.17, \eta_p^2 = 0.03$).

**Effort.** The self-reported effort was not significantly different in the 2D desktop condition compared to the 3D VR condition (2D desktop: $\mu = 5.8, \sigma = 2.09 \sim 3$-D VR: $\mu = 5.73, \sigma = 1.94; F(1, 59) = 0.016, p = 0.89, \eta_p^2 = 0.01$). The self-reported effort was not significantly different when novice training was given first compared to when novice training was given second (Novice First (N1): $\mu = 5.7, \sigma = 2.23 \sim$ Novice Second (N2): $\mu = 5.96, \sigma = 1.58; F(1, 58) = 0.288, p = 0.6, \eta_p^2 = 0.003$). The effort was also not significantly different when expert training was given first compared to when expert training was given second (Expert First (E1): $\mu = 5.4, \sigma = 2.45 \sim$ Expert Second (E2): $\mu = 5.83, \sigma = 2.21; F(1, 58) = 0.51, p = 0.47, \eta_p^2 = 0.08$). The interaction between the type of virtual environments and the order of training did not significantly influence the self-reported effort ($F(1, 56) = 0.79, p = 0.38, \eta_p^2 = 0.014$).

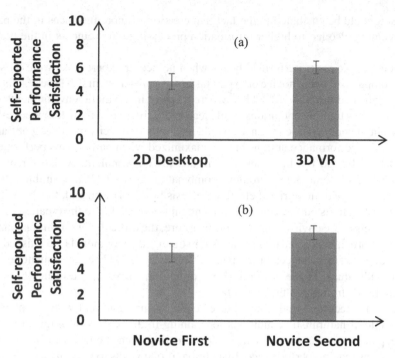

**Fig. 11.** Means and 95% confidence intervals of the self-reported performance satisfaction (a) across different types of virtual environments (b) across different novice training conditions. The error bars show 95% CI across point estimates.

## 5 Discussion and Conclusions

In the current study, we evaluated the performance and cognitive implications of training in 2D desktop and 3D VR environments under varying order of task difficulty in dynamic search-and-shoot scenarios. Results suggested that the participants performed better in the 3D VR environment compared to the 2D desktop environment. Participants performed better when novice or expert tasks were presented second compared to when novice or expert tasks were presented first. Also, results indicated that participants performed the best in the VR environment when the novice training was followed by the expert training. These results could be explained on the basis of Instance-based Learning Theory (IBLT; [10]) and the desirable difficulty framework [11].

First, participants performed better in the 3D VR environment compared to the 2D desktop environment. These results are consistent with the broader literature [9, 12–15], where the immersive 3D VR environment has been shown to enhance performance compared to the 2D desktop environment. In fact, the VR environment offers a more comprehensive knowledge of the 3D structures, textures, and the general aesthetics of the environment. These features may lead individuals to create a better mental model or instances of the scene and its constituents in VR environments compared to desktop environments [9]. However, results also indicated that mental demand requirement in the 3D VR environment was significantly higher compared to the 2D desktop environment.

This result could be attributed to the fact that creation of more instances in the participant's memory also led to higher information processing requirements in the brain [9, 10].

Second, participants performed better when novice or expert tasks were presented second compared to when novice or expert tasks were presented first. These results could also be explicated through IBLT [10], where training in synthetic environments under difficulty or easy training conditions would lead to the creation of more knowledge-based instances in memory. These instances would lead to better decision-making at transfer.

Third, the performance at transfer was maximized when participants performed the novice task followed by the expert task in the VR environment. A likely reason for this result could be elucidated through a combination of IBLT [10], desirable difficulty framework [11], and the retrieval effort hypothesis [17]. As discussed, the 3D VR environment, owing to its enhanced resolution and immersivity, led to the creation of better mental models of the environment [9]. Furthermore, the difficult conditions encountered first by the participants acted as a stimulus for strategy development [17]. This led to the creation of superior instances in the person's memory [10]. The subsequent decisions taken by individuals in the novice task were the derivatives of the difficult instances recorded during training in the expert task.

The results recorded in this novel research have important real-world implications. For example, if naturalistic conditions for training in the real world might turn out be either expensive or dangerous or both, then creating such scenarios in VR would be effective in enhancing performance. In addition, if one wants to maximize performance during training in VR, then it would be ideal to train people first in difficult tasks and then on easier tasks.

To the best of the author's knowledge, this study is the first of its kind to investigate the effects of 2D desktop and 3D VR environments under varying orders of task difficulty in a dynamic decision-making context. Even though the results derived from this study are believed to be ecologically valid, cross-validation of these results in the real-world in different domains (including military) and individuals with varying demographic profiles may be carried out as a part of future work. The conclusions obtained from this experiment may be utilized to design a framework for personnel-specific decision-making assessment and enhancement.

**Acknowledgments.** This research was supported by a grant from Defence Research and Development Organization (DRDO) titled "Development of a human performance modeling framework for visual cognitive enhancement in IVD, VR and AR paradigms" (IITM/DRDO-CARS/VD/110) to Prof. Varun Dutt.

# References

1. Jenkins, D.P., Stanton, N.A., Salmon, P.M., Walker, G.H.: A formative approach to developing synthetic environment fidelity requirements for decision-making training. Appl. Ergon. **42**(5), 757–769 (2011)
2. Gonzalez, C., Fakhari, P., Busemeyer, J.: Dynamic decision making: learning processes and new research directions. Hum. Factors **59**(5), 713–721 (2017)

3. Dreyfus, H., Dreyfus, S.E., Athanasiou, T.: Mind over machine. Simon and Schuster (2000)
4. Donovan, S.L., Triggs, T.: Investigating the effects of display design on unmanned under-water vehicle pilot performance (No. DSTO-TR-1931). Defense Science and Technology Organization, Maritime Platforms Division, Victoria, Australia (2006)
5. Donovan, S., Wharington, J., Gaylor, K., Henley, P.: Enhancing Situation Awareness for UUV Operators. ADFA, Canberra (2004)
6. McCarley, J.S., Wickens, C.D.: Human factors implications of UAVs in the national airspace (2005)
7. ter Haar, R.: Virtual reality in the military: present and future. In: 3rd Twente Student Conference IT (2005)
8. Adhikarla, V.K., Wozniak, P., Barsi, A., Singhal, D., Kovács, P.T., Balogh, T.: Freehand interaction with large-scale 3D map data. In: 2014 3DTV-Conference: The True Vision-Capture, Transmission and Display of 3D Video (3DTV-CON), pp. 1–4. IEEE, July 2014
9. Rao, Akash K., Chahal, J.S., Chandra, S., Dutt, V.: Virtual-Reality Training Under Varying Degrees of Task Difficulty in a Complex Search-and-Shoot Scenario. In: Tiwary, U.S., Chaudhury, S. (eds.) IHCI 2019. LNCS, vol. 11886, pp. 248–258. Springer, Cham (2020). https://doi.org/10.1007/978-3-030-44689-5_22
10. Gonzalez, C., Dutt, V.: Instance-based learning: Integrating sampling and repeated decisions from experience. Psychol. Rev. **118**(4), 523 (2011)
11. Bjork, R.A.: Memory and metamemory considerations in the, p. 185. Knowing about knowing, Metacognition (1994)
12. Srivastava, P., Rimzhim, A., Vijay, P., Singh, S., Chandra, S.: Desktop VR is better than nonambulatory HMD VR for spatial learning. Front. Robot. AI **6**, 50 (2019)
13. Parmar, D., et al.: A comparative evaluation of viewing metaphors on psychophysical skills education in an interactive virtual environment. Virt. Real. **20**(3), 141–157 (2016). https://doi.org/10.1007/s10055-016-0287-7
14. Murcia-López, M., Steed, A.: The effect of environmental features, self-avatar, and immersion on object location memory in virtual environments. Front. ICT **3**, 24 (2016)
15. Rao, A.K., Pramod, B.S., Chandra, S., Dutt, V.: Influence of indirect vision and virtual reality training under varying manned/unmanned interfaces in a complex search-and-shoot simulation. In: Cassenti, D.N. (ed.) AHFE 2018. AISC, vol. 780, pp. 225–235. Springer, Cham (2019). https://doi.org/10.1007/978-3-319-94223-0_21
16. Schneider, V.I., Healy, A.F., Bourne Jr., L.E.: What is learned under difficult conditions is hard to forget: contextual interference effects in foreign vocabulary acquisition, retention, and transfer. J. Mem. Lang. **46**(2), 419–440 (2002)
17. Pyc, M.A., Rawson, K.A.: Testing the retrieval effort hypothesis: Does greater difficulty correctly recalling information lead to higher levels of memory? J. Mem. Lang. **60**(4), 437–447 (2009)
18. For Building Fun: Groovy Little Games Quickly. Packt Publishing Ltd., Birmingham (2010)
19. Roosendaal, T., Selleri, S. (eds.): The Official Blender 2.3 Guide: Free 3D Creation Suite for Modeling, Animation, and Rendering, vol. 3. No Starch Press, San Francisco (2004)
20. Rao, A., Satyarthi, C., Dhankar, U., Chandra, S., Dutt, V.: Indirect visual displays: Influence of field-of-views and target-distractor base-rates on decision-making in a search-and-shoot task. In: 2018 IEEE International Conference on Systems, Man, and Cybernetics (SMC), pp. 4326–4332. IEEE, October 2018
21. Hart, S.G., Staveland, L.E.: Development of NASA-TLX (Task Load Index): results of empirical and theoretical research. In: Advances in Psychology, vol. 52, pp. 139–183. North-Holland, Amsterdam (1988)

# Computer-Based PTSD Assessment in VR Exposure Therapy

Leili Tavabi, Anna Poon, Albert Skip Rizzo, and Mohammad Soleymani(✉) iD

Institute for Creative Technologies, University of Southern California,
12015 Waterfront Drive, Los Angeles, CA 90094, USA
{ltavabi,soleymani}@ict.usc.edu
http://ict.usc.edu

**Abstract.** Post-traumatic stress disorder (PTSD) is a mental health condition affecting people who experienced a traumatic event. In addition to the clinical diagnostic criteria for PTSD, behavioral changes in voice, language, facial expression and head movement may occur. In this paper, we demonstrate how a machine learning model trained on a general population with self-reported PTSD scores can be used to provide behavioral metrics that could enhance the accuracy of the clinical diagnosis with patients. Both datasets were collected from a clinical interview conducted by a virtual agent (SimSensei) [10]. The clinical data was recorded from PTSD patients, who were victims of sexual assault, undergoing a VR exposure therapy. A recurrent neural network was trained on verbal, visual and vocal features to recognize PTSD, according to self-reported PCL-C scores [4]. We then performed decision fusion to fuse three modalities to recognize PTSD in patients with a clinical diagnosis, achieving an F1-score of 0.85. Our analysis demonstrates that machine-based PTSD assessment with self-reported PTSD scores can generalize across different groups and be deployed to assist diagnosis of PTSD.

**Keywords:** Post-traumatic stress disorder · Virtual reality · Exposure therapy · Human behavior · Multimodal machine learning

## 1 Introduction

Post-traumatic stress disorder (PTSD) is a psychiatric disorder that can occur in people who have experienced or witnessed a traumatic event such as a natural disaster, a serious accident, a terrorist act, war/combat, rape or other violent personal assault [19]. Despite the growing number of people experiencing mental health disorders including PTSD, there is still a large gap between available clinical resources and the need for accurate assessment, detection, and treatment. This problem highlights the importance for devising automated methods for assessing mental health disorders in order to augment clinical resources. As a result, there is growing interest in using automatic human behavior analysis for

C. Stephanidis et al. (Eds.): HCII 2020, LNCS 12428, pp. 440–449, 2020.
https://doi.org/10.1007/978-3-030-59990-4_32

computer-aided mental health diagnosis. Such computer-based diagnosis methods are based on the detection, quantification, and analysis of behavioral cues such as verbal content, facial and body expressions and speech prosody [27].

In this work, we trained a machine learning model for PTSD assessment by leveraging data collected during semi-structured interviews between a Virtual Human (VH) agent and subjects from the general population. We then evaluate the model on the clinical data collected from PTSD patients as a result of sexual trauma. We demonstrate how the weak labels that are based on self-reported questionnaire assessments can be used to train a classifier to inform the diagnosis of PTSD due to sexual trauma using behavioral data from patients. During the test phase, we evaluate our model's performance on real patients with expert clinical diagnosis. To this end, we use verbal and nonverbal data obtained from human-VH agent interactions during semi-structured interviews with probing cues relevant to mental health disorders. The semi-structured interviews consist of either a wizard-controlled or AI-controlled agent, asking a set of predefined questions along with follow-up questions and continuation prompts from the VH depending on the user's responses. We use verbal, speech and visual data from the user during the interactions for performing machine learning with the goal of PTSD detection.

The remainder of this paper is organized as follows. The related work on computational analysis of PTSD is given in Sect. 2. The datasets are described in Sect. 3. The feature extraction and machine learning methods are given in Sect. 4. Experimental results are presented and discussed in Sect. 5. Finally, the work is concluded in Sect. 6.

## 2   Related Work

Verbal and nonverbal communicative behaviors have been shown to be effective indicators for assessing mental health disorders [7,13,25]. Previous studies indicate that mental health disorders can result in consistent changes in patterns of behavior, including reduced vocal pitch, slower speaking rate or less intense smiles [9,23].

Compared to automated depression diagnosis, for example [1,16,18], there have been fewer studies on automatic detection of PTSD. Recent research with this purpose has leveraged the data recorded during interaction with the VH agent interviewer that is part of the SimSensei Kiosk [21]. Stratou et al. [27], analyzed the association between behavioral responses related to affect, expression variability, motor variability and PTSD. They found that correlations were gender dependent. For example, expression of disgust was higher in men with PTSD while the opposite was true for women. These findings motivated additional attention investigating gender-based PTSD recognition, which resulted in higher accuracy compared to the gender-independent models. Scherer et al. [22] reported that subjects with PTSD and depression had a significantly lower vowel space in their speech. In a previous study Scherer et al. [24], investigating voice quality features, identified that PTSD patients spoke in a more tense voice.

More recently, Marmar *et al.* [17] studied audio features for automatic recognition of PTSD using a dataset of patients' voices, recorded during psychotherapy sessions, with PTSD patients without comorbidity. After analyzing audio descriptors they found that PTSD patients spoke more slowly in a more affectively flat, monotonous tone with less activation.

## 3   Data

We used two datasets consisting of clinical interviews from research participants and a VH agent interviewer. The first dataset is the Extended Distress Analysis Interview Corpus (E-DAIC), a subset of the larger DAIC database [12], which contains data captured during clinical interviews with the SimSensei Kiosk VH agent, referred to as "Ellie". The second dataset also includes similar interviews with Ellie, recorded from patients who participated in a virtual reality (VR) exposure therapy program to address PTSD due to military sexual trauma (MST) [15].

### 3.1   Distress Analysis Interview Corpus

The Distress Analysis Interview Corpus is an audiovisual dataset of human-agent interactions captured during a VH (Ellie) clinical interview [12]. The interviews were designed to investigate the occurrence of user's behavioral signals that were hypothesized to be related to psychological distress conditions such as anxiety, depression, and PTSD. DAIC was recorded using SimSensei Kiosk [21], an interactive system with multimodal behavior recognition and generation components, including dialogue management, body tracking, facial expression analysis, agent visualization, and behavior realization. In this paper, we use a portion of DAIC dataset, *i.e.*, E-DAIC, that was used in the Audio/Visual Emotion Challenge and Workshop (AVEC 2019), depression detection sub-challenge [20]. The Extended Distress Analysis Interview Corpus (E-DAIC) [10] is an extended version of a dataset developed in former AVEC challenges, called DAIC-WoZ [12].

For the purpose of the challenge, the E-DAIC dataset was partitioned into training, development, and test sets while preserving the overall speaker diversity – in terms of age, gender distribution, and the eight-item Patient Health Questionnaire (PHQ-8) scores – within the partitions. Whereas the training and development sets include a mix of WoZ and AI scenarios, the test set is solely constituted from the data collected by the autonomous AI. E-DAIC data also includes PCL-C (PTSD Checklist-Civilian version) self-reported questionnaire scores that are indicative of PTSD [4]. The E-DAIC dataset includes recordings from 275 subject, from the general population (Fig. 1).

**Fig. 1.** A participant and the virtual agent, Ellie, during a clinical interview.

## 3.2 PTSD Patients

The data used for evaluation consists of verbal and nonverbal data collected from PTSD patients participating in a clinical trial investigating the safety and efficacy of VR exposure therapy for PTSD due to MST [14]. The patients are all military veterans who were victims of sexual assault. Patients went through the same interview as in DCAPS with the same virtual agent (Fig. 2).

For more information regarding the VR exposure therapy, experimental protocol and details of clinical intervention, we refer the reader to [14].

## 4　Method

We analyzed three modalities for automatic behavioral assessment of PTSD, namely, language (verbal content), vision (face and head behavior) and voice (speech prosody). A machine learning model was trained to recognize PTSD from each modality and their results were fused at decision-level to perform multimodal classification.

**Textual Features.** The textual data is transcribed from the recorded audio and using the Google Cloud's speech recognition service. Google's pretrained Universal Sentence Encoder [5] is then used to extract sentence embeddings (feature vectors) of size 512. We input every speech turn from the participant to the encoder and as a result, a sequence of $N \times 512$ vectors was generated for each participant, where $N$ is the number of subject's speech turns.

**Fig. 2.** Snapshots of content in the Bravemind Military Sexual Trauma (MST) exposure therapy.

**Audio Features.** We use VGG-16 DEEP SPECTRUM[1] features as our audio feature representation [2]. These features are inspired by deep representation learning paradigms common in image processing: spectrogram images of speech frames are fed into pretrained CNNs and the resulting activations are extracted as feature vectors.

To extract deep spectrum features, the audio signals were transformed to mel-spectrogram images with 128 mel-frequency bands with a temporal window of 4 s and hop size of 1 s.

This feature set is extracted from a VGG-16 pretrained CNNs [26] to obtain a sequential representation of vector size 4096. As a result, for each participant, we obtain a sequence of $T \times 4096d$ vectors as speech features, where $T$ is the duration in seconds.

---

[1] https://github.com/DeepSpectrum/DeepSpectrum.

**Visual Features.** For the visual features, we extract action units and head pose angles per frame using OpenFace [3], a computer-vision based open-source software for head pose tracking and facial action unit detection. OpenFace is used to extract the intensity of 17 facial action units (FAU), based on the Facial Action Coding System (FACS) [11] along with 3d head pose angles per frame, therefore providing a $25 \times T \times 20$ representation, where $T$ is the duration in seconds and 25 is the video frame rate.

### 4.1   Model Architecture

Our multimodal learning model relies on fusing the three mentioned modalities, namely, vision (facial expressions and head pose), speech prosody and verbal content for PTSD recognition. We use deep encoders to map voice and verbal modalities to embeddings. Visual features were extracted per-frame by Open-Face [3]. Sequences of features for all three modalities were fed to a one-layer recurrent neural network followed by a linear layer for classification. The results of unimodal classifiers are in turn fused for multimodal PTSD recognition.

Since all the embeddings retain the temporal dimension of the modalities, we use a single-layer Gated Recurrent Unit (GRU) that maps each representation to a fixed-size embedding, keeping only the last state. GRU is a variant of recurrent units able to capture long and short term dependencies in sequences [6]. The dimensionality of the hidden layer is adjusted based on the size of the representation for each modality. Therefore the input feature space for the text, audio and visual representations are mapped to 128-d, 256-d and 8-d embedding space respectively. The resulting embedding space is then mapped to a single output using a linear layer. We use a sigmoid activation function for the last layer and train the model with a binary cross entropy loss. To alleviate the class imbalance problem in E-DAIC (large number of healthy vs PTSD), we use a weighted loss whose weights for each class is inversely proportional to the number of samples in that class, in each training batch.

## 5   Experimental Results

The neural networks are implemented in PyTorch. The network training is optimized using Adam, with a batch size of 15 and a learning rate of $10^{-4}$. The models are trained for 80 epochs and the best performing model on the validation set is selected. The evaluation results are computed using the area under the curve (AUC) of the Receiver Operating Characteristics (ROC) curve. The training and validation sets are fixed and obtained from the E-DAIC dataset, consisting of 219 and 56 subjects respectively. The test set is from the VR exposure therapy dataset and consists of 14 sessions (from seven unique patients). The results of the binary classification from each modality are fused together using a majority-vote late fusion to obtain the multimodal fusion results.

PTSD recognition is evaluated using AUC of ROC curves and F1-score. F1-score is a harmonic mean of recall and precision scores. The recognition performances are given in Table 1. Unimodal results demonstrate that features extracted from nonverbal behavior, including head movement, facial action units and speech prosody perform slightly better than verbal features for this task. Moreover, multimodal fusion achieve superior results compared to unimodal results.

The recognition performance with a relatively simple machine learning pipeline is promising. The results demonstrate how the machine learning models trained on a limited experimental dataset, from the general population and with self-reported scores can generalize to a different population with clinically validated labels.

**Table 1.** Recognition performance for multimodal PTSD classification on clinical data. AUC stands for area under curve for ROC.

| Modality | Features | AUC | F1-score |
|----------|----------|-----|----------|
| Verbal | USE | 0.53 | 0.67 |
| Vocal | DS-VGG | 0.68 | 0.67 |
| Visual | AU+Pose | 0.54 | 0.76 |
| Multimodal fusion | - | 0.70 | 0.85 |

Even though the obtained results are fairly accurate, there is still room for improvement. Marmar *et al.* [17] argued that since a large number of PTSD patients are also diagnosed with other mental health disorders, such as depression, that can alter behavior, the training samples should not be from comorbid patients. We did not follow this approach, since we did not have a large enough sample to do so. The data recorded form patients were also of much lower quality, as both video and audio were captured with a camcorder, as opposed to high quality webcams and wearable microphones, in DAIC. We believe this might have also had a negative effect on the performance.

## 6   Conclusions

In this paper, we reported on our efforts for detecting behaviors relating to PTSD using verbal and audiovisual data captured during a clinical interview between a patient and a VH agent. Our analysis shows that our model trained on a diverse population can be applied to a population of actual patients for automatic clinical behavioral assessment to inform diagnosis. We trained and evaluated PTSD recognition models on verbal, audio and visual modalities. The results show that audio has the highest ROC score, outperforming verbal and visual modalities and the visual modality appears to have a higher recall on PTSD patients. Late fusion of the three modalities results in improved F1-score of the PTSD patients, as well as the AUC score.

In the future, we will perform domain adaptation [8] to reduce the effect of using different types of recording apparatus. Training gender-based models, such as the one proposed in [27], and training samples without comorbidity, as proposed in [17] will be also explored in our future work.

Automated mental health assessment methods have the potential to augment and improve mental health care. We hope the advancement of such technologies combined with novel treatment methods, such as VR exposure therapy, improves access and quality of care for mental health disorders.

**Acknowledgments.** The work of Poon was supported by the National Science Foundation under award 1852583, "REU Site: Research in Interactive Virtual Experiences" (PI: Ron Artstein). This work was supported in part by the U.S. Army. Any opinion, content or information presented does not necessarily reflect the position or the policy of the United States Government, and no official endorsement should be inferred.

# References

1. Alghowinem, S., Goecke, R., Wagner, M., Epps, J., Breakspear, M., Parker, G.: Detecting depression: a comparison between spontaneous and read speech. In: 2013 IEEE International Conference on Acoustics, Speech and Signal Processing, pp. 7547–7551. IEEE (2013)
2. Amiriparian, S., et al.: Snore sound classification using image-based deep spectrum features. In: Proceedings of INTERSPEECH 2017, 18th Annual Conference of the International Speech Communication Association, pp. 3512–3516. ISCA, Stockholm, Sweden, August 2017
3. Baltrusaitis, T., Zadeh, A., Lim, Y.C., Morency, L.P.: Openface 2.0: facial behavior analysis toolkit. In: 2018 13th IEEE International Conference on Automatic Face & Gesture Recognition (FG 2018), pp. 59–66. IEEE (2018)
4. Blanchard, E.B., Jones-Alexander, J., Buckley, T.C., Forneris, C.A.: Psychometric properties of the PTSD checklist (PCL). Behaviour Res. Therapy **34**(8), 669–673 (1996)
5. Cer, D., et al.: Universal sentence encoder. arXiv preprint arXiv:1803.11175 (2018)
6. Chung, J., Gulcehre, C., Cho, K., Bengio, Y.: Empirical evaluation of gated recurrent neural networks on sequence modeling. arXiv preprint arXiv:1412.3555 (2014)
7. Cohn, J.F., et al.: Detecting depression from facial actions and vocal prosody. In: Proceedings of the 3rd International Conference on Affective Computing and Intelligent Interaction and Workshops. IEEE, Amsterdam, Netherlands (2009). 7 pages
8. Csurka, G.: Domain adaptation for visual applications: a comprehensive survey. arXiv preprint arXiv:1702.05374 (2017)
9. Cummins, N., Scherer, S., Krajewski, J., Schnieder, S., Epps, J., Quatieri, T.F.: A review of depression and suicide risk assessment using speech analysis. Speech Commun. **71**, 10–49 (2015)
10. DeVault, D., et al.: SimSensei Kiosk: a virtual human interviewer for healthcare decision support. In: Proceeding of the International Conference on Autonomous Agents and Multi-Agent Systems, AAMAS 201, pp. 1061–1068. ACM, Paris, France (2014)
11. Ekman, P., Friesen, W.: The Facial Action Coding System (FACS). Consulting Psychologists Press, Stanford University, Palo Alto (1978)

12. Gratch, J., et al.: The Distress Analysis Interview Corpus of human and computer interviews. In: Proceeding of the 9th International Conference on Language Resources and Evaluation, LREC 2014, pp. 3123–3128. ELRA, Reykjavik, Iceland, May 2014

13. Joshi, J., et al.: Multimodal assistive technologies for depression diagnosis and monitoring. J. Multimodal User Interfaces **7**(3), 217–228 (2013). https://doi.org/10.1007/s12193-013-0123-2

14. Loucks, L., et al.: You can do that?!: Feasibility of virtual reality exposure therapy in the treatment of PTSD due to military sexual trauma. J. Anxiety Disorders **61**, 55 – 63 (2019). https://doi.org/10.1016/j.janxdis.2018.06.004, http://www.sciencedirect.com/science/article/pii/S0887618517304991, virtual reality applications for the anxiety disorders

15. Loucks, L., et al.: You can do that?!: Feasibility of virtual reality exposure therapy in the treatment of PTSD due to military sexual trauma. J. Anxiety Disorders **61**, 55–63 (2019)

16. Low, L.S.A., Maddage, N.C., Lech, M., Sheeber, L.B., Allen, N.B.: Detection of clinical depression in adolescents' speech during family interactions. IEEE Trans. Biomed. Eng. **58**(3), 574–586 (2010)

17. Marmar, C.R., Brown, A.D., Qian, M., Laska, E., Siegel, C., Li, M., Abu-Amara,D., Tsiartas, A., Richey, C., Smith, J., Knoth, B., Vergyri, D.: Speech-based markers for posttraumatic stress disorder in us veterans. Depression Anxiety **36**(7), 607–616 (2019). https://doi.org/10.1002/da.22890, https://onlinelibrary.wiley.com/doi/abs/10.1002/da.22890

18. Meng, H., Huang, D., Wang, H., Yang, H., Ai-Shuraifi, M., Wang, Y.: Depression recognition based on dynamic facial and vocal expression features using partial least square regression. In: Proceedings of the 3rd ACM international workshop on Audio/visual emotion challenge. pp. 21–30 (2013)

19. Organization, W.H., et al.: Guidelines for the management of conditions that are specifically related to stress. World Health Organization (2013)

20. Ringeval, F., et al.: Avec 2019 workshop and challenge: State-of-mind, detecting depression with AI, and cross-cultural affect recognition. In: Proceedings of the 9th International on Audio/Visual Emotion Challenge and Workshop. AVEC 2019, pp. 3–12. Association for Computing Machinery, New York, NY, USA (2019). https://doi.org/10.1145/3347320.3357688

21. Rizzo, A., et al.: Detection and computational analysis of psychological signals using a virtual human interviewing agent. J. Pain Manag. **9**, 311–321 (2016)

22. Scherer, S., Lucas, G.M., Gratch, J., Rizzo, A.S., Morency, L.P.: Self-reported symptoms of depression and PTSD are associated with reduced vowel space in screening interviews. IEEE Trans. Affect. Comput. **7**(1), 59–73 (2015)

23. Scherer, S., et al.: Automatic behavior descriptors for psychological disorder analysis. In: Proceedings of the 10th IEEE International Conference and Workshops on Automatic Face & Gesture Recognition (FG). IEEE, Shanghai, P. R. China, April 2013. 8 pages

24. Scherer, S., Stratou, G., Gratch, J., Morency, L.P.: Investigating voice quality as a speaker-independent indicator of depression and PTSD. In: Interspeech, pp. 847–851 (2013)

25. Scherer, S., et al.: Automatic audiovisual behavior descriptors for psychological disorder analysis. Image Vis. Comput. **32**(10), 648–658 (2014)

26. Simonyan, K., Zisserman, A.: Very deep convolutional networks for large-scale image recognition. https://arxiv.org/abs/1409.1556 (2014). 14 pages
27. Stratou, G., Scherer, S., Gratch, J., Morency, L.: Automatic nonverbal behavior indicators of depression and PTSD: exploring gender differences. In: 2013 Humaine Association Conference on Affective Computing and Intelligent Interaction, pp. 147–152, September 2013. https://doi.org/10.1109/ACII.2013.31

# Text Entry in Virtual Reality: A Comparison of 2D and 3D Keyboard Layouts

Caglar Yildirim[1](✉) and Ethan Osborne[2]

[1] Northeastern University, Boston, MA 02115, USA
c.yildirim@northeastern.edu
[2] University of California Santa Cruz, Santa Cruz, CA 95064, USA
etosborn@ucsc.edu

**Abstract.** Text entry is an important task in most interactive technologies in use today. Virtual Reality (VR) is becoming increasingly popular and is used in a variety of contexts, including tasks that involve text entry. With this being the case, it has become increasingly important to determine what the best keyboard layout is for text entry tasks in VR environments. To address this need, the current study compared two keyboard layouts, 2D (flat UI) and 3D (curved UI), with respect to text entry performance in VR. Results indicated that, compared to the 3D keyboard layout, using the 2D keyboard layout for the text entry task led to a greater number of words per minute, fewer corrections, and fewer redundant key presses while typing. These results indicate that the 2D keyboard layout was more efficient in VR text entry performance, compared to the 3D keyboard layout. Implications for the design and development of VR text entry tasks are discussed.

**Keywords:** Virtual reality · Text entry in VR · 3D text entry · Keyboard layout · Flat UI · Curved UI

## 1 Introduction

Text entry is an integral task afforded by most of the interactive technologies that users commonly use. As an interaction style, text entry allows the user to perform various tasks: users can look up information, they can communicate with other users, and they can even play games using text entry. Due to the pervasiveness of text entry tasks in modern technologies, text entry has been of particular interest to researchers in human computer interaction (HCI) circles [1–4], partly because of the increasing variety of different devices (e.g., desktop computers, hand-held devices, head-mounted displays, etc.) and input methods (e.g., touch, stylus, swipe, etc.).

While HCI researchers have studied text entry in desktop computers, hand-held mobile devices, and voice-based interfaces, research into text entry in virtual reality (VR) environments is in its infancy [5, 6], which is partly attributable to the recent proliferation of relatively affordable VR headsets and resurgence of academic and industrial interest. VR technology affords the ability to immerse users in computer-generated three-dimensional (3D) environments in which they can control and interact with 3D objects.

© Springer Nature Switzerland AG 2020
C. Stephanidis et al. (Eds.): HCII 2020, LNCS 12428, pp. 450–460, 2020.
https://doi.org/10.1007/978-3-030-59990-4_33

This also means that 3D user interfaces (UI) can be adopted for interaction purposes. While it would be reasonable to utilize this affordance of 3D UI layouts and metaphors for most VR tasks on the grounds that 3D representations would be perceived as more natural, many existing VR applications rely on 2D UI layouts and interaction metaphors for common interaction tasks, including text entry. Specifically, most VR applications adopt the familiar mental model of 2D keyboards, which are displayed on a flat UI layout in the VR environment. This begets the question of whether directly appropriating this metaphor for VR tasks is preferable to using a 3D representation of keyboard layouts for text entry tasks.

The current study was designed to address this question and to compare two keyboard layouts, 2D (flat UI) and 3D (curved UI), with respect to text entry performance in VR. We designed two keyboard layouts and conducted a user study in which we evaluated the performance of these two layouts in enabling users to complete text entry tasks. More specifically, we measured text entry performance in terms of words per minute (WPM), number of corrections, and redundant key entry/deletion rate. We found that the 2D keyboard layout significantly outperformed the 3D layout in all these metrics.

With the current study, our goal was to empirically evaluate the performance of two keyboard layouts. Another aim of the current study was to call for future research into the design, implementation, and evaluation of text entry tasks and methods for VR environments. The contributions of the current study are as follows:

- User study of 3D text entry in VR,
- Empirical evaluation of two keyboard layouts for text entry tasks in VR,
- Design guidelines for text entry tasks in VR.

## 2  Related Work

While other forms of text entry in VR (e.g., speech-based) has been of interest to VR researchers [7, 8], in this paper we particularly focus on studies that utilize keyboard-based (either physical or virtual) text entry in VR.

In one such study, Grubert et al. examined the effects of hand representation on text entry performance in VR [9]. Participants used a physical keyboard while typing. The virtual representation of the keyboard was 2D with a flat panel. The researchers were interested in comparing four hand representations: no hand representation, inverse kinematic model (virtual simulation of 3D hand models displayed based on users' hand/finger movements in the physical environment), fingertip visualization using spheres and video inlay of users' own hands in the virtual environment. Text entry speed ranged from 34.4 WPM to 38.7 WPM, with no significant differences among the four groups.

Knierim et al. studied the effect of hand avatars and transparency of these avatars on text entry performance in VR [10]. Participants used a physical keyboard, and both their hands and the keyboard were tracked using special sensors and were simulated in the virtual environment. Compared to no hand visualization, inexperienced typists benefited significantly from having their hands visualized in the virtual environment while typing, which translated into faster text entry. This, however, was not the case for experienced typists. Their performance under the no-hand and hand-avatars conditions was comparable. As for the effect of the realism of hand representations and transparency of these hand models, the researchers found no difference in text entry performance. This study highlights the importance of providing hand visualizations during text entry tasks in VR, especially when a physical keyboard is being used. In another study that utilized a physical keyboard for text entry tasks in VR, Grubert et al. compared touchscreen and standard, desktop keyboards on text entry performance in VR and found that using touchscreen keyboard was slower than using standard, desktop keyboard [11].

In addition to studying the use of physical keyboards for text entry tasks in VR, previous research has also explored the feasibility of using alternate text entry methods in VR. For instance, Rajanna et al. investigated the viability of gaze typing in VR in sitting and biking conditions [6], while examining the effect of selection method (dwelling vs. button click). They utilized a commercial HMD that supported eye-tracking in VR, using which they enabled users to select keys with gaze movements. In Study 1, the researchers used a 2D representation of a virtual QWERTY keyboard and found that gaze typing speed ranged from 8.1 WPM to 10.15 WPM. While the researchers found a significant difference across the four conditions, they failed to follow up the significant result with a post hoc analysis, which would reveal which conditions were different from which conditions. In Study 2, the researchers used a curved UI for the keyboard layout, which was larger than users' field-of-view (FOV) afforded by the VR headset. With the curved UI, text entry speed went down, ranging from 6.8 WPM to 9.2 WPM. The researchers attributed this performance decrement to the fact that participants needed to make more head movements to see all the keys on the virtual keyboard (because the keyboard was larger than users FOV). Because the researchers did not directly compare the flat UI used in Study 1 to the curved UI used in Study 2, it is not possible to conclude from these two separate studies that the curved UI should lead to slower text entry performance in VR. If the curved UI was within-view as well, the results would probably have been different, for instance. That said, one implication of the Rajanna et al. study is that virtual keyboards, whether 2D or 3D, should be displayed within the FOV of participants, so as to minimize redundant head movements, which might slow the participants down while typing.

In an attempt to develop a set of guidelines for text entry tasks in VR, Speicher et al. [5] investigated selection-based text entry in VR by comparing six different methods of text entry: head pointing, controller pointing, controller tapping, freehand selection, discrete cursor selection, and continuous cursor selection. Participants were presented with a virtual 2D keyboard and were asked to enter short phrases. Text entry speed ranged from 5.3 WPM to 15.4 WPM, with the controller pointing selection method outperforming all the other five methods. This study indicates that for selection-based text entry tasks utilizing a virtual keyboard the best method of key selection is controller pointing, in which a raycasting metaphor is used to enable participants to point to a key and select the key by pressing the trigger or selection button on the controller.

Based on the foregoing review of related work, the comparison of 2D and 3D keyboard layouts on text entry performance in VR still remains understudied, which provided the impetus for the current study. We sought to investigate the effect of keyboard layouts on text entry performance when using native VR controllers that implement a raycasting metaphor for selection tasks. This decision was made on the basis of Speicher et al.'s findings [5], as discussed earlier. We also decided to implement both 2D and 3D virtual keyboard layouts so that both layouts would be displayed within users' FOV. This way the performance of the two layouts could be compared, which was not possible in Rajanna et al. [6].

## 3  Method

### 3.1  VR Keyboard

We developed two keyboard layouts in Unity 3D (Fig. 1). The keyboard layouts were identical in that both were built to mimic a virtual representation of standard QWERTY layout. The two layouts were different in that one had was a 2D keyboard using a flat UI, whereas the other one was a 3D keyboard using a curved UI.

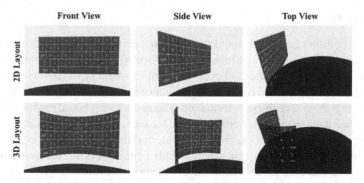

**Fig. 1.** Screenshots of the keyboard layouts from different angles. 2D keyboard layout uses a flat panel, while 3D keyboard layout uses a curved user interface panel.

Regardless of the keyboard layout, the strings were displayed in one of the horizontal bars placed above the keyboard (Fig. 2). The top bar represented the submit button the users pressed to submit their text entry. The label bar, placed below the top bar, showed the phrase the participants were to enter. The bottom bar was the input field and showed the letters typed as participants selected keys.

### 3.2  Text Entry Task

Users performed a simple text entry task in which they were asked to enter the short strings (sentences or phrases) presented to them as quickly and accurately as possible. For this purpose, we designed two keyboard layouts, one with a flat UI and another with curved UI. Users were asked to enter five such strings using each of the keyboard layouts. The strings were randomly selected from the memorable mobile emails dataset provided by Vertanen and Kristensson [12]. For key selection, based on [5], we used a raycasting metaphor in which participants used the Oculus Touch controller to point to the key they wanted to select.

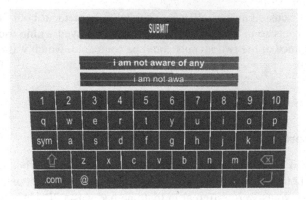

**Fig. 2.** VR keyboard screen for 2D layout

### 3.3  Study Design

To examine the effect keyboard layout on text entry performance in VR, we conducted a within-subjects experiment in which users performed the same text entry task (with different strings) using both the 2D keyboard layout and 3D keyboard layout. This was done to eliminate the potential confounding effect of individual differences on the effect of keyboard layout on text entry performance. The order of keyboard layout presentation was counterbalanced to minimize order effects. As such, half of the users first completed the text entry task using the 2D keyboard layout, and then moved on to completing the task using the 3D keyboard layout, whereas half of the users started out with the 3D keyboard layout and then proceeded to the 2D keyboard layout.

The independent variable manipulated in the experiment was the keyboard layout with two levels: 2D keyboard layout, which used a flat UI design, and 3D keyboard layout, which used a curved UI design, as shown in Fig. 1. We hypothesized that the 3D keyboard layout would lead to better text entry performance, because it would provide a more natural representation of the keyboard in a 3D virtual environment.

The dependent variable was text entry performance operationally defined in terms of the average number of words per minute (WPM), average number of corrections, and average redundancy rate. These are described in greater detail below.

**Words per Minute.**  To standardize the WPM metric, we defined one word as a string with five characters. We computed the WPM for each trial by dividing the number of words entered during the trial by the number of seconds taken to enter the string presented in each trial. The average WPM across the five trials was used as the dependent variable for each user.

**Number of Corrections.**  A correction refers to a key press performed to correct a typo during text entry. The number of corrections was defined in terms of the number of key presses required to correct typos during each trial. Again, we used the average number of corrections across all trials.

**Redundancy Rate.** Redundancy rate was defined by the proportion of the number of redundant key presses performed to the total number of key pressed required, where the number of redundant key presses was computed by subtracting the total number of key pressed required (the length of a given string) from the total number of key pressed actually performed by the user during each trial. Similar to the previous two metrics, we used the average redundancy rate across all trials.

## 3.4  Apparatus

We used the Oculus Rift headset in this experiment. Oculus Rift is a head-mounted display (HMD) that track the users' position around the virtual environment. The HMD features a display with a $1080 \times 1200$ resolution and a 90 Hz refresh rate [13]. The HMD also tracks the user's head motion using a gyroscope and an accelerometer. The Oculus Rift also has a field of view (FOV) of 110° [13]. The Oculus Rift comes equipped with two motion controllers, only one of which was used during this experiment, specifically the right Touch controller. The VR Keyboard application was run on a VR-compatible computer with i7 CPU and 16 GB RAM.

## 3.5  Participants

Upon arrival in the experimental room, participants read and signed an informed consent form. Then the experimenter explained the experimental procedures. Specifically, participants were told that they were going to complete the experiment in two parts and that they were going to complete a short text entry task in each part. They were also told that they could take a break at any point, and to take off the headset should they ever feel uncomfortable.

## 3.6  Procedures

Upon arrival in the experiment room, participants read and signed an informed consent form. Then the experimenter explained the experimental procedures. Specifically, participants were told that they were going to complete the experiment in two parts and that they were going to complete a short text entry task in each part. They were also told that they could take a break at any point, and to take off the headset should they ever feel uncomfortable.

Before participants donned the Oculus Rift headset, the experimenter explained and demonstrated how to use the right-handed motion controller to perform a selection by pressing the trigger key on the controller. After the experimenter ensured the participant was comfortable with how to use the controller, the participant put on the VR headset. When the headset was donned properly and comfortably, the experimenter opened up the tutorial file where the participant would get an opportunity to practice inputting a phrase and understanding how the keyboard was laid out. When the participant indicated that they felt comfortable, the experimenter had them begin the first part of the experiment in their randomly assigned starting condition, which was the 2D keyboard layout for half of the participants and the 3D keyboard layout for the other half. When the participants were done with all the five phrases in the starting condition, the instructions panel in the virtual environment indicated that the first part was completed and that they should remove the headset.

Following the completion of the first part, participants took a five-minute break. After the break, they were told that they were going to perform the same text entry task (with different strings) again. In this second part, those participants who started with the 2D layout used the 3D layout, and those who started with the 3D layout used the 2D layout. Once the second part was completed, participants were debriefed about the study and were encouraged to ask any questions they may have.

# 4  Results

To compare the two keyboard layouts on the performance metrics described earlier, we conducted several descriptive and inferential statistics tests. Table 1 presents a summary of these tests for each of the dependent variables of the experiment.

**Table 1.** Summary of descriptive and inferential statistics

|  | $M (SD)$ |  |
| --- | --- | --- |
| **WPM** |  | $t(31) = 11.57, p < .001$ |
| 2D Layout | 17.35 (.59) | Cohen's $d = 2.05$ |
| 3D Layout | 11.81 (.45) |  |
|  |  |  |
| **Corrections** |  | $t(31) = 4.14, p < .001$ |
| 2D Layout | 6.30 (1.11) | Cohen's $d = .73$ |
| 3D Layout | 12.34 (1.55) |  |
|  |  |  |
| **Redundancy Rate** |  | $W = 128, p = .01$ |
| 2D Layout | 14.99 (3.05) | Cohen's $d = .41$ |
| 3D Layout | 23.88 (2.91) |  |

Alpha level set at .05 for all hypothesis tests.

The assumption of normality, as assessed by Shapiro-Wilk's test, ($p > .05$) was met for the average WPM and average number of corrections variables, but not for the redundancy rate variable. Therefore, a paired-samples $t$ test was conducted to compare the average WPM and number of corrections between the 2D and 3D keyboard layouts, and a Wilcoxon rank test was conducted for the average redundancy rate variable.

As seen in Table 1, results revealed a significant difference between the 2D keyboard layout and 3D keyboard layout in the average number of WPM, with a greater number of average WPM in the 2D keyboard layout (Fig. 3).

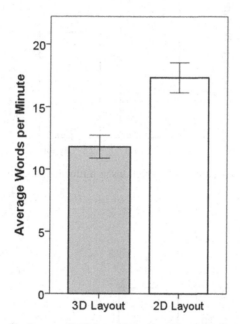

**Fig. 3.** Average WPM as a function of keyboard layout

Regarding the effect of keyboard layout on the number of corrections made during text entry, results showed that participants made more corrections in the 3D keyboard layout compared to the 2D keyboard layout (Fig. 4).

As for the effect of keyboard layout on redundancy rates, results revealed a statistically significant difference between the two keyboard layouts, with increased redundancy rates in the 3D keyboard layout (Fig. 5).

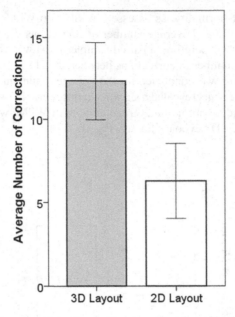

**Fig. 4.** Average number of corrections as a function of keyboard layout

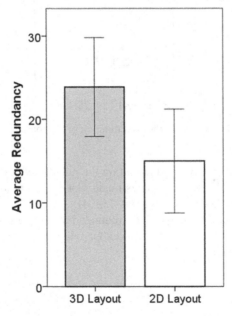

**Fig. 5.** Average redundancy rate as a function of keyboard layout

# 5  Discussion

The purpose of the current study was to investigate the effect of keyboard layout on text entry performance in VR. We set out to compare a 2D keyboard layout with flat UI to a 3D keyboard layout with curved UI with respect to efficiency. More specifically, we focused on the average WPM, a standard metric to evaluate text entry speed, average number of corrections made to edit/correct the text entry, and averaged redundancy rate.

In relation to task entry speed, our results indicate that compared to the 3D layout, the 2D keyboard layout was more efficient in terms of enabling users to enter text faster in VR. In fact, users entered an average of 17.4 WPM using the 2D keyboard layout, in contrast to an average of 11.8 WPM using the 3D keyboard layout. The stark difference between two keyboard layouts was unexpected because we hypothesized the 3D keyboard layout would be a more natural choice in a 3D virtual environment, compared to the 2D keyboard layout. One potential explanation for this finding is that users' familiarity with flat UIs might be the reason why their performance was better in the 2D keyboard layout. Due to the purported novelty of curved UI, the users might have found it more challenging to adjust to this new layout, resulting in slower text entry. While this finding is inconsistent with our initial prediction, the average WPM using the 2D keyboard layout is congruent with a prior study on text entry in VR [5]. In comparing multiple input methods, Speicher et al. found that participants could enter an average of 15.4 WPM (SD = 2.7) on a 2D keyboard layout, using a controller with the same raycasting metaphor as used in our study.

Regarding the number of corrections made during text entry, results indicated that the 3D keyboard layout led to an increased number of corrections than did the 2D keyboard layout, a finding incongruent with our initial prediction. This means that users made a greater number of key selection errors in the 3D keyboard layout condition than in the 2D keyboard layout condition. The increased number of corrections in the 3D keyboard layout condition could also partially explain why users were slower in this condition: The greater the number of corrections made during text entry, the longer it takes to complete the text entry task, leading to fewer WPM in the 3D keyboard layout condition. Closely tied to the number of corrections is the redundancy rate, because an increase in the number of corrections translates to increases in the redundancy rate. Results indicated users needed to make more redundant key selections in the 3D keyboard layout, compared to the 2D keyboard layout.

Given the nascent nature of the literature on VR interaction techniques, it is important to develop empirically-driven guidelines for the design, development, and implementation of effective, efficient, and satisfactory VR systems. One direct implication of the current findings is that a 2D keyboard layout should be used for text entry tasks in VR if task performance and efficiency are prioritized over the visual aesthetics and cohesiveness of the environment. It can be argued that flat UI panels hovering in the virtual environment are not as natural and well-integrated as curved UIs, the difference in performance levels is too stark to overlook.

While the current study provides a first step in the direction of better understanding the effectiveness and efficiency of 2D and 3D layouts for VR interaction tasks, future studies are warranted to corroborate, challenge, and/or expand on the findings from the current study. To the best of our knowledge, the current study is the first to compare 2D and 3D keyboard layouts on text entry performance in VR. Therefore, future research should attempt to replicate the findings from the current study. It would also be particularly useful to extend the current study to investigations of menu design for VR environments and examine the effect of layout on menu selection tasks. Furthermore, in the current study, we focused on objective, task-based metrics when comparing the 2D and 3D layouts, but it would be useful to incorporate user preference as a metric into the comparison of these two layouts.

# References

1. Kristensson, P.O.: Next-generation text entry. Computer **48**(7), 84–87 (2015)
2. Zhai, S., Sue, A., Accot, J.: Movement model, hits distribution and learning in virtual keyboarding. In: Proceedings of the SIGCHI Conference on Human Factors in Computing Systems, pp. 17–24. ACM (2002)
3. Zhai, S., Kristensson, P.O., Smith, B.A.: In search of effective text input interfaces for off the desktop computing. Interact. Comput. **17**(3), 229–250 (2005)
4. MacKenzie, I.S., Soukoreff, R.W.: Phrase sets for evaluating text entry techniques. In: CHI 2003 extended abstracts on Human factors in computing systems, pp. 754–755. ACM (2003)
5. Speicher, M., Feit, A.M., Ziegler, P., Krüger, A.: Selection-based text entry in virtual reality. In: Proceedings of the 2018 CHI Conference on Human Factors in Computing Systems, p. 647. ACM (2018)
6. Rajanna, V., Hansen, J.P.: Gaze typing in virtual reality: impact of keyboard design, selection method, and motion. In: Proceedings of the 2018 ACM Symposium on Eye Tracking Research & Applications, p. 15. ACM (2018)
7. Bowman, D.A., Rhoton, C.J., Pinho, M.S.: Text input techniques for immersive virtual environments: an empirical comparison. In: Proceedings of the Human Factors and Ergonomics Society Annual Meeting, vol. 46, no. 26, pp. 2154–2158. Sage CA: Los Angeles, CA: SAGE Publications (2002)
8. Pick, S., Puika, A.S., Kuhlen, T.W.: SWIFTER: design and evaluation of a speech-based text input metaphor for immersive virtual environments. In: 2016 IEEE Symposium on 3D User Interfaces (3DUI), pp. 109–112. IEEE (2016)
9. Grubert, J., Witzani, L., Ofek, E., Pahud, M., Kranz, M., Kristensson, P.O.: Effects of hand representations for typing in virtual reality. In: 2018 IEEE Conference on Virtual Reality and 3D User Interfaces (VR), pp. 151–158. IEEE (2018)
10. Knierim, P., Schwind, V., Feit, A.M., Nieuwenhuizen, F., Henze, N.: Physical keyboards in virtual reality: analysis of typing performance and effects of avatar hands. In: Proceedings of the 2018 CHI Conference on Human Factors in Computing Systems, p. 345. ACM (2018)
11. Grubert, J., Witzani, L., Ofek, E., Pahud, M., Kranz, M., Kristensson, P.O.: Text entry in immersive head-mounted display-based virtual reality using standard keyboards. In: 2018 IEEE Conference on Virtual Reality and 3D User Interfaces (VR), pp. 159–166. IEEE (2018)
12. Vertanen, K., Kristensson, P.O.: A versatile dataset for text entry evaluations based on genuine mobile emails. In: Proceedings of the 13th International Conference on Human Computer Interaction with Mobile Devices and Services, pp. 295–298. ACM (2011)
13. Oculus Rift Specs (2018). https://www.cnet.com/products/oculus-rift/specs/

# Author Index

Printed in the United States
By Bookmasters